THE
SCIENCE OF
PHOTOBIOLOGY

THE
SCIENCE OF
PHOTOBIOLOGY

Edited by Kendric C. Smith

Stanford University School of Medicine
Stanford, California

PLENUM PRESS · NEW YORK AND LONDON

Library of Congress Cataloging in Publication Data

Main entry under title:
The Science of photobiology.

Includes bibliographical references and index.
1. Photobiology. I. Smith, Kendric C.

QH515.S37	574.1'9153	77-22130

ISBN 0-306-31051-1

© 1977 Plenum Press, New York
A Division of Plenum Publishing Corporation
227 West 17th Street, New York, N.Y. 10011

Printed in the United States of America

Foreword

Although there are several excellent books covering a few of the specialized areas of photobiology, at the present time there is no book that covers all areas of the science of photobiology. This book attempts to fill this void. The science of photobiology is currently divided into 14 subspecialty areas by the American Society for Photobiology. The first 14 chapters of this book deal with those subspecialty areas, each written by a leader in the field. Chapter 15, entitled "New Topics in Photobiology," highlights areas of research that may be designated subspecialties of photobiology in the future.

This book has been written as a textbook to introduce the science of photobiology to advanced undergraduate and graduate students. The chapters are written to provide a broad overview of each topic. They are designed to contain the amount of information that might be presented in a one- to two-hour general lecture. The references are not meant to be exhaustive, but key references are included to give students an entry into the literature. Frequently a more recent reference that reviews the literature will be cited rather than the first paper by the author making the original discovery. Whenever practical, a classroom demonstration or simple laboratory exercise has been provided to exemplify one or more major points in a chapter. The chapters are not meant to be a repository of facts for research workers in the field, but rather are concerned with demonstrating the importance of each specialty area of photobiology, and documenting its relevance to current and/or future problems of man.

A great deal of research has been performed upon the effects of temperature (arctic and desert), of pressure (deep-sea diving and space flights), and of gravity (space flights) on the physiology of man, but until recently very little effort has been expended on studying the effects of light on man. Yet, light is one of the most important elements of our environment.

It is hoped that this book will serve as a basic text for introductory courses in photobiology, and as a vehicle for encouraging students to enter the field of photobiology. It is also expected that this book will be of interest to scientists outside of the area of photobiology and to interested laypersons, since it is now becoming more apparent, even to the general public, that light, both natural and artificial, has important consequences to man other than just as an aid to vision. Photobiology appears to have come of age as a major scientific discipline.

Kendric C. Smith

September 1977

Preface

The reader should be aware of the major sources of literature and information relevant to the science of photobiology.

For review articles, there are two major sources: *Photophysiology,* Volumes 1–8 (A. C. Giese, ed.), Academic Press, New York, covering the years 1964–1973; and *Photochemical and Photobiological Reviews,* Volumes 1*ff.* (K. C. Smith, ed.), Plenum Press, New York, that was begun in 1976.

For research papers, the major source is the international journal *Photochemistry and Photobiology* (Pergamon Press, London). This journal was inaugurated in 1962, and is now the official organ of the American Society for Photobiology.

The American Society for Photobiology was started in 1972 (1) to promote original research in photobiology, (2) to facilitate the integration of different disciplines in the study of photobiology, (3) to promote the dissemination of knowledge of photobiology, and (4) to provide information on the photobiological aspects of national and international problems. Membership in the Society is open to persons who share the stated purpose of the Society and who have educational, research, or practical experience in photobiology or in an allied scientific field.

The name of the Society was chosen to encompass both North and South America, but members from other parts of the world are also welcome. The Journal is included in the membership dues. The American Society for Photobiology holds an annual scientific meeting (usually in June), and publishes frequent Newsletters of interest to photobiologists. Further information may be obtained by writing to the Executive Secretary, American Society for Photobiology, 4720 Montgomery Lane, Suite 506, Bethesda, MD 20014.

The Association International de Photobiologie sponsors an international congress on photobiology every four years. The next congress will be held in France in 1980.

The Editor

Contents

1

Phototechnology and Biological Experimentation

1.1. INTRODUCTION[1]

A great number of experiments in photobiology can be done with remarkably simple and inexpensive equipment. Nevertheless, the experimenter usually finds, after completing his initial "rough" experiments, that he must concern himself more and more with details such as the relative effectiveness of different wavelengths and the exact energies required. Consequently, there are needs for both simple and complex apparatus. One usually tries to compromise by using equipment that is only as sophisticated as is necessary for sound experimentation.

John Jagger • Biology Programs, The University of Texas at Dallas, Richardson, Texas

The selection of equipment to be used depends, of course, on the nature of the problem. In deciding among the enormous variety of equipment available, considerable trouble and expense may be spared if one keeps in mind the following generalizations concerning photobiological (as opposed to photochemical or photophysical) experimentation: (1) high intensities of light are usually required, (2) narrow bandwidths (high monochromaticities) are usually not required, and (3) irradiation times shorter than a few seconds are usually not required. Thus, the extremely high monochromaticity of a laser and the extremely short (nanosecond range) pulse of a pulsed laser are usually not required in photobiological (as opposed to photochemical) work.

A few conventions should be stated at the outset. Figure 1-1 illustrates the wavelengths, photon energies, and colors in the ultraviolet (UV), visible, and near-infrared regions. For purposes of this chapter, we consider far-UV radiation to lie in the range of 210–300 nm, and near-UV radiation to lie in the band 300–380 nm. We shall not be concerned with the vacuum-UV or the near-infrared regions. The width of a band of radiation will sometimes be characterized as the "half-maximum bandwidth" (bandwidth at one-half the peak of intensity), usually when discussing filters or where the transmission curve has a Gaussian form, and sometimes as the "total bandwidth" (bandwidth at ~1% of the peak of intensity), usually in discussing the output of a dye laser or the complete wavelength spread in the output of a monochromator. Finally, the SI (Système International) units for spectral energy and power will be used (Table 1-1).

TABLE 1-1. **International System of Units (Système International d'Unités; SI)[a]**

ampere	A	lux	lx
calorie	cal	meter	m
candela	cd	milligram	mg
coulomb	C	million electron volts	MeV
curie	Ci	minute	min
degree Celsius	°C	mole	mol
degree Kelvin	K	nanometer	nm
degree (angle)	°	newton	N
electron volt	eV	ohm	Ω
erg	erg	ohm-centimeter	Ω-cm
farad	F	parts per million	ppm
gauss	G	percent	%
gram	g	radian	rad
henry	H	roentgen	R
hertz	Hz	second	s
hour	h	steradian	sr
joule	J	tesla	T
kilogram	kg	torr	Torr
lambert	L	volt	V
liter	liter	watt	W
lumen	lm	weber	Wb

[a]Note that units named for persons are not capitalized except when abbreviated. All abbreviations stand for the plural as well as the singular forms.

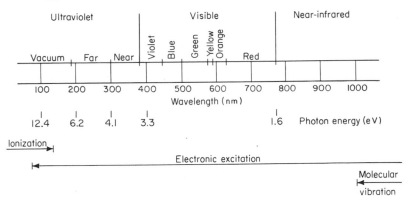

Fig. 1-1. Ultraviolet, visible, and near-infrared regions of the electromagnetic spectrum. Beneath the spectrum are indicated major physical events that occur in various regions. The limits are not sharp, as shown. Furthermore, any photon can cause events generally associated with photons of lower energy.

1.2. CHOOSING THE SOURCE[1-5]

1.2.1. Visible Radiation

Both the sun and high-wattage incandescent lamps are good bright* sources of visible radiation. They are not well suited for experimental work, however, since the sun's radiation at the surface of the earth is highly variable and difficult to get into the laboratory, and both solar and incandescent sources produce too much undesired heat and infrared radiation.

Incandescent lamps are nevertheless occasionally used, especially in work with plants. Figure 1-2 shows the spectrum of solar radiation at the surface of the earth, the spectrum of normal human visual sensitivity (which defines visible light), and the emission spectrum of an incandescent lamp with a color temperature of 3475 K (the spectrum of an incandescent lamp is determined by its filament temperature). Incandescent lamps produce so much heat that they are usually used in conjunction with a heat-absorbing filter, such as a 10% solution of copper sulfate (Section 1.3.2.). A particularly useful form of the incandescent lamp is the General Electric[6] colored (Dichro-color) spot lamp, which has an interference filter built into the lens; these lamps emit relatively narrow bands of light (half-maximum bandwidth of 50–80 nm), but also emit in the far-red; they come in five colors and are useful for obtaining rough action spectra at low cost.

Fluorescent lamps are the most useful sources of visible light for photobiological purposes, having relatively high intensity* but low heat output. They are not very bright, but this is balanced by the fact that they are good for irradiating large areas. Fluorescent lamps are mercury discharge lamps that operate at *low pressure* (a few millimeters of argon containing about 10^{-3} mm of mercury

*For terminology of photometry and radiometry, see Section 1.4.1. and Appendix C of reference 1.

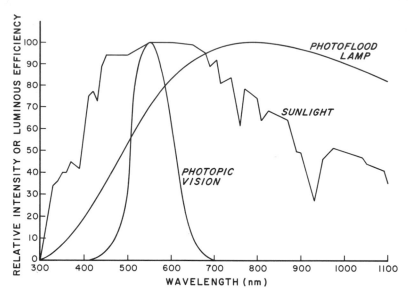

Fig. 1-2. Spectrum of (a) sunlight from the zenith at the surface of the earth. The dips in the red and infrared regions are due to absorption by water vapor [300–650 nm from R. D. Cadle and E. R. Allen, Science **167**, 243–249 (1970); 650–1100 nm from R. B. Withrow and A. P. Withrow, Chap. 3 in *Radiation Biology,* Vol. III (A. Hollaender, ed.), McGraw–Hill, New York (1956)]. (b) 250-W incandescent photoflood lamp (color temperature, 3475 K). [From R. B. Withrow and A. P. Withrow, Chap. 3 in *Radiation Biology,* Vol. III (A. Hollaender, ed.), McGraw–Hill, New York (1956)], and (c) average human visual sensitivity.[2]

vapor). Electrons from the hot filament, and positive mercury ions produced by them, excite mercury atoms by collision. The excited electrons in the mercury atoms tend to lose energy and to collect in a low-lying metastable (triplet) state, from which they fall back to the ground state emitting radiation at 254 nm. Because of the low mercury vapor pressure, this radiation reaches the surface of the tube. Figure 1-3c shows that about 85% of the total energy emitted by the lamp may lie at 254 nm, while small amounts of radiation at longer wavelengths, representing transitions from higher energy states down to the metastable state, represent some 15% of the total energy. In a *germicidal lamp,* all of this radiation is permitted to leave the fluorescent tube, but, in a *fluorescent lamp,* phosphors (such as calcium halophosphate) that coat the inside wall of the tube absorb the far-UV radiation and reemit it in a smooth continuum of visible fluorescent light. Thus, the output of a fluorescent lamp (Fig. 1-3a,b) consists of this fluorescent continuum superimposed upon the long-wavelength mercury lines. The envelope of a fluorescent lamp is made of glass, which cuts off all residual far-UV radiation, and lowers the intensity of the near-UV components.

The fluorescent lamps of a given type (e.g., "daylight," "cool-white") produced by the three major American companies (General Electric, GTE-Sylvania, Westinghouse) are virtually identical. Figure 1-4 shows that they may be obtained with various spectral biases. Some of these lamps are useful for general room illumination when one wishes to avoid particular wavelengths of

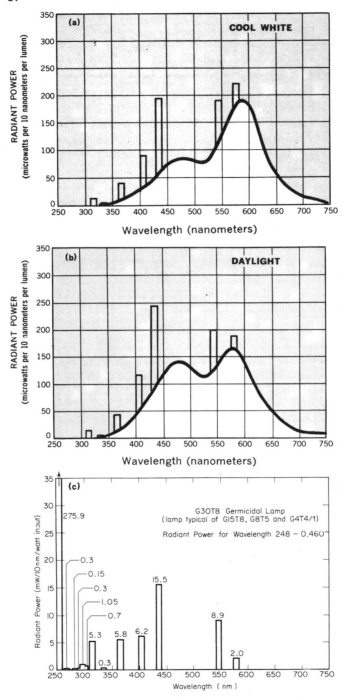

Fig. 1-3. Spectral output after 100 h of operation of (a) 40-W T12 "cool white" fluorescent lamp, (b) 40-W T12 "daylight" fluorescent lamp, and (c) 30-W T8 germicidal lamp. Note that the same mercury lines above 300 nm appear in all three spectra. For the germicidal lamp, about 85% of the total radiant power is emitted at 254 nm, but this percentage will decrease with age of lamp and nearby light shielding that raises lamp temperature. (Courtesy of General Electric Co.)

light [for example, the use of gold fluorescent lamps to avoid bacterial photoreactivation (Sections 1.5.3. and 5.4.2.), which occurs only below 500 nm]. Fluorescent lamps with an output optimized for plant growth are produced by several companies (Fig. 1-5).

It is important to recognize that there may be very considerable differences in the spectral output of various "white" fluorescent lamps. Thus, the emission spectrum of a "daylight" lamp is quite different from that of a "cool white" lamp (Fig. 1-3).

Germicidal and fluorescent lamps come in a variety of sizes and shapes; the 15-W size is convenient to use in an ordinary desk-lamp holder. The lamps require a transformer-ballast and a starter, both of which are usually built into the lamp fixture.

The output of a germicidal or fluorescent lamp changes considerably during the first 100 h of operation, and lamps should be "aged" for this period of time before use. In addition, the output changes considerably during the first few minutes of operation, and the lamps should be "warmed up" for about 10 min before use. Lamp output varies considerably with the temperature of the lamp, which in turn is influenced by how closely it is surrounded by light shields—a confined lamp with no ventilation will operate at a higher temperature, and will have a lower output and a shift in its emission spectrum.

Fluorescent lamps are versatile, stable, cheap, and easy to use. They are clearly the source of choice for work in the visible region. For some applications, however, a source of greater brightness is needed. One will then usually turn to a high-pressure xenon or mercury arc lamp. These sources and their power supplies, however, are expensive and can be dangerous to use (Section 1.2.4.).

The *high-pressure xenon arc lamp* is the most useful high-intensity source of visible light. Typically, the luminous arc itself is only 1 mm wide by 3 mm high, virtually a point source. The source is very stable, and has a lifetime of the order of 1000 h. It has an almost constant emission spectrum in the visible region, except for a few small peaks around 400–500 nm (Fig. 1-6).

The *high-pressure mercury arc lamp,* unlike the xenon lamp, produces a discrete spectrum in the visible region that is superimposed upon a lower-level continuous spectrum (Fig. 1-6). The high gas pressure (~100 atm) results in total absorption of the 254 nm line, leaving the spectrum *dark* at this wavelength, but at none of the other emission wavelengths of mercury. This source is useful in conjunction with interference filters or a monochromator, where one wishes to obtain high-intensity monochromatic light, but is not too particular about the exact wavelength being used. There are really only four major mercury lines above 380 nm, and this limits the value of the mercury arc as a source of visible light. High-pressure mercury arcs are harder to use than xenon arcs. They are more unstable, and the mercury, usually contained in a capillary about 2×12 mm, must be carefully redistributed before each ignition. Another complication is that mercury arcs are usually water-cooled, whereas xenon arcs are air-cooled. The lifetime of a high-pressure mercury arc source averages about 100 h but may be twice this or as little as zero (i.e., the lamp may not light at all). In spite of these moderate difficulties, the lamps are rather widely used. Philips produces the most convenient source, and supplies the appropriate water jacket and power supply. Oriel supplies high-intensity xenon and mercury lamps of various manufacturers and provides a variety of lamp housings, which may include condensing-lens assemblies (see also ref. 6).

Medium-pressure mercury arc lamps provide a discrete spectrum with no continuum throughout the UV and visible regions. They are used primarily with monochromators, but the intensity is much lower than that obtainable with the high-pressure sources. One advantage for UV work is that they provide both the 254-nm line and the other lines. Medium-pressure mercury vapor lamps, which provide a more extended source, are available from Hanovia, and are useful as immersion sources for irradiation of chemical solutions or cell suspensions (see also ref. 6).

A *metallic vapor arc* unit (300-W, General Electric Marc 300), supplying very high intensity, with a spectrum similar to that of the sun, has been described.[7]

Lasers alone are not very useful for most photobiological (as opposed to photochemical) work in the visible region. However, *dye lasers,* which may be excited either by a flash lamp or by another laser, are becoming popular. Such lasers excite various dyes (typically oxazines, xanthenes, or coumarins) to a single-wavelength (linewidth ~0.1 nm) fluorescence. A single dye may be "tuned" to laser action over a band of about 10 nm for far-UV dyes, 20 nm for near-UV dyes, and 40 nm for visible dyes. Although very high intensities may be

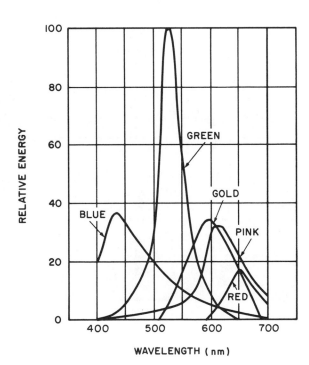

Fig. 1-4. Spectral output of various colored fluorescent lamps of equal wattage. The mercury spectral lines are omitted. (Courtesy of Westinghouse Electric Corp.)

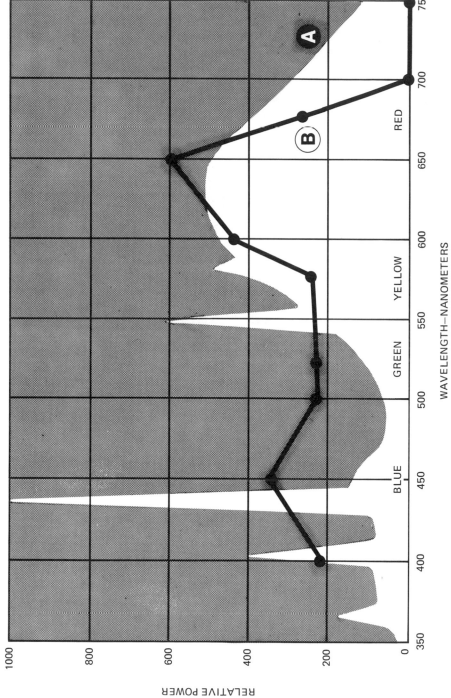

Fig. 1-5. (A) Spectral output of Agro-lite fluorescent lamp compared with (B) action spectrum for optimum plant growth ("saturation"). (Courtesy of Westinghouse Electric Corp.)

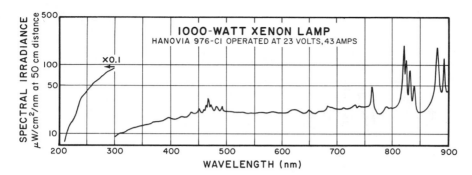

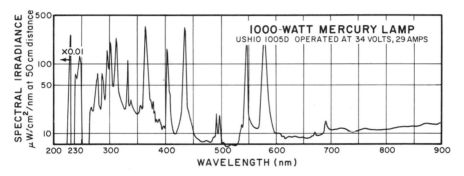

Fig. 1-6. Spectral output of 1000-W xenon and mercury arc lamps. (Courtesy of Oriel Corp.)

obtained with a laser alone (e.g., the ruby laser, which emits at 694 nm), the dye lasers do not generally provide a higher output than can be obtained with a monochromator fitted with a high-pressure mercury or xenon arc. Since the monochromator is more versatile and cheaper, the usefulness of the dye laser in experimental photobiology is still somewhat limited. They have been very useful, however, for the microirradiation of cells,[5] and in photochemical studies.

1.2.2. Near-Ultraviolet Radiation

Incandescent lamps are very inefficient in the near-UV region, and are generally not used.

Some fluorescent lamps (e.g., the "daylight"; Fig. 1-3) provide usable near-UV output, but the intensity is low. Much more appropriate are the "black-light" lamps, fluorescent lamps with phosphors that emit in the near-UV region (Fig. 1-7). The regular "black-light" (BL) lamp has some mercury lines in the visible region, but the "black-light-blue" (BLB) lamp has a quite narrow spectrum with relatively little line emission. The output of the 40-W T12 BLB lamp peaks at about 355 nm (the 15-W BLB lamp spectrum is shifted slightly to longer wavelengths and peaks at 366 nm).

High-pressure mercury arcs are very useful in the near-UV region, but

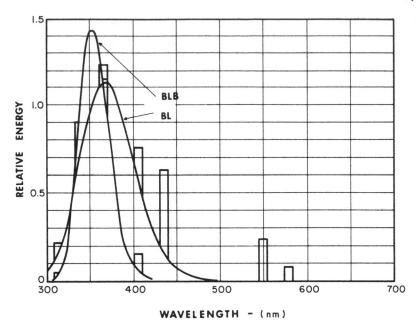

Fig. 1-7. Spectral output of 40-W T12 black-light (BL) and black-light-blue (BLB) lamps after 100 h of operation. (Courtesy of General Electric Co.)

xenon lamps are less useful because their intensity falls off rapidly below 400 nm (Fig. 1-6).

A useful laser for the near-UV region is the pulsed nitrogen laser, which emits at 337 nm. The output of this laser is about 10 times what one can expect from a monochromator with a high-pressure mercury arc, and its relatively simple design and low cost make it a reasonable competitor. Some manufacturers also supply dye heads for laser emission at longer wavelengths (the emission intensity is much lower with the dye head, however).[6]

1.2.3. Far-Ultraviolet Radiation[1,2]

There are only two sources commonly used for the far-UV region. These are the germicidal lamp, which is virtually monochromatic at 254 nm, and the high-pressure mercury arc, which emits some moderately strong lines in the far-UV region (Section 1.2.1.). It must be remembered that one is virtually constrained to using quartz optics when working below 300 nm, which makes work in this region more complicated and more expensive.

1.2.4. Hazards

Because visible and UV radiation are relatively easily obtained, the dangers of these radiations are often ignored. While it is perfectly true that we live in an environment of "natural" visible light, it is also true that a person may permanently blind himself by staring at the sun for a period as short as 1 min. A high-pressure mercury arc lamp has a brightness approaching that of the sun, and similar consequences would result from looking directly at such a source with no eye protection.

In addition, radiations in the region 280–330 nm are very damaging to human epithelium, and can produce skin cancers and keratosis of the cornea. Radiations below 280 nm are absorbed largely in the dead layers of the skin, and are thus less dangerous to the living cells of skin than slightly longer wavelengths. Therefore, except for tissues with live cells near the surface, such as the lips and cornea, a black-light lamp is more dangerous than a germicidal lamp.

Thus, one must consider two important hazards of photobiological light sources: (1) very-high-intensity light damage to the retina, and (2) UV radiation damage to the skin, lips, and cornea. The only good protection from very-high-intensity light is not to expose oneself to it. If exposure is necessary, it should be done through a thick glass filter that transmits only a narrow band of the light, preferably in some region away from the wavelength maximum of the source. Protection from the carcinogenic UV region is provided by thick window glass, but generally not with plastics. UV-absorbing plastics are available, however. Eyeglasses provide fairly good protection for the cornea from frontal UV radiation, but one must beware of radiation entering from the sides.

Finally, it should be noted that high-pressure mercury and xenon arc lamps operate at high voltages, and at currents of several amperes, which can be lethal if the operator is grounded.

1.3. NARROWING THE SPECTRAL OUTPUT[1-4]

Most sources provide a very broad spectrum of light, and it is usually necessary, after the most preliminary biological experiments have been done, to narrow this spectrum as much as possible to determine what wavelengths are most effective. There are two reasons for this: (1) to ensure that light sources are used more efficiently in producing an effect (or certain wavelengths may be avoided when one does not wish to produce the effect), and (2) to gain some information concerning the nature of the molecules responsible for the effect being studied, i.e., to obtain an action spectrum (Section 1.5.4.).

As noted in Section 1.1., extremely narrow bands of light are not needed in most biological work because effects dependent upon narrow-wavelength bands are rarely observed. Bands 5–10 nm wide represent a good compromise between the requirements of specificity of absorption, and the high fluences needed to produce most biological effects. Narrow bands of light can be obtained by using (1) monochromatic sources, (2) filters, or (3) monochromators. In general, the cost of apparatus and the difficulties in experimentation also increase in this order (lasers are an exception, being expensive monochromatic sources).

1.3.1. Monochromatic Sources

Use of a monochromatic source is clearly the easiest way to obtain a narrow band of radiation, but it is very limiting, since the wavelength cannot be changed.

The most monochromatic source is the laser, which typically provides a single-wavelength emission that is only ~0.01 nm in width (although certain lasers, such as the pulsed nitrogen laser, produce a number of closely spaced wavelengths of light). As noted in Section 1.1., it is the high intensity available from lasers that makes them useful for biological work, rather than their extreme monochromaticity.

The germicidal lamp emits up to 85% of its radiation in a single mercury resonance line at 254 nm, thus making this virtually a monochromatic source. The BLB black-light lamp also can be considered to be virtually monochromatic, since its half-maximum bandwidth is only about 40 nm (Fig. 1-3).

1.3.2. Filters[6]

Filters provide a simple way to limit a broad spectrum. They may be classified as cutoff filters that pass all radiation on one side of a certain wavelength and none on the other side, or band-pass filters that exclude wavelengths on both sides of a transmitted band.

Glass filters of the cutoff type are available throughout the UV and visible regions, but only one far-UV glass band-pass filter is available, and it includes the near-UV region (Corning 9863 filter). Glass band-pass filters can result in a reasonable narrow band pass, although the peak transmittance of band-pass filters or combinations of filters is generally lower the narrower the band pass. Some very useful combinations of glass filters with high transmission are shown in Fig. 1-8.

One should note that transparent materials typically reflect about 4% of the radiation incident on their surfaces. Since a glass filter has two surfaces, one expects about an 8% loss of transmitted light through reflection; this explains why the transmittance curves of glass filters level off at about 92%. It should also be noted that manufacturers' transmission curves are often not reliable; one should check the transmission of a commercial filter in a spectrophotometer if it is possible. In the far-UV region, many common materials serve as good cutoff filters (Fig. 1-9). Note the extremely sharp cutoff of Mylar polyester film at about 300 nm; this material is very useful in biological work for eliminating inactivating wavelengths that are absorbed by protein or nucleic acid.

Chemical filters, usually aqueous solutions, may be quite effective as band-pass filters.[2,9] Two of the most useful chemical filters in the UV region are aqueous solutions of copper sulfate (half-maximum transmission ~330–550 nm) and nickelous-cobaltous sulfate (half-maximum transmission ~235–330 nm).[1] Chemical filters are usually good absorbers of infrared radiation, since 10 cm of water will absorb virtually all radiation above 1200 nm. The major problem with most chemical filters is that they decompose rapidly upon irradiation. Combinations of chemical and glass filters are also very useful.[4]

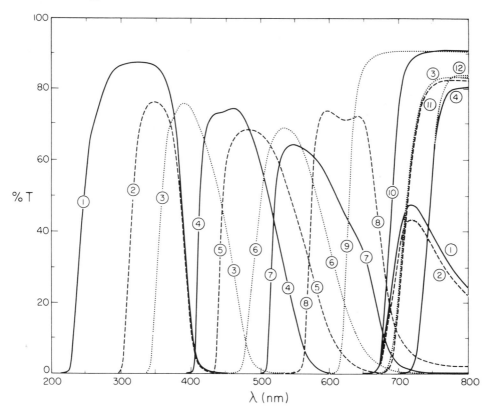

Fig. 1-8. Percent transmittance of some high-transmission combinations of glass filters[8]. (1) Corning 9863; (2) Corning 9863 + 0160; (3) Kodak Wratten 39 + Corning 7380; (4) Corning 5031 + 3391; (5) Schott BG-14 + Corning 9788 + 3387; (6) Corning 9788 + 3385; (7) Schott BG-38 + Kodak infrared cutoff, No. 301 + Corning 3486; (8) Kodak No. 301 + Corning 3480; (9) Schott RG630; (10) Schott RG695; (11) Kodak 39 + Schott RG695; (12) Corning 5031 + Schott RG695.

Fig. 1-9. Transmittance of some common materials that may serve as cutoff filters in the UV region (see also Appendix A of reference 1). (A) window glass, 2.5 mm; (B) Pyrex #774, 1 mm; (C) Pyrex #9741, 1 mm; (D) clear fused quartz, 1 cm; (E) distilled water, 15 cm; (F) polystyrene film, 0.17 mm initial; (G) polystyrene film, 0.17 mm, after 150 h exposure to S-1 lamp at 15 cm distance; (H) Mylar (50A), 0.13 mm. (Courtesy of GTE-Sylvania Lighting Center.)

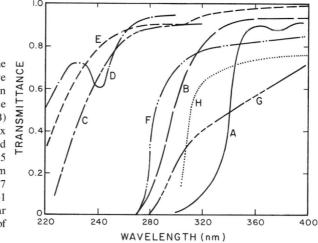

Interference filters are the best band-pass filters available, short of a monochromator. They are expensive, but a set of six or eight will cost much less than a good monochromator. In an interference filter, radiation undergoes multiple reflection between semitransparent surfaces that are very close together, such that multiply-reflected radiation of the desired wavelength emerges from the filter in phase with radiation directly transmitted by the filter. The desired wavelength thus passes straight through the filter, while other wavelengths interfere destructively, and leave the filter at various angles. Although very narrow bandwidths are available (as low as 5–10 nm), the peak transmission, as with glass filters, decreases as the bandwidth decreases. Interference filters have the additional advantage that they are easy to handle (compared with a monochromator), and their characteristics do not alter appreciably with use (as do those of plastic and chemical filters).

Neutral-density filters provide the best way to lower the intensity of a beam without changing its spectral composition or geometry. Such filters are available for the visible or the UV regions, and may provide intensity reduction by factors of 2–100. Other ways of altering the intensity of a beam of light, such as moving the source farther from the sample or changing the slit widths on a monochromator, usually alter the geometry and/or spectral composition of the beam, and are thus less desirable. (The fluence rate falls off as the inverse square of the distance from a point source; however, a line source—such as a fluorescent lamp—will approximate a point source at distances greater than twice the length of the source.) Wire-mesh screens make good neutral-density filters, but they create intensity variations across the sample that may not be desirable. Oriel[6] makes a variable neutral filter; although it costs more than a set of half a dozen individual filters, its versatility and convenience are assets.

1.3.3. Monochromators[1,4,10]

The most versatile and sophisticated way to provide narrow bands of light in different regions of the spectrum is to use a diffraction-grating or prism monochromator. It is also often the most expensive and difficult way.

Monochromators have a dispersing element, usually a prism or a diffraction grating, that breaks a beam of white light down into its colored components (Fig. 1-10). By rotating the dispersing element, different colors can be made to fall upon the exit slit of the monochromator. An important feature of a monochromator is the *reciprocal linear dispersion* at the exit slit, which indicates the spread of wavelengths that one will have for a given slit width. Monochromator slits are typically set at widths from 1 to 5 mm, and a typical reciprocal linear dispersion would be about 3 nm/mm. The radiant flux transmitted by a monochromator is a function of the area of the dispersing element; thus, all high-power monochromators have large prisms or diffraction gratings. For a monochromator in which the distance from the entrance slit to the dispersing element is the same as that from the dispersing element to the exit slit (the usual case, but not that shown in Fig. 1-10), a *doubling of slit width (both entrance and exit) will double the bandwidth*

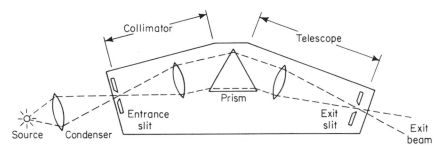

Fig. 1-10. Schematic diagram of a prism monochromator. In the instrument shown, the collimator lens has a shorter focal length than the telescope lens.

and quadruple the transmitted power, provided that the monochromator is optimally illuminated, and the source emits a continuous spectrum.

Prism monochromators are the classical type, but they have many disadvantages. They typically contain lenses, whose focal lengths change with wavelength, and this means that for each wavelength used all of the lenses inside the monochromator, as well as the prism, must be set to a new position. In addition, the linear dispersion changes rapidly as a function of wavelength.

Normal-incidence *grating monochromators* with reflection optics (focal length independent of wavelength) are the most useful machines for biological work. Such monochromators do not require internal adjustments as the wavelength is changed (except for rotating the diffraction grating), and the linear dispersion is virtually constant with wavelength. One of the most useful of such instruments is the Bausch & Lomb 33-86-45-49, which has a 10 × 10 cm grating (high power transmission) with 1200 grooves/mm (low reciprocal linear dispersion). Used in conjunction with a high-pressure mercury or xenon arc, it provides very high power transmission with narrow bandwidths. (One should note that the so-called "high-intensity" monochromators of Bausch & Lomb actually provide lower power transmission than this machine.)

High-intensity monochromators are usually rather expensive, and should only be used if one cannot use simpler apparatus. A number of cheaper medium-power monochromators are available.[6]

The operation of any monochromator requires some care. One should avoid letting more light enter the monochromator than is necessary; this implies using filters, or even a prism, between the source and entrance slit of the monochromator, and keeping the monochromator shutter (placed in front of the entrance slit) closed as much as possible. Most monochromators have a lens at the exit slit that focuses an image of the dispersing element in some plane beyond the exit slit. The light in this image should show a *uniform intensity;* if it does not, adjustments should be made in the optics of the system; such a nonuniform field indicates that the monochromator is not being illuminated optimally. Finally, one should note that optical elements will retain their properties for a long time if they are kept enclosed or covered so that dust cannot reach them. Tobacco smoke will rapidly produce a film on such surfaces that will decrease the transmission, especially in the UV region. The optics of high-intensity water-cooled sources may develop mineral deposits, which will also rapidly decrease transmission; these can usually be removed by mild acid washing.

1.4. MEASURING THE RADIATION ENERGY[1–4]

1.4.1. Terminology[1,11]

In addition to the SI units used throughout this book (Table 1-1), a variety of *dosimetry* terms are also used.

The most common dosimetry term is *energy fluence* (F), which is the *amount of energy crossing a unit area normal to the direction of propagation.** It is measured ideally in "joules per square meter (J/m^2)." The unit "ergs per square millimeter ($ergs/mm^2$)," used extensively in the past, is now discouraged ($1\ J/m^2 = 10\ ergs/mm^2$).

The *energy fluence rate* is the *power crossing a unit area normal to the direction of propagation*, or the *energy per unit area per unit time* ($J\ m^{-2}s^{-1} = W \cdot m^{-2}$).

If the light quantity is expressed in terms of numbers of photons, one then uses the expressions *photon fluence* and *photon fluence rate*. When the term "fluence" is used alone, it usually implies energy fluence. The reader is reminded that the energy (E, in joules) of a photon is given by

$$E = h\nu = hc/\lambda$$

where h is Planck's constant ($6.624 \times 10^{-34} J \cdot s$), ν is the frequency (per second), λ is the wavelength (m), and c is the velocity of light ($3 \times 10^8\ m \cdot s^{-1}$).

The terms *absorbed dose* and *absorbed dose rate* pertain to the energy and power, respectively, absorbed by the irradiated object. While these are common terms in dealing with ionizing radiation, they are rarely used in photobiology.[11]

All of the above terms pertain to the object being irradiated. Other terms pertain to a source of light. The *intensity* of a source is the power emitted per unit solid angle, and the *radiance* of a source is the intensity per unit area of the source. The *brightness* is the radiance expressed in terms of human visual sensitivity. Thus, a 1000-W light blub (frosted) is very intense but not very bright

---

*The term *fluence* (originally invented by the International Commission of Radiation Units and Measurements for ionizing radiations) has been objected to by many persons because its Latin root (fluens = flowing) seems to suggest a time rate of—rather than an accumulated—energy passage. The consequent tendency to use the term where flux per unit area is intended represents an entrapment which a better name might avoid. Unfortunately, no other internationally agreed upon name for this quantity now exists. Readers should be alert, however, to a possible change of nomenclature in the future. The older literature uses "*D*" for "dose" (see reference 11).

(or radiant), while the filament of a flashlight bulb is very bright but not very intense. High-radiance sources are needed to provide the small images that must be focused on the entrance slits of monochromators. High-intensity sources are generally required in photobiological work to provide the large fluences usually needed.

1.4.2. The Thermopile[6]

The thermopile (usually 4 to 12 thermocouples in series) is the most fundamental device for the measurement of UV, visible, or near-infrared radiation. *All other radiant energy measurement systems are, at some point, calibrated with thermopiles.* When a circuit is composed of two dissimilar metals, a contact potential (voltage) is present at each of the two junctions. If the two junctions are at the same temperature, the potentials are equal and opposite, and no current flows. However, if one of the junctions is covered with carbon black and is then illuminated, the carbon black absorbs all the incident radiation and converts it to heat. This causes the temperature of that junction to rise, which increases the voltage across the junction, thus leading to a current flow through the circuit. Such a circuit is called a thermocouple. Thermopiles contain a number of thermocouples, alternate junctions being exposed to the light, and the others shielded from the light. The voltage produced is very low (of the order of 0.1 μV), and must be measured by a very sensitive microvoltmeter.

Thermopiles measure the fluence rate in a parallel beam. If the beam is highly convergent or divergent, and is considerably larger than the thermopile surface, then an error proportional to the cosine of the angle of incidence will be introduced, since the fluence includes all rays crossing a unit area *normal* to the direction of propagation of the ray.

Thermopiles are rather fragile, and they measure any source of heat, so one must be careful not to stand too close to a thermopile during use, or to have other heat sources nearby. In spite of these difficulties, thermopiles are often used for routine measurements. Their great virtue is that *the response is independent of wavelength,* and this is one reason why they are used as standards for all other measurement devices. They are particularly useful in measuring the output of a monochromator, which is usually in a small beam and may on different occasions involve widely differing bandwidths.

Thermopile calibration is essential for absolute dosimetry. This can be done with the classical 50-W carbon-filament standard lamp, developed by the National Bureau of Standards. However, for most photobiological work, where high fluence rates are used, it is better to use one of the new standards, such as the 1000-W tungsten-filament lamp. Both lamps are sold by Eppley.[6] Complete instructions come with these lamps. Calibrations should be done with the greatest care and, with normal use, should be repeated about every three years.

Yellow Springs Instrument Co.[6] makes a complete radiometer measuring system that consists of a breaker amplifier and a small thermistor probe (a thermistor is a semiconductor whose electrical resistance is very sensitive to temperature). The system costs several hundred dollars, is a little less sensitive than a thermopile, but can be used in confined spaces because of the small probe (0.9 × 1.2 × 3.8 cm).

1.4.3. Electronic Photocells

Photovoltaic cells make very handy dosimeters for particular sources of light, such as a germicidal lamp or a fluorescent lamp. In a photovoltaic cell, such as the selenium barrier-layer type, light falling on a metal-coated semiconductor releases electrons from the semiconductor into the metal coating, and this sets up a current through an external circuit that connects to the other side of the semiconductor. The device is completely self-powered and, at the usual light fluxes, produces enough current to actuate a microammeter. Like all electronic devices, photovoltaic cells must be calibrated against a thermopile for each source of light, or each wavelength being measured. Jagger[12] describes a photovoltaic cell that has been adapted for use in the UV region by covering the cell with a Corning 7-54 filter.

Phototubes are electron vacuum tubes in which the cathode is coated with a photoemissive metal such as cesium. When light strikes the cathode, electrons are emitted, and are then accelerated by an electric field to the anode, thus producing a current in an external circuit that connects the anode and cathode. This current may then be amplified electronically. For low fluence rates, often encountered when working with a monochromator, the output of a phototube becomes too small to be conveniently amplified, and one must use a photomultiplier tube.

Photomultiplier tubes produce most of the needed amplification in the tube itself. In addition to a photosensitive cathode, a photomultiplier tube has several (usually 5 to 10) anode–cathodes *(dynodes)*. Electrons emitted by the cathode are accelerated by a voltage gradient to the first dynode, where each bombarding electron will cause the emission of several electrons, which then travel through a voltage gradient to a second dynode, etc. Amplification by a factor of a million is not unusual with a 10-stage photomultiplier tube. For use below 300 nm, phototubes and photomultipliers must of course have quartz windows.

Although thermopiles are the most accurate detectors one can use, they are rather fragile, are sensitive to temperature, and may require several minutes for the taking of a measurement. Photomultipliers are more rugged and easier to use, and measurements may be made very quickly, but they must be *calibrated for every wavelength* to be measured.

1.4.4. Chemical Systems (Actinometers) and Biological Systems

The physical devices discussed above usually measure the *fluence rate in a parallel beam*. Chemical and biological systems do the same, but they may also provide a direct measure of the *average fluence within the sample,* regardless of divergence or convergence of the beam, provided that the dosimeter solution has the same absorption as the experimental sample. Chemical and biological dosimeters are particularly useful where the fluence rate fluctuates, the radiation field is not uniform, or the sample has an unusual geometry or a geometry that changes with time (e.g., by sampling).

Actinometry is the measurement of radiation by the change produced in a chemical, and is usually done with aqueous solutions.[1,4] If one uses an actinometer solution that absorbs the radiation very effectively, e.g., within the first millimeter, one obtains a measure of the fluence at the sample surface, which then, by dividing by the time, may be readily converted into fluence rate. On the other hand, most biological samples transmit a good deal of the light incident upon the irradiation cuvet,* and one can arrange to have the actinometer concentrations such that the absorption by the actinometer will be the same as that by the sample; one then obtains a measure of the *fluence received by individual cells,* which again may be easily converted into *fluence rate for individual cells.* The best liquid actinometer for UV work is the potassium ferrioxalate type. Its advantages are; (1) the actinometer solution absorbs light completely throughout the UV region; (2) the quantum yield (Section 1.5.2.) is practically constant; (3) the yield is independent of fluence rate and virtually independent of temperature; and (4) the quantum yield is not critically dependent on the composition of the solutions. Ferrioxalate actinometry must be conducted under red or yellow light, since the actinometer is sensitive well into the visible region. Purified crystals of potassium ferrioxalate for actinometry may now be purchased from Duke Standards Co.[6]

Liquid-phase actinometers for use in the visible range usually depend on the measurement of oxygen uptake by dye solutions. Porphyrins are sometimes used. Pheophytin- and pheophorbid-containing actinometers have a quantum yield independent of wavelength from 440 to 640 nm. One of the difficulties with these systems is maintenance of oxygen saturation.[4]

Biological dosimetry can be relatively easy and quite accurate. As with chemical actinometry, a highly absorbing suspension will provide a measure of fluence at the sample surface, while a transparent suspension will indicate the fluence upon individual cells, regardless of beam geometry. Good substrates for UV work are bacteriophage, preferably those that show exponential inactivation, in which the fluence will be proportional to the logarithm of the percentage survival of the phage. Although workers often obtain different results in different laboratories, the use of a given bacteriophage and a given host in the same laboratory under the same conditions usually provides highly reproducible results. The system must of course be calibrated. Biological dosimetry has a peculiar advantage not shared by any of the other measuring systems, namely, that the biological dosimeter may often be mixed with the biological sample being studied, and both irradiated simultaneously; this provides a highly accurate measure of the average fluence incident on each cell in a suspension.

---

*The vessel that holds the sample is called a ''cuvet'' rather than a ''cell'' to avoid confusion with the living cell.

1.4.5. The Fluence "Seen" by a Cell

When irradiating a monolayer of cells, it is easy to determine the fluence rate "seen" by each cell, which will be simply the fluence rate in the surface plane of the cells. A liquid suspension of cells, however, may absorb some of the light, so that cells near the back of the cuvet receive less radiation than those near the front. Even if the sample is stirred vigorously, the average fluence rate per cell will be lower than the fluence rate at the front surface of the cuvet. A correction for this effect can be made according to a procedure devised by Morowitz (Table 1-2).

One must also remember that the front surface of the front wall of the cuvet will normally reflect about 4% of the incident radiation. The back surface of the front wall, however, will reflect very little, because the index of refraction of water is similar to that of glass. If one makes a measurement behind a water-filled cuvet, then reflection losses at two surfaces must be taken into account. If the cuvet is empty, there will be reflection losses at four surfaces.

Finally, the *light scattering* by suspensions of small cells, such as bacteria, can introduce large errors in estimations of fluence incident upon individual cells. The Morowitz correction, for example, depends upon a measurement of the optical density of the suspension, but, if this is made in the ordinary spectrophotometer, there can be an error as great as a factor of 10, owing to the fact that most of the apparent absorption of the sample is caused by light being scattered by the sample so widely that it completely misses the photoreceiver. Corrections can be made for this effect.[14]

TABLE 1-2.   Incident Energy Correction for Nonscattering Particles in a Dense Suspension[13]

Transmittancy (%)	Optical density	Correction factor[a]
100	0	1.00
90	0.046	0.95
80	0.097	0.90
70	0.155	0.84
60	0.222	0.78
50	0.301	0.72
40	0.398	0.66
30	0.523	0.58
20	0.699	0.50
10	1.000	0.39
5	1.301	0.32

[a]This factor, applied to the energy entering the sample, gives the average energy incident upon particles in the sample, under conditions of *vigorous stirring*. One must also consider apparent absorption by the front wall of the cuvet. For applications to scattering bacterial suspensions, see reference 14.

1.5. DOING THE EXPERIMENT[1,15]

1.5.1. Absorption by the Sample

Absorption of light by a biological system is the fundamental event in all photobiology. A photon produces chemical (and therefore biological) effects only if it is absorbed. The absorbed photon raises a valence electron of a molecule or of part of a molecule (the chromophore) to an excited state. Both the molecule as a whole and the electron are then considered to be excited. The excited molecule may lose energy and go back to the ground state by several pathways, chiefly: (1) by converting its electronic excitational energy into vibrational energy of itself or of other molecules (heat); (2) by emitting a photon (fluorescence or phosphorescence); or (3) by producing a chemical reaction. In the usual biological system (liquid state at room temperature), most of the absorbed energy is dissipated as heat, some of it as fluorescence or phosphorescence, and relatively little as chemical reaction. The chemical reaction may involve a rearrangement of the absorbing molecule to a new stable form, or it may involve a transfer of energy (electronic or vibrational) to another molecule, which then undergoes some stable change in structure. (Further details are given in Chapters 2 and 3.)

Although it seems a very obvious thing, some workers do not take into account the absorption properties of their biological substrate. If, for example, one wishes to produce an effect that is probably mediated by carotenoids, it would not be wise to use a low-wattage incandescent lamp, which emits most of its radiation in the red and infrared regions where the carotenoid does *not* absorb; a blue light would be far more effective.

Absorption effects become very important in estimating the fluence received by a cell or part of a cell. If one is dealing with a thick suspension, absorption by cells near the surface may shield much of the light from cells farther back in the cuvet. This is one reason for stirring cell suspensions whenever possible during irradiation. However, although stirring will ensure that each cell receives the same *average* fluence, this average fluence will be lower than the fluence incident on the cuvet if the suspension absorbs appreciably; in this event, the average fluence can be calculated if one knows the optical density of the sample (Section 1.4.5. and Table 1-2). A more drastic circumstance occurs if one is irradiating a large biological object, such as an insect egg, in which light may not even reach the desired target (e.g., the DNA) because of extensive shielding by the surface layers of the cell.

Absorption problems are greatest in the far-UV region, where proteins and nucleic acids absorb very strongly. A bacterial cell (e.g., *Escherichia coli*) is only about 1 μm in diameter, yet it absorbs about 25% of radiation at 254 nm. In thicker tissues, one can expect 254 nm radiation to be attenuated to 10% of its initial intensity after passing through about 10 μm of tissue. Thus, experiments in the far-UV region are usually done only with small microorganisms; work with anything larger than a yeast cell requires a sophisticated approach, and may yield results that are difficult to analyze.

In the near-UV and visible regions, absorption is much less of a problem, since the proteins and nucleic acids absorb very little. Absorption properties in the near-UV and visible regions will vary tremendously with the organism. Above 350 nm, an *E. coli* cell is almost transparent, but blue-green algae, which may be as small as *E. coli,* absorb strongly. One may make a few generalizations about expected absorption in the near-UV and visible regions: (1) photosynthetic organisms will always have a lot of chlorophyll and carotenoid and may have a lot of associated photosynthetic pigments, such as xanthophylls or phycobilins, and thus may be expected to absorb strongly throughout this region; (2) all organisms with oxidative metabolism will have moderate concentrations of cytochromes (absorbing in the violet and the red), flavines (absorbing below 500 nm), and quinones (absorbing below 400 nm); (3) many other strongly absorbing molecules, such as the vitamins, are generally relatively unimportant, since they are usually present at low concentration; and (4) animal cells are usually rather transparent, and plant cells usually rather opaque throughout this region.

Scattering of light can seriously affect both the measurement of absorption spectra, and the estimation of average fluence at a depth in a sample. Scattering decreases as the wavelength increases (for bacteria it varies as the inverse first or second power of the wavelength), so that measurements with the UV region are complicated by scattering much more than with red light. The effects of scattering on absorption spectra can be alleviated in a variety of ways,[1] the most effective being the use of a spectrophotometer (such as the Aminco-Chance Duochromator) that has a large end-window photomultiplier placed very close to the sample cuvet such that it intercepts a large fraction of the scattered radiation. In typical spectrophotometers, however, the phototube intercepts very little of the scattered radiation. Thus, *E. coli* absorption is negligible at 500 nm, but one may follow its growth very effectively by measuring in a spectrophotometer the "turbidity," which is almost exclusively due to scattering. Effects of scattering on the estimation of average fluence in a sample has been discussed in Section 1.4.5.

The absorption of a biological system is of crucial importance if one is determining an action spectrum (Section 1.5.4.).

1.5.2. Survival Curves

Many of the biological effects studied by photobiologists involve the inactivation of a cell, a subcellular organelle, a biological property, or a biological molecule. Such inactivation processes may be analyzed in terms of "survival curves," normally plotted with the fractional survival on a logarithmic ordinate, and the fluence on a linear abscissa (see, e.g., Fig. 5-12). A straight line of negative slope on such a graph indicates that the rate of decrease of the number (N) of active particles (molecules, organelles, or cells) with respect to fluence (F) is proportional (k) to the number of active particles remaining at that fluence level:

$$-dN/dF = kN \qquad \text{or} \qquad N/N_0 = e^{-kF}$$

This *exponential inactivation* is known as *single-hit kinetics,* and indicates that a single photon is *capable of* (although it usually does not succeed in) inactivating a biological particle. At a survival of 37%, $kF = 1$ and this means that there is an average of one "hit" per particle. The fluence at this level is called the F_{37}, a number useful in characterizing the rate of inactivation. The kinetics of survival curves, or *target theory,* has been treated in a simple and elegant fashion by Clayton.[15]

The *quantum yield* (Φ) for an event is defined as

$$\Phi = \frac{\text{number of events}}{\text{number of photons absorbed}}$$

Except for chain reactions, the quantum yield is always $\leqslant 1$. It represents the probability that an absorbed photon will produce an event of interest, such as inactivating a molecule or a cell. A typical quantum yield for enzyme inactivation would be 0.01, which means that 100 photons are absorbed by the molecule before the molecule is inactivated. However, it is important to recognize that *only one of these photons does the entire job,* as is reflected in the single-hit inactivation kinetics for most enzymes.

One should not fail to observe that Φ is defined in terms of number of photons *absorbed,* not number of *incident* photons, the quantity usually measured. Furthermore, since photons are the active agents, fluences expressed in J/m^2 must be converted to photons/m^2. The quantum yield may vary with wavelength, temperature, or a variety of other parameters.

The *photosensitivity* can be defined as

$$\sigma = \frac{\text{number of events}}{\text{number of photons incident}}$$

where σ is called the "inactivation cross section." The "absorption cross section" can similarly be defined as

$$s = \frac{\text{number of photons absorbed}}{\text{number of photons incident}}$$

$s = 3.82 \times 10^{-21}\,\epsilon$, where ϵ is the molar extinction coefficient measured in liters mol^{-1} cm^{-1}. It should then be clear that the photosensitivity

$$\sigma = s\Phi$$

or *the true photosensitivity of a system is the product of its absorption cross section and its quantum yield.*

Accurate survival curves will be obtained only under conditions where the sample absorbs very little of the incident light. It is not hard to see that an unstirred sample that is inactivated exponentially at low concentration will show a concave-upward curve at high concentration: cells near the back of the cuvet will receive relatively little of the incident radiation and will thus be inactivated much more slowly; therefore, the inactivation curve will represent an initial rapid inactivation followed by progressively slower and slower inactivation. If the suspension is stirred rapidly, exponential inactivation will be observed, but the slope will be shallower than at low concentrations, since the average fluence will be lower. For the same reasons, suspensions of cells ideally should be irradiated in a medium that does not absorb the incident light (e.g., for the UV region, use phosphate buffer or a saline solution).

1.5.3. UV Recovery Phenomena

The history of a biological system after UV irradiation can greatly influence the response, since conditions after irradiation may either encourage or discourage cellular systems for repairing UV-induced damage to DNA. Such recovery effects do not appear to operate to any great extent after inactivation by near-UV and visible irradiation. Therefore, this discussion concerns only UV inactivation. These repair phenomena and their mechanisms are treated in detail in Chapter 5.

Photoreactivation is a recovery from far-UV radiation damage that is usually mediated by radiation in the range of 300–500 nm. Undesired photoreactivation is usually a serious problem only if UV-irradiated cells are in a room lighted by sunlight, or if one leaves such systems exposed to fluorescent room light for periods of more than half an hour. Some UV workers equip their laboratories with "gold" fluorescent lamps, which emit only above 500 nm, in order to avoid undesired bacterial photoreactivation.

Dark repair of UV damage has at least two mechanisms (Section 5.4.2.), one of which *(excision–resynthesis repair)* occurs primarily before DNA replication, and the other *(postreplication repair)* after DNA replication. Excision–resynthesis repair tends to be more effective if the cells are kept in a less-than-optimal growth medium for an hour or two after irradiation; the resultant recovery is known as *liquid-holding recovery*. It is avoided by plating or incubating the cells immediately after irradiation in complete growth medium. One should recognize that a great deal of dark repair occurs normally, even with immediate incubation, the only purpose of the precautionary measures noted here being to avoid additional repair that may lead to variable results.

Some strains of bacteria lack both of these dark-repair mechanisms and are thus extremely sensitive to UV radiation (see Fig. 5-12). Such strains are so sensitive that even the small amount of radiation at 313 nm present in the light from fluorescent room lamps may inactivate them in a few minutes. A further hazard is that this inactivation will quickly select for more resistant cells in the population, and may result in loss of the sensitive strain.

It may seem curious that extensive repair phenomena appear to occur mainly for UV-radiation-induced damage. This is undoubtedly because the UV-induced damage usually occurs in DNA, a molecule unique in that it is composed of genes that are generally present in only a single copy, and whose loss or damage can be lethal. In contrast, targets for near-UV and visible light are usually present in the cell in many copies, and are replaceable by further cell metabolism, thus obviating the need for repair. Near-UV and visible light can also damage DNA, but at low efficiency and by indirect mechanisms (see Chapter 4 and Section 15.7.1.).

1.5.4. Action Spectra[1,15]

An action spectrum (see Section 3.7.) is a plot of the reciprocal of the number of *incident* photons required to produce *a given effect* vs. wavelength (i.e., a plot of the photosensitivity = $s\Phi$). Peaks in the action spectrum represent the most efficient wavelengths, which require the fewest incident photons to produce the effect. Since photons are usually much more energetic than is necessary to produce a given chemical effect, the same effect will often be produced by electronic excitations of different energy. For example, photosynthesis is produced by absorption of photons of either red or violet light in chlorophyll, although the photons of violet light have about twice as much energy, and initially produce a different electronic excitation. Consequently, the action spectrum for an effect usually resembles the absorption spectrum of the molecule (the chromophore) that absorbs the radiation responsible for the effect.

To get a good match between an action spectrum and an absorption spectrum requires that many conditions be fulfilled.[1] These include:

1. The fluence–effect curves should be similar at all wavelengths (superimposible if the fluence is multiplied by a constant factor), which implies that the mechanism of action is the same at all wavelengths.
2. The quantum yield should be the same at all wavelengths.
3. The absorption spectrum of the chromophore must be the same *in vivo* and *in vitro*.
4. Absorption and scattering of light within a cell by an inactive substance in front of the chromophore (intracellular screening) must be either negligible or constant at all wavelengths.
5. At no wavelength of interest should a large fraction of the radiation be absorbed by the sample.
6. Reciprocity of time and fluence rate must hold under the conditions of the determination (e.g., if the time is doubled and the fluence rate halved, the effect should be the same).

Since action spectra are based on the effectiveness of photons, it is necessary to make a "quantum correction," which may be easily done by multiplying the energy fluence required at a given wavelength by the ratio of that wavelength to any reference wavelength, since the energy of the photon is inversely proportional to wavelength (Section 1.4.1.).

It should be noted that an action spectrum represents the response of a biological system under a given set of conditions: one may get a different spectrum under different conditions. When describing an action spectrum, one should therefore specify the organism, the effect, and the conditions. For example, one might speak of "an action spectrum for photosynthesis in *Chlorella* at room temperature," rather than "*the* action spectrum for photosynthesis."

Action spectroscopy has been an exceedingly useful tool in photobiology. The technique has been used for about a century in the identification of chromophores for photosynthesis.[16] Action spectra also supplied strong evidence for the crucial genetic role of DNA in biological systems[17] more than a decade before the experiments of Avery and co-workers on transforming principle.

1.6. REFERENCES

1. J. Jagger, *Introduction to Research in Ultraviolet Photobiology*, Prentice–Hall, Englewood Cliffs, N.J. (1967). [This is the primary reference for the present chapter, which is drawn largely from Chapter 2. The Appendixes contain useful material.]
2. L. R. Koller, *Ultraviolet Radiation*, Wiley, New York (1965). [An excellent, elementary, highly readable book on techniques. The tabular and graphical data are most useful.]
3. H. H. Seliger and W. D. McElroy, *Light: Physical and Biological Action*, Academic Press, New York (1965). [This was the first textbook on photobiology. The discussion of equipment and physical techniques (seven appendixes) are rather advanced.]
4. J. G. Calvert and J. N. Pitts, *Photochemistry*, Wiley, New York (1966). [Chapter 7 is an extensive discussion of sources, filters, monochromators, measurement techniques, and experimental setups. Highly recommended.]
5. M. W. Berns, *Biological Microirradiation*, Prentice–Hall, Englewood Cliffs, N.J. (1974).
6. Guide to Scientific Instruments, *Science* **194** (1976) [Complete, up-to-date listing of suppliers of specific items. Published every November.]
7. D. A. LeBuis, J. R. Lorenz, W. J. Eisler, Jr., and R. B. Webb, High-intensity vapor lamp for biological research, *Appl. Microbiol.* **23**, 972–975 (1972).
8. W. G. Herkstroeter, A series of sharp-cut and band-pass glass filters for the range 252–800 nanometers, *Mol. Photochem.* **4**, 551–557 (1972).
9. M. Kasha, Transmission filters for the ultraviolet, *J. Opt. Soc. Am.* **38**, 929–934 (1948).
10. H. E. Johns and A. M. Rauth, Theory and design of high intensity monochromators for photobiology and photochemistry: Comparison of spectral purity and intensity of different monochromators, *Photochem. Photobiol.* **4**, 673–707 (1965).
11. C. S. Rupert, Dosimetric concepts in photobiology, *Photochem. Photobiol.* **20**, 203–212 (1974).
12. J. Jagger, A small and inexpensive ultraviolet dose-rate meter useful in biological experiments, *Radiat. Res.* **14**, 394–403 (1961).
13. H. J. Morowitz, Absorption effects in volume irradiation of microorganisms, *Science* **111**, 229–230 (1950).
14. J. Jagger, T. Fossum, and S. McCaul, Volume irradiation of microorganisms: errors involved in the estimation of average fluence per cell, *Photochem. Photobiol.* **21**, 379–382 (1975).
15. R. K. Clayton, *Light and Living Matter.* Vol. 1: *The Physical Part*, McGraw–Hill, New York (1970); Vol. 2: *The Biological Part* (1971).
16. C. S. French and V. M. K. Young, The absorption, action, and fluorescence spectra of photosynthetic pigments in living cells and in solutions, in: *Radiation Biology* (A. Hollaender, ed.), Vol. 3, Chap. 6, McGraw–Hill, New York (1956).
17. F. L. Gates, A study of the bactericidal action of ultraviolet light. III. The absorption of ultraviolet light by bacteria, *J. Gen. Physiol.* **14**, 31–42 (1930).

2

Spectroscopy

2.1. INTRODUCTION

Central to photochemistry is the principle, first stated by Grotthus and Draper in 1818, that *only absorbed light can produce a chemical change*. Thus, molecular photobiology is concerned with the sequence of molecular events that begins with the absorption of light by some part of what is considered to be the biological system, and terminates with some observed biological response. It is,

Angelo A. Lamola • Bell Laboratories, Murray Hill, New Jersey   **Nicholas J. Turro** • Department of Chemistry, Columbia University, New York, New York

therefore, not only reasonable but necessary to begin a course in photobiology with a treatment of the laws governing the absorption of light by molecules.

Spectroscopy refers to that science originally concerned with quantitative measurements of the spectral distribution of light, but which now also encompasses much of what is learned from interpretations of such measurements. For our purposes, spectroscopy can be taken to refer to that science concerned with the measurement of light to gain an understanding of the interactions of light with molecular systems, the structures of the various electronic states of molecules, and the relaxation processes they undergo.

2.2. PRELIMINARIES: ELECTROMAGNETIC RADIATION AND THE NATURE OF LIGHT

Electromagnetic radiation is that form of energy that can be transmitted through a vacuum, and which, when present in sufficient quantity, can be described as transverse electromagnetic waves, i.e., a series of rapid alterations of electric and magnetic fields perpendicular to each other, and both transverse to the direction of propagation. As a wave, the radiation can be characterized by its wavelength λ or its frequency ν, which are related by the simple equation

$$\lambda\nu = c \qquad (2\text{-}1)$$

where c is the propagation velocity (speed of light) that has the value 3×10^8 m s^{-1}. Electromagnetic radiation of a single wavelength is called monochromatic.

Electromagnetic radiation is classified according to various wavelength ranges. The basis for classification was at first empirical, and arose from the fact that very different methods for generation and detection are required for radiation in the different wavelength ranges. One classification scheme is given in Table 2-1.

What is commonly called light is that radiation which elicits a visual sensation in man; this radiation is in the wavelength range of 380–800 nm. More specifically this wavelength range is called the *visible* portion of the electromagnetic spectrum. Photobiology is concerned predominantly with radiation of wavelengths from 200 to 800 nm, which is capable of causing chemical changes when absorbed by molecules. Generally, radiation of longer wavelengths is ineffective unless it is very intense. Radiation of much shorter wavelengths generally ionizes molecules indiscriminantly, and is referred to as ionizing radiation (e.g., X- and γ-radiation).

The description of light as an electromagnetic wave is not adequate to explain phenomena such as the absorption and emission of light by molecules. The central notion that is required is that electromagnetic radiation is quantized, i.e., it is composed of discrete packets of energy called photons. The wave description of light works for phenomena in which high numbers of photons are involved, e.g., in classical optics. However, light must be considered as individ-

**TABLE 2-1. The Electromagnetic
Spectrum**

Range	Wavelength
γ rays	10^{-4} to 10^{-1} nm
X rays	10^{-2} to 10 nm
Vacuum ultraviolet	10 to 200 nm
Ultraviolet	200 to 300 nm
Near-ultraviolet	300 to 380 nm
Visible	380 to 800 nm
Infrared	0.8 to 1000 μm
Microwaves	1 mm to 100 cm

ual photons in addressing phenomena that involve systems, such as molecules, which themselves must be treated within the quantum theory.

As first proposed by Einstein, the energy per quantum, E, is related to the wavelength and frequency of the radiation by the equation

$$E = h\nu = hc/\lambda \tag{2-2}$$

where h is Planck's constant (6.63×10^{-34} J s). Thus, the energy value of a photon is lower at the longer wavelengths.

The energy content of a mole of photons (6×10^{23} photons, sometimes called an einstein) is then (6×10^{23})$hc\lambda^{-1}$ or, in units of kcal/mol, $E = 2.86 \times 10^4/\lambda(nm)$ (1 cal = 4.184 J). Energies of photons of various wavelengths in the range of interest here are listed in Table 2-2. A "red" photon of wavelength 700 nm has about half the energy content of a "violet" one of wavelength 400 nm. The energy of the photon is of utmost importance in some instances. For example, one system may absorb violet light but be completely transparent to red light; another system may absorb both wavelengths. If the latter is the case, then about

TABLE 2-2. The Ultraviolet and Visible-Light Spectrum

λ (nm)	ν (s^{-1})	Color	Energy (kcal/mol)	(eV)
200	1.5×10^{15}	Invisible ultraviolet	143	6.2
250	1.2×10^{15}	Invisible ultraviolet	114	4.9
300	1.0×10^{15}	Invisible ultraviolet	102	4.4
380	7.9×10^{14}	Edge of visible ultraviolet	76	3.3
400	7.5×10^{14}	Violet	72	3.1
470	6.4×10^{14}	Blue	60	2.6
530	5.8×10^{14}	Green	54	2.3
580	5.2×10^{14}	Yellow	49	2.1
620	4.9×10^{14}	Orange	46	2.0
700	4.3×10^{14}	Red	41	1.8

twice as many "red" photons must be absorbed than "violet" ones to supply equal amounts of energy to the system. This may be an important consideration, e.g., if one is concerned with heating with sunlight. However, in consideration of photomolecular events, while the energy of the photon may dictate whether or not it is absorbed at all, the energy of the absorbed photon may not be critical to the subsequent occurrence of the event of interest. In this case, the number of photomolecular events is proportional to the number of photons absorbed, and not to the total light energy absorbed. It is important to understand this difference from the beginning, and it is important that one be able to convert from light intensity measurements made with an energy-measuring device (such as a thermopile) to units based upon numbers of photons.

2.3. LIGHT ABSORPTION[1-3]

2.3.1. Empirical Considerations

Bouguer (1729) and Lambert (1760) discovered that the fraction of incident light absorbed by a medium is proportional to the thickness of the medium traversed, and is independent of the intensity of the light. That is,

$$-dI = aI \ dx \tag{2-3}$$

where I is the intensity of the light at a distance from where it enters the medium and a is the absorption coefficient (extinction coefficient). The Bouguer–Lambert law holds only if: (1) the incident light is monochromatic and collimated, (2) the absorbing medium is homogeneous, and (3) the absorbing centers act independently of each other.

Beer (1852) showed, for many solutions of absorbing solutes in virtually transparent solvents, that the absorption coefficient, a, was proportional to the number or concentration, c, of the solute (Beer's law). Incorporation of this into the Bouguer–Lambert law gives

$$-dI = -acI \ dx \tag{2-4}$$

which upon integration gives

$$\ln(I_0/I) = acx \tag{2-5}$$

where I_0 and I are the incident and transmitted intensities, respectively, and x is the thickness of the absorbing medium. For solutions this law is usually expressed in terms of logarithms to the base 10 such as

$$\log_{10}(I_0/I) = \epsilon cl \tag{2-6}$$

where, if c is the concentration in moles/liter, and l is the optical pathlength in centimeters, the constant ϵ, in units of liter mol^{-1} cm^{-1} is called the *molar extinction coefficient*. Equation (2-7) defines the *absorbance A*, frequently called the optical density (O.D.):

$$A = \log_{10}(I_0/I) = \text{O.D.} = -\log_{10} T \qquad (2\text{-}7)$$

The fraction of incident radiation transmitted is called the transmittance, T.

It should be pointed out that the absorbance A, which specifically refers to a reduction in transmitted light because of absorption, is not identical to O.D. or $-\log_{10} T$, if the Bouguer–Lambert conditions are not met. Reflection or scattering of incident light contributes to the reduction of transmitted light, insofar as fewer incident photons reach a detector on the other side of the medium. Transmittance (T) and optical density (O.D.) are defined by Eq. (2-7) without specification as to the reason for loss of transmitted intensity.

Exceptions to Beer's law are sufficiently frequent that one should assume that it does not apply unless experimentally verified. Molecular aggregation and complexation, partial ionization (dissociation), and other phenomena that are concentration dependent can lead to deviations from Beer's law. Luminescence of the sample can lead to deviations from the Lambert–Bouguer–Beer law. The law is frequently violated when high light intensities, such as obtained from a laser, are involved due to the operation of two-photon processes of various kinds and other "nonlinear" effects.

2.3.2. Absorption Spectra and Spectrophotometers[4,5]

Absorption spectra can be represented in various ways. Although there is no standard representation, the most common is to plot wavelength λ along the abscissa and the absorbance A along the ordinate, with a description of the sample included in the text (see Fig. 2-1).

The choice of wavelength as a parameter is convenient because grating instruments disperse the spectrum linearly with respect to wavelength. The energy, of course, decreases with increasing wavelength and is proportional to the frequency and inversely proportional to the wavelength. Most instruments plot either transmittance (I/I_0) or optical density [$-\log(I_0/I)$]. When the Lambert–Bouguer–Beer law holds, optical density is identical to absorbance, and it is valid to plot the ordinate as ϵ, the extinction coefficient. When ϵ (or A) varies over a large range, the ordinate is usually plotted as $\log \epsilon$ (or $\log A$).

The measurement of an absorption spectrum consists of obtaining a number of closely spaced wavelength values of I and I_0. Ideally I_0 is obtained by using a reference for zero absorbance that is identical to the sample to be examined, except for the presence of the absorbing material of interest. Double-beam spectrophotometers allow automatic scanning and recording of spectra. In such instruments, which are the most common in use now, light from a source with a continuous spectrum (all wavelengths in the range are present) is collimated and

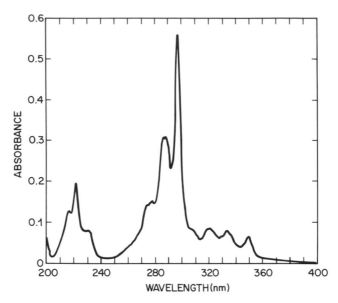

Fig. 2-1. The absorption spectrum of 1,2,5,6-dibenzanthracene in ethanol. The path length of the solution was 1 cm and the concentration of the dibenzanthracene was 4 μM. Thus, the molar extinction coefficient at 297 nm, the maximum of the most intense band, is 1.4×10^5. The weak absorption between 360 and 400 nm is genuine, and represents a molar extinction coefficient on the order of 5×10^2. At least four transitions can be observed in this spectrum. They represent promotions from the ground state to four excited singlet states (see Sections 2.3.5. and 2.3.6.). The onsets of the transitions are at about 400, 355, 300, and 235 nm, respectively (see Section 2.3.7.).

split into two beams, one of which passes through the sample and the other through the reference sample. The light is monochromated either before or after passing through the samples. The beam intensities are translated into electrical signals through the use of some photoelectric detector such as a photomultiplier and then electronically processed and fed to a strip chart recorder that is driven in synchrony with the wavelength drive of the monochromator. A schematic of such an instrument is shown in Fig. 2-2.

2.3.3. Absorption Spectra and Molecular Structure[1-4]

The color of a pure organic compound was recognized by the earliest organic chemists to be an important characteristic property. The absorption spectrum, an objective criterion, replaced color once reliable instruments became available. Absorption spectra in the UV range, of course, allowed for characterization of many colorless compounds, and so measurements of such spectra became part of the chemist's daily activities. Once a sufficient library of spectra was obtained, empirical rules were discovered that related the molecular structures of organic compounds to their absorption spectra. It was quickly realized that saturated compounds (e.g., cyclohexane) do not absorb light in the

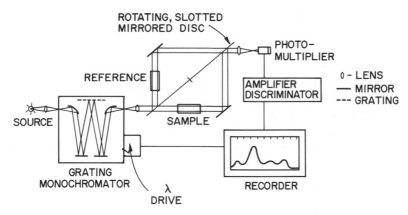

Fig. 2-2. A schematic representation of a dual-beam recording spectrophotometer as described in Section 2.3.2.

visible and UV ranges, but compounds containing unsaturated groups or conjugated combinations of these groups always absorb UV radiation and sometimes visible light. Such an unsaturated molecular center is referred to as a *chromophore*. The chromophore comprises all the conjugated multiply bonded (unsaturated) atoms in the molecule. Generally the higher the number of conjugated unsaturated centers in the chromophore the further toward longer wavelengths does the compound absorb. *Auxochromes* are groups, such as —OH, whose presence in a saturated molecule do not endow the molecule with the ability to absorb visible or UV light, but bring about large changes in the absorption spectra when bound to chromophores.

2.3.4. Absorption Spectra and Molecular Electronic States[1,2]

Quantum mechanics tells us that the energy (kinetic plus potential) of the electrons of a molecule with fixed nuclei can only be of certain discrete values corresponding to certain distributions of the electrons about the nuclei. Each of these distributions corresponds to an electronic state of the molecule. At room temperature almost all molecules in equilibrium with the surroundings are found in the electronic state of lowest energy, i.e., the ground state. States of higher energy are referred to as excited electronic states. The energy difference between the ground state and the excited state of lowest energy is usually so large that significant population of the excited state cannot be achieved by heating the molecule. This is because bond dissociations or rearrangements would occur first due to the population of vibrational modes before the population of electronic states could occur.

Promotion of a molecule from its ground electronic state to one or another excited electronic state occurs upon absorption of a photon by the ground-state molecule. For a photon to be absorbed (leading to electronic excitation) there must be an interaction between the molecule and the light. For molecules of interest to photobiologists, the chief interaction is that between the electrons and the electric field of light. Only photons of energies corresponding to the energy differences, ΔE, between the lower and upper electronic states of the molecules can be absorbed. Since ΔE is related inversely to the wavelength λ of the light [Eq. (2-2)], the longer the wavelength of the light that is absorbed the smaller must be E. Thus, the difference between the energy of the ground state and the lowest excited electronic state of an organic molecule containing a chromophore must be much smaller than the corresponding difference in saturated molecules. The molecular orbital model for molecular electronic states easily accounts for this observation.

2.3.5. Molecular Electronic States, Orbitals, and Energy Levels[6–9]

Each bound electronic state of a molecule is characterized by a particular lowest-energy geometry or arrangement of atomic nuclei. The electrons occupy specific regions of space around the nuclei. These regions are called orbitals. Electrons in dissimilar orbitals possess different energies. Each orbital can be occupied by not more than two electrons, and these must have opposite spins (Pauli exclusion principle).

In a molecule the electrons (orbitals) of lowest energy are the inner core, essentially atomic $1s$ orbitals, because they are closest on the average to nuclei. Taking ethylene as an example, the lowest-lying orbitals are the carbon $1s$ orbitals. In molecules containing carbon, nitrogen, or oxygen, the $2s$ and $2p$ electrons of the separated atoms are employed in the interactions that bond the atoms together in the molecule. These bonding electrons occupy molecular orbitals that are associated with two or more nuclei. The electron density in the region between the nuclei shield the nuclear charges, thereby, in concert with other electrons, balancing the repulsive interaction between the nuclei.

In ethylene there are two types of molecular orbitals, called σ and π, which are differentiated on the basis of the symmetry with respect to the internuclear axes (or molecular plane). By analogy with the atomic s and p orbitals, a σ orbital has cylindrical symmetry with respect to the bond axis, and a π orbital is antisymmetric with respect to a 180° rotation about the bond axis (Fig. 2-3). A σ orbital is essentially localized between two nuclei and represents single bonding. The bonds between the carbon and hydrogen atoms in ethylene are effected by electrons in σ orbitals, as is one of the carbon–carbon bonds. The other carbon–carbon bond is a π bond. In molecules with conjugated double bonds (e.g., butadiene, benzene) there are π orbitals that are delocalized over (associated with) several nuclei.

The orbitals that are occupied (each by two electrons) in the ground or lowest electronic state of ethylene have been described so far. In the molecular orbital model there is an antibonding molecular orbital corresponding to each bonding molecular orbital. Thus, for ethylene one considers two antibonding orbitals, π^* and σ^*, where the asterisk (called star) denotes the antibonding character. These antibonding orbitals have a nodal plane perpendicular to the internuclear axis (Fig. 2-3). In such orbitals the electron density in the region

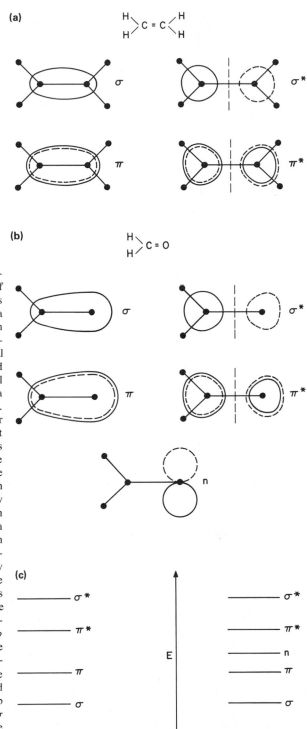

Fig. 2-3. (a) Schematic representation of some molecular orbitals of ethylene. The molecular plane is the plane of the page. Ethylene is a planar molecule (all six atoms in the molecular plane). The carbon–carbon σ orbital has cylindrical symmetry with respect to the bond axis. The corresponding σ^* orbital has the same symmetry, but has a node halfway between the atoms. The σ and σ^* orbitals for the four carbon–hydrogen bonds are not shown. The π orbital nodal plane is in the molecular plane; one lobe (outer solid line) extends above the plane, and the lobe of opposite sign (inner broken line) extends below the plane. The π^* orbital has an additional node halfway between the carbon atoms. (b) In addition to the orbitals that must be considered for ethylene, an essentially atomic orbital on oxygen must be considered for formaldehyde. This n orbital lies in the molecular plane and is perpendicular to the carbon–oxygen bond axis. It is a p atomic orbital with a node at the atom. (c) The energies of an electron placed in each of the ethylene and formaldehyde orbitals pictured in (a) and (b) have the relationship shown. The carbon–hydrogen σ and σ^* energies (not shown) lie below and above, respectively, the levels shown.

between the nuclei is low and, depending on the distribution of other electrons, may not be sufficient to shield the nuclear charges, and a net repulsive interaction may obtain.

In order of the increasing energy of an electron in the orbital, the orbitals of ethylene may be written $2 \times$ (C)$1s$, $4 \times$ (C—H)σ,(C—C)σ,π,π^*,(C—C)σ^*,$4 \times$ (C—H)σ^* (Fig. 2-4a). To generate the ground-state or lowest-energy electronic configuration (distribution) two electrons are placed in each orbital starting with the lowest-energy orbital (Aufbau principle). Ethylene has 16 electrons and its ground-state configuration is $2 \times$ (C)$1s^2$,$4 \times$ (C—H)σ^2,(C—C)σ^2,π^2; i.e., all the bonding orbitals are filled and none of the antibonding orbitals is occupied (the occupancy number of the orbital, which may be 0, 1, or 2, is indicated by a superscript after the orbital). To generate electronic configurations or states of higher energies, one or more electrons are promoted from orbitals occupied in the ground state to an empty orbital(s). In the one-electron molecular orbital model the energies of the orbitals are independent of the configuration or orbital occupancy, i.e., certain electron–electron interactions are neglected. Thus, the energy difference between two electronic states is simply the sum of the energy differences between the orbitals occupied in the two states (with the occupancy number, 1 or 2, taken into account). Furthermore, one has only to consider, of course, those orbitals whose occupancy is different in the two states.

In the process of light absorption, the energy difference between the lower and higher molecular state is equal to the energy of the photon absorbed. When the lower state is the ground state and the molecule contains only low-atomic-number elements, the energies required to promote an electron out of core orbitals or out of σ orbitals, or into σ^* orbitals corresponds to wavelengths below about 220 nm. These energies are generally outside the range of interest of photobiology, and for most purposes such orbitals can be ignored. If this simplification is applied to ethylene, then the only orbitals that need be considered are the π and π^* orbitals. The ground-state configuration for ethylene is simply π^2. The only one-electron excited state that need be considered is $\pi^1\pi^{*1}$. The change (transition) from ground state to excited state may be written in a way that refers only to the electron whose orbit was changed, that is, $\pi \rightarrow \pi^*$ (π to π^*). The excited state may be named by reference to the same orbitals, e.g., in this case, π,π^*.

Unshared electrons in the valence shell have energies comparable to electrons in π orbitals and must be considered in light absorption processes in the wavelength range of interest. Such electrons are referred to as nonbonded electrons or n electrons and the orbitals they occupy are called n orbitals. An example of such electrons is the nonbonded, essentially $2p$, electron pair on oxygen in formaldehyde. In formaldehyde the bonding between carbon and oxygen involves electrons in σ and π orbitals as in ethylene (Fig. 2-3). The ground-state configuration of formaldehyde may be written $\pi^2 n^2$. Now two low-lying excited states must be considered. These are generated by the one-electron transitions $n \rightarrow \pi^*$ and $\pi \rightarrow \pi^*$. Since the n orbital lies higher in energy than does the π orbital (Fig. 2-4b) the transition $n \rightarrow \pi^*$ requires less energy than does the $\pi \rightarrow \pi^*$ transition, i.e., the n,π^* excited state is the lowest-energy excited state.

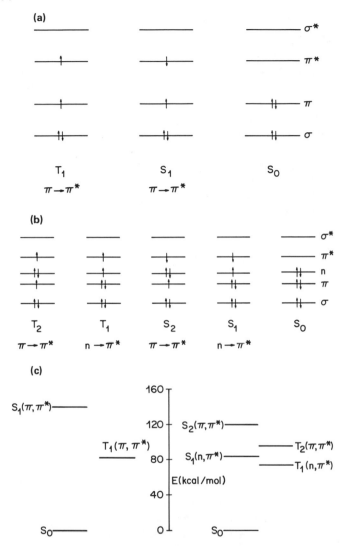

Fig. 2-4. (a) A schematic representation of the configurations of the ground state (S_0), and lowest excited singlet (S_1) and triplet (T_1) states of ethylene. Atomic core electrons, and electrons in carbon–hydrogen bonds are not depicted. (b) Configurations of the ground state and low-lying excited states of formaldehyde. (c) The energies above the ground state for the low-lying excited states of ethylene (left) and formaldehyde (right). This scheme is called a state energy diagram or a Jablonski diagram. The triplet states are displaced horizontally from the singlet states to make the representation of interconversion processes clearer (see Fig. 2-8b).

When all occupied orbitals contain two electrons there are, by the Pauli principle,[10] equal numbers of electrons of opposite spin ($+\frac{1}{2}, -\frac{1}{2}$), or a net zero electron spin. The state multiplicity is given by $2S + 1$, where S is the total electron spin. A state of multiplicity $=1$ ($S = 0$) is a singlet state, and a state of multiplicity $=3$ ($S = 1$) is a triplet state. The ground states of ethylene and formaldehyde are singlet states. In the discussion thus far, the spin of the promoted electron has not been mentioned. If the spin of the promoted electron is conserved on going from the ground state to the excited state, then the latter is also a singlet state. Since the Pauli principle allows either spin direction for electrons in singly occupied orbitals, the one-electron excited states can obtain a total spin of 1, e.g., if the spin of the promoted electron is reversed from the direction it had in the ground state. Thus, the excited configurations can be singlet or triplet spin states (Fig. 2-4a,b). In the one-electron model, singlet and triplet states of the same configuration are degenerate (have equal energy content). However, because there is actually more than one electron in each molecule, one must consider interactions among the electrons. It turns out that the motions of the electrons in singlet and triplet states are correlated differently (as required by the laws of quantum mechanics) such that the electron–electron repulsion energy is lower in the triplet state than it is in the singlet state.[10] Thus, the triplet state always lies lower in energy than the singlet state of the same orbital configuration. This means that the lowest-energy excited state is a triplet state. This is true for all diamagnetic non-metal-containing compounds encountered in biological systems.

In a state energy level diagram the states, represented by horizontal lines, are vertically displaced according to the energy differences between them (ground state set at zero energy). Singlet and triplet states are grouped and displaced horizontally. Such a diagram for the low-lying electronic states of ethylene and formaldehyde is given in Fig. 2-4c. The arrangement of the various states is based on various kinds of spectroscopic data including absorption spectra, emission spectra, and photoelectron spectra. Note that the difference (splitting) in energy between the singlet and triplet π, π^* states is larger than the difference between the singlet and triplet n, π^* states. Singlet–triplet splittings can vary widely, and for many molecules there is more than one triplet level below the lowest-excited singlet level.[11] Furthermore, the configuration of the lowest-lying triplet state need not correspond to that of the lowest-lying singlet state. Molecular electronic states observed spectrally or by some indirect method are often not assigned, i.e., the electron configuration and concomitant properties, are not known. When this is the case, the states are simply numbered S_1, S_2, etc., and T_1, T_2, etc., in order of increasing energy, it being understood that the configurations of S_1 and T_1 may not correspond.

To summarize, organic molecules absorb light in the wavelength range above 200 nm because there exist electronic states for those molecules that lie at energies less than about 140 kcal/mol above the ground states. These excited electronic states are generated by the promotion of an electron from an n or π orbital to a π^* orbital. The energy gap (excitation energy) between the highest-energy n or π orbital and the lowest-energy π^* orbital in conjugated systems decreases with the extent of conjugation. Thus, ethylene absorbs light at 200 nm, benzene at 280 nm, carotene at 450 nm, and some porphyrins at 600 nm. These absorptions (see below) are due to promotions to excited singlet states (zero net electron spin). There is a triplet state corresponding to each of the excited singlet states and lying at lower energies, making the lowest electronic excited state a triplet state. There are both theoretical and many experimental approaches to the identification and location (energy above the ground state) of the low-lying excited states of an organic molecule.

Before proceeding from this discussion it should be pointed out that molecules in excited electronic states can have quite different chemical and physical properties than the ground-state molecules. This is one of the reasons why photochemical reactions generally give different products than do reactions that involve only ground-state molecules. For example, it is easy to understand why the charge distribution and equilibrium geometry of the $S_{n,\pi*}$ state of formaldehyde are different than they are for the ground state. The promotion of an n electron to a π^* orbital weakens the bonding between the carbon and oxygen. Thus, the C—O bond is longer and has a lower stretching frequency in the excited state molecule compared to the ground state. With respect to charge, the oxygen center is considerably more positive and the carbon more negative in the excited state compared with the ground state, because an electron associated with only the oxygen atom in the ground state is associated with both the carbon and oxygen in the excited state.

2.3.6. Optical Transitions between Electronic States and Rules for Light Absorption[2,8,9,12,13]

An optical transition from one electronic state of a molecule to another involves the loss or gain of energy due to the emission or absorption of a photon by the molecule. To satisfy energy conservation, the energy gained or lost by the molecule on account of the transition is equal to the energy of the photon, $\Delta E = h\nu$, i.e., an optical transition is a resonance process. This energy conservation rule can be stated another way: photons of a particular energy (light of a particular wavelength) may be absorbed by a system of molecules only if there exists an excited state of the system that lies higher in energy by an amount equal to the energy of the photons. This energy matching is a necessary but not a sufficient condition for efficient light absorption.

The probability of light absorption, assuming the resonance condition is met, is referred to as the oscillator strength f. The experimental measure of the probability of light absorption (absorption intensity) is related to the molar extinction coefficient, ϵ [Eq. (2-6)]. Because the spectra of polyatomic molecules are broad, i.e., spread over a range of wavelengths (Fig. 2-1), the measure of the oscillator strength is the integrated intensity of the absorption band,

$$f \sim \int_{\bar{\nu}_1}^{\bar{\nu}_2} \epsilon \, d\bar{\nu} \qquad (2\text{-}8)$$

where $\bar{\nu}_1$ and $\bar{\nu}_2$ are the wavenumbers ($\bar{\nu} = 1/\lambda$) between which the absorption band lies.

The oscillator strength may be related to the electron distributions in the two molecular electronic states between which the transition occurs, in that the oscillator strength is proportional to the square of a quantity called the transition moment:

$$f \sim |\mathbf{M}|^2 \tag{2-9}$$

For molecules of interest to biologists, the transition moment is given by†

$$\mathbf{M} = <\Psi_f| e\mathbf{r}|\Psi_i> \tag{2-10}$$

where Ψ_f and Ψ_i are the electronic wavefunctions of the final and initial states, respectively (i.e., the transition is $\Psi_i \rightarrow \Psi_f$), e is the electronic charge, and $\mathbf{r}$ is the position vector ($\mathbf{r} = \mathbf{x} + \mathbf{y} + \mathbf{z}$ in Cartesian coordinates). The transition moment is a measure of the linear displacement of electronic charge in the molecule as a consequence of the electronic transition. The interaction that leads to the absorption of a photon is essentially that between the electric field of the light and the electrons in the molecule. This interaction, and hence the probability of the absorption of a photon, increases with transition moment.[10]

In the one-electron approximation, Ψ_f and Ψ_i, the wavefunctions describing the electrons in the molecule, are replaced by the final and initial "one-electron" orbitals (wavefunctions) occupied by the electron that is promoted. For example, ignoring electron spin, the transition moment for the $n \rightarrow \pi^*$ transition in formaldehyde, referred to in the last section is given by

$$\mathbf{M}_{n,\pi^*} = e <\pi^*|\mathbf{r}|n> \tag{2-11}$$

A further shortcut in estimating $\mathbf{M}$ is provided by the selection rules that can tell for mathematical reasons whether the one-electron transition moment is zero. For example, the *spin-selection rule* requires that the transition moment be zero if the multiplicities of the initial and final states are different, i.e., if one is a singlet state and the other is a triplet state. The change in multiplicity may be associated with a change in the spin state of the promoted electron. The total wavefunction is a product of spatial and spin functions and, e.g., one may express the total wavefunction for an n electron as $n\alpha$, where α is one of the two spin states. Consider an electron in a π^* orbital with the opposite spin state β, i.e., $\pi^*\beta$. For the transition $n\alpha \rightarrow \pi^*\beta$, $\mathbf{M}$ is given by $e <\pi^*\beta|\mathbf{r}|n\alpha>$. This expression is equivalent to $e <\pi^*|\mathbf{r}|n><\beta|\alpha>$. Since $<\beta|\alpha> = 0$‡, $\mathbf{M} = 0$. When $\mathbf{M} = 0$ the transition is said to be forbidden. In this case the radiative (involving a photon) transition is said to be *spin-forbidden*. The physical reason for this may be viewed as follows: The interaction between the electric field of light and an electron is ineffective in changing the spin of the electron; i.e., it is forbidden to go from one spin state to another spin state in a molecule by the absorption of a photon.

The *symmetry-selection rules* tell if the transition moment is zero because

---

†The bracket <> is a mathematical notation that denotes an integration over the proper space.
‡The integrals $<\alpha|\alpha>$ and $<\beta|\beta> = 1$, while $<\alpha|\beta>$ and $<\beta|\alpha> = 0$.[10]

terms cancel out each other due to the symmetries of the wavefunctions. A transition may be spin-allowed but symmetry-forbidden. There are mathematical shortcuts, e.g., group theory, that facilitate the consideration of symmetries.[2,6-9,12,13]

The transition probability may be low, i.e., the absorption may be weak, even if it is not forbidden by spin or symmetry, but because the magnitude of the transition moment is small due to a small overlap in space of the final and initial one-electron wavefunctions. This is the case for n, π^* transitions, and is one reason for the low extinction coefficients ($\epsilon = 10$ to 100) for n, π^* absorption bands.

In reality the magnitude of the transition moment is never zero even if the transition is forbidden by the first-order selection rules mentioned above. This is because the approximate ("one-electron") wavefunctions employed do not include electron–electron interactions, and do not include certain features of the interactions of the electrons with the nuclei. The first-order selection rules are also based on the Born–Oppenheimer approximation, which assumes independent motion of the electrons and nuclei. In fact, the nuclear motions (vibrations of the molecule) can greatly affect the magnitude of the transition moment.

Transitions formally forbidden because of spin considerations have finite strengths because the spin states are not "pure," i.e., singlet states possess some triplet character and triplet states possess some singlet character. This "mixing" of singlet and triplet states is a consequence of *spin–orbit coupling*, the interaction due to the magnetic moment associated with the electron spin, and the magnetic moment due to the orbital motion of the electron. Wavefunctions that take spin–orbit coupling into account can be constructed by mixing together the approximate wavefunctions that ignored this interaction. Consider that the lowest triplet state, T_1, of a molecule has a small admixture of an excited singlet state, S_x. The triplet state may be written $(T_1 + \delta S_x)$, where S_x is the excited singlet state mixed in, and δ is the coefficient of mixing. The transition moment for the transition to this triplet state from the ground state S_0 is given by

$$\mathbf{M}_{S_0 \to T_1} = e <(T_1 + \delta S_x)|\mathbf{r}|S_0> \qquad (2\text{-}12)$$
$$= e <T_1|\mathbf{r}|S_0> + \delta e <S_x|\mathbf{r}|S_0> \qquad (2\text{-}13)$$

The first term is zero because of the spin-selection rule, so that

$$\mathbf{M}_{S_0 \to T_1} = 0 + \delta \mathbf{M}_{S_0 \to S_x} \qquad (2\text{-}14)$$

That is, the $S_0 \to T_1$ transition derives its intensity from the $S_0 \to S_x$ transition. The strength of the $S_0 \to T_0$ transition depends upon δ, the coefficient of mixing, which reflects the spin–orbit coupling, and $\mathbf{M}_{S_0 \to S_x}$, the strength of the transition from which intensity is "borrowed." Spin–orbit coupling is weak for compounds containing elements of low atomic number, but increases dramatically when heavy atoms, e.g., bromine, are incorporated. There are symmetry selection rules that help to find which singlet and triplet states can mix by spin–orbit coupling.

Transitions that are formally symmetry-forbidden are not strictly so because the motions of the electrons are not, in fact, independent of the motions of the nuclei. In a formalism similar to that used above for singlet–triplet mixing, the various electronic states are not of pure symmetry. Mixing of states of different symmetry can be thought to occur because of the vibrations of the molecule; the states are mixed due to what is called vibronic coupling. The symmetry-forbidden transition "borrows" intensity from symmetry-allowed transition(s). There are selection rules that tell which states can mix by vibronic coupling.

2.3.7. Vibrational Energy Levels and Absorption Bandwidths[1,2,11–13]

For each electronic state of a molecule there exists a set of vibrational modes ($3n - 6$ for nonlinear polyatomic molecules, where n is the number of atoms in the molecule). Each of these modes is associated with a set of energy levels whose spacing depends upon the frequency of the vibration. The spacings are small compared to the spacings between electronic levels. Superimposed upon the electronic energy levels, depicted as bold lines in Fig. 2-5, are levels representing the vibrational states (only one vibrational mode is depicted). The total energy of the molecule is the sum of its electronic, vibrational, rotational (connected with the rotations of the molecule) and translational (connected with the linear motion of the molecule) energies. The latter two energies are ignored here, i.e., the total energy is taken to be the electronic energy plus the vibrational energy.

At physiological temperatures most molecules in equilibrium with the surroundings are in the lowest ($v = 0$) vibrational level of the ground electronic state. With the appropriate wavelength of light, the absorption process can promote the molecule to some combination of electronic state and vibrational level. If transitions can occur to a number of vibrational levels of the upper electronic state, the absorption spectrum is broad (spread over energy). Whether or not the individual electronic plus vibrational (vibronic) transitions are observed depends upon the number of vibrational modes that can be excited and the spacings between the vibrational levels, among other things. Most times, the absorption spectrum of biomolecules in solution consists of broad bands without much "structure." These unstructured bands are the envelopes of many overlapping vibronic transitions (Fig. 2-5).

The purely electronic transition is from the $v = 0$ level of the ground state to the $v = 0$ level of the excited state. The 0–0 band, as it is called, is the longest wavelength (lowest energy) vibronic band in the absorption spectrum for the transition. Thus, the electronic energy difference between the ground state and the various excited states observed by absorption spectroscopy is obtained from the red edges or onsets of the various absorption bands.

The oscillator strength (probability) of the electronic transition is distributed among the various vibronic bands [Eq. (2-8)]. The relative strengths of the individual vibronic bands depend upon several factors, among which is the difference in the equilibrium geometry of the molecule in the excited state compared to the ground state. The Franck–Condon principle states that the redistribution of the electrons (orbital changes) occurs much faster than the

Fig. 2-5. (a) Atomic-like electronic
energy levels. The optical transitions are very narrow in bandwidth, i.e., only light of wavelengths corresponding to the
electronic energy difference can be
absorbed. The absorption spectrum is a "line" spectrum, with the
width of the absorption bands, as
shown here, due to limitations of
the spectrometer. (b) Molecular-
like energy level scheme. Superimposed upon the electronic levels
(bold lines) are levels (lighter lines)
depicting vibrational states (only
one vibrational mode is shown). In
addition to the purely electronic
transition (0–0′ band), the ground-
state molecule may absorb light

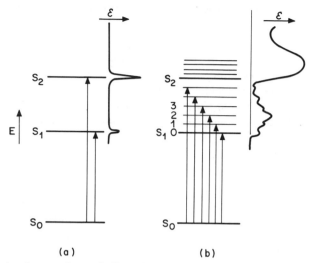

corresponding to promotion of the molecule to an electronically excited state that is also vibrationally
excited, i.e., 0–1′, 0–2′, etc. The individual vibronic transitions may be observed (the spectrum for
$S_0 \rightarrow S_1$), or, in the case where the vibrational spacings are close, only the envelope of them is
seen (the spectrum for $S_0 \rightarrow S_2$).

motions of the nuclei. Thus, the geometry of a molecule immediately before and
after the absorption of a photon is unchanged; light absorption is a "vertical"
process. As exemplified by the case of diatomic molecules (Fig. 2-6), the 0–0
band will be intense if there is little or no difference in the equilibrium geometry
of the two states (and the transition is allowed), but some higher-energy vibronic
transition will be the more intense if there is a significant difference in equilibrium
geometry. The relative probabilities of the individual vibronic transitions are
called the Franck–Condon factors.

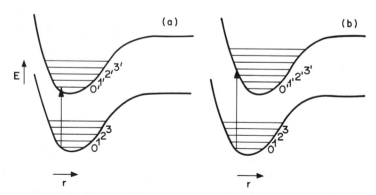

Fig. 2-6. The potential energy of a diatomic molecule as a function of its interatomic distance r for the
ground state and an excited electronic state. In (a) the minimum energy interatomic distance
(equilibrium bond distance) is identical in both electronic states. The most probable vibronic transition
is the 0–0′ band. In (b) the excited state has a larger equilibrium bond distance. As shown here, the
most probable vibronic transition is the 0–3′ band.

2.4. PATHS AND RATES OF RELAXATION OF ELECTRONICALLY EXCITED MOLECULES[13-19]

Consider a molecule in some upper vibrational level, v_x, of some excited electronic state, S_x or T_x, lying in the visible or UV energy range, and that the excited molecule is part of some system in the condensed phase in thermal equilibrium at some not-too-high temperature, e.g., the excited molecule is part of some biological system. It is generally expected that the molecule will undergo vibrational relaxation, i.e., it will lose vibrational energy to its surroundings, and within a short time (10^{-13} to 10^{-12} s) reach the $v = 0$ vibrational level of S_x or T_x. For the molecule to relax further, there must be a loss of electronic energy, i.e., electronic relaxation. Electronic relaxation may occur by radiative means whereby the molecule emits a photon and is demoted to a lower electronic state, the opposite of absorption. As will be shown below, when molecules in biosystems are considered, radiative electronic relaxation is only important for S_1 and under some circumstances T_1. Alternatively, electronic relaxation may occur by radiationless means whereby electronic energy is converted into vibrational energy within the molecule, followed by vibrational relaxation. Both electronic and vibronic relaxation eventually return the molecule to its ground state, and are referred to as photophysical processes. Relaxation may occur by means of chemical processes whereby the molecule undergoes some reaction, bond breaking and/or making. Such processes are called photochemical processes (see Chapter 3).

2.4.1. Radiative Relaxation, Fluorescence, and Phosphorescence[13-18]

Spontaneous emission of a photon from a molecule in its lowest excited singlet state or lowest triplet state demotes the molecule to the ground state. The $S_1 \rightarrow S_0$ emissive process is called *fluorescence*, and the $T_1 \rightarrow S_0$ emissive process is called *phosphorescence*.

The selection rules governing the emission probabilities are the same as those for the absorption probabilities, since the processes are just the reverse of each other. The higher the transition probability, the faster is the rate of emission and the shorter the lifetime of the excited state. If the excited molecules relax solely by photon emission, the excited state lifetime is governed solely by the emission rate (k_{emission}), and this is proportional to the oscillator strength,

$$k_{\text{emission}} = 1/\tau^0 \propto f \qquad (2-15)$$

where τ^0 is called the radiative lifetime. The oscillator strength, f, and τ^0 can be estimated from the longest wavelength transition in the absorption spectrum. Radiative lifetimes for S_1 states range from 10^{-9} s for fully allowed transitions to 10^{-5} s for very weak $S_1 \rightarrow S_0$ transitions. Spin forbiddenness leads to long radiative lifetimes for T_1 states ranging from 10^{-3} to 10^2 s.

Because vibrational relaxation is rapid compared with the emission processes, the emission originates from the $v = 0$ vibrational level of the excited state. Transitions can occur to various vibrational levels of the ground state. The

0–0′ band is, therefore, the highest energy (shortest wavelength) band in the emission spectrum. The absorption and emission spectra corresponding to the same electronic transition overlap at the 0–0′ bands, but then go in opposite directions: the absorption spectrum to shorter wavelengths and the emission spectrum to longer wavelengths. The absorption and emission spectra are very often mirror images of each other (Fig. 2-7). The energy of the excited state

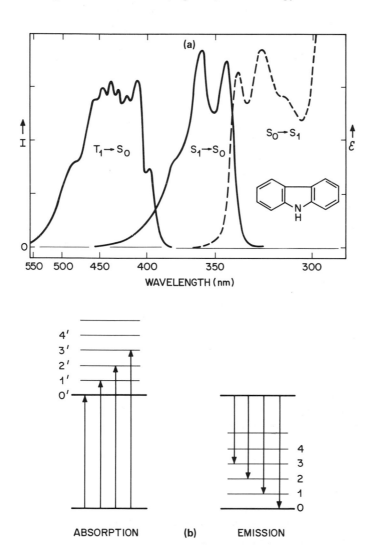

Fig. 2-7. (a) Absorption and emission spectra of carbazole in ethanol. The intensities of the various spectra are not to scale. The spectra are portrayed on an energy scale with the energies expressed in wavelength units. This display best demonstrates the mirror-image relationship between the $S_0 \rightarrow S_1$ absorption spectrum and the fluorescence ($S_1 \rightarrow S_0$) spectrum. Note the position of the phosphorescence spectrum to the red (lower energies) of the fluorescence spectrum. The $S_0 \rightarrow T_1$ absorption, too weak to be observed by ordinary absorption spectrophotometry, is expected to commence near 400 nm. (b) Schematic representation of the spectral relationship between absorption and fluorescence when both transitions originate from vibrationally relaxed states.

above the ground state may be obtained from the blue (highest-energy) edge of the emission spectrum.

2.4.2. Radiationless Relaxation, Internal Conversion, and Intersystem Crossing[17−19]

Internal conversion is that radiationless relaxation process in which a molecule in some excited electronic state converts to a state of the same multiplicity but of lower electronic energy ($S_x \rightsquigarrow S_y$, $T_x \rightsquigarrow T_y$, where $x > y$) and in which the lost electronic energy is converted initially into vibrational energy. *Intersystem crossing* is the analogous relaxation process in which a molecule in some excited electronic state converts to a state of different multiplicity, e.g., $S_1 \rightsquigarrow T_1$, $T_1 \rightsquigarrow S_0$.

That internal conversion occurs between excited electronic states is made evident by the observation that, with few exceptions, fluorescence is observed only from S_1 even when the molecules are initially excited to higher excited singlet states, i.e., the fluorescence spectrum is independent of the wavelength of the light used to excite the molecules. It can also be shown that some molecules do not fluoresce because the internal conversion of S_1 to S_0 is much faster than the fluorescence process.

That intersystem crossing occurs is made evident by the observation that many molecules exhibit phosphorescence upon excitation into excited singlet states. It can also be shown that some molecules excited to T_1 do not phosphoresce because intersystem crossing from T_1 to S_0 is much faster than the phosphorescence process.

In both internal conversion and intersystem crossing the converted electronic energy initially appears as vibrational energy within the molecule, i.e., the total energy of the molecule immediately before and after the process occurs is essentially the same. The rate of the radiationless relaxation depends upon the electronic and vibronic interactions that couple the initial and final states, and upon the difference in their electronic energy. Generally, the greater the electronic energy to be converted, the slower is the process; i.e., the closer in electronic energy the initial and final states are, the faster and more likely it is for radiationless relaxation to occur. For most molecules of concern in photobiology, the electronic energy differences between excited states are much smaller than the energy gap between S_1 or T_1 and S_0. Radiationless relaxation from higher excited states to S_1 or T_1 is usually much faster than that from S_1 or T_1 to S_0. Generally, internal conversion between excited states is fast compared to radiative transitions or photochemical processes from those states. However, it is often the case that radiationless relaxation from S_1 or T_1 to S_0 is comparable in rate to radiative (fluorescence or phosphorescence) and photochemical processes from S_1 or T_1. Spin restrictions for radiationless processes are similar to those for radiative processes. Thus, for the same electronic energy gap, intersystem crossing is usually slower than internal conversion.

In summary, what is generally expected for the unimolecular paths of relaxation of a chromophore excited to S_x ($x > 1$) by the absorption of light are the following (refer to Fig. 2-8). Fast internal conversion and vibrational relaxation processes quickly bring the molecule to its vibrationally equilibrated ($v = 0$) S_1 state. Excluding photochemistry, the relaxation path has three branches at S_1, fluorescence ($S_1 \rightarrow S_0$), internal conversion ($S_1 \rightsquigarrow S_0$), and intersystem crossing ($S_1 \rightsquigarrow T_x$). The relative numbers of molecules that relax by each of these paths depends upon the relative rates of the three competing processes but is generally independent of which singlet state is initially excited. Intersystem crossing from S_1 is the major pathway by which molecular triplet states are populated. The fraction of S_1 states that undergo intersystem crossing (yield of triplet states)

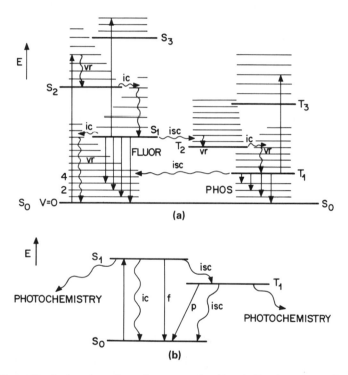

Fig. 2-8. (a) Generalized relaxation scheme for an organic molecule (Section 2.4.2.). Electronic levels and vibrational levels are shown as heavy and light horizontal lines, respectively. Only one vibrational mode is portrayed for each electronic state. Radiative transitions are shown as straight arrows. Radiationless relaxation processes are shown as wavy arrows: vr, vibrational relaxation; ic, internal conversion; isc, intersystem crossing; FLUOR, fluorescence; PHOS, phosphorescence. In this scheme the molecule is excited to S_2 by absorption of a photon. After fast internal conversion and vibrational relaxation the molecule finds itself in S_1 from which it either converts radiationlessly to S_0, fluorescences, or undergoes intersystem crossing, in this case, to T_2. An excited state molecule may absorb a photon and be promoted to a still higher level. Two such absorptions, $S_1 \rightarrow S_3$ and $T_1 \rightarrow T_3$, are shown. Both are spin-allowed and may be strong. They provide a means of detecting the S_1 and T_1 states (see Section 2.6.). (b) For most purposes only the processes shown in this simplified relaxation scheme need be considered in relating kinetics and quantum yields of excited state processes. The various relaxation processes are labeled to correspond to the rate constants considered in Section 2.4.3. Thus, f, p, ic, and isc, refer to fluorescence, phosphorescence, internal conversion, and intersystem crossing, respectively.

varies widely. However, there are many molecules, e.g., aryl ketones like acetophenone, in which fluorescence is slow (S_1 is an n, π^* state) and intersystem crossing is relatively fast for several reasons, and consequently almost all the molecules cross from S_1 to a triplet state. Intersystem crossing may take place from S_1 to T_1 directly, or to some higher triplet state followed by rapid internal conversion to T_1. Excluding photochemistry, the relaxation path has two branches at T_1, phosphorescence ($T_1 \rightarrow S_0$) and intersystem crossing ($T_1 \rightsquigarrow S_0$). Since intersystem crossing from T_1 to S_0 is slow because of spin restrictions, the forbidden radiative process phosphorescence can compete in many instances, and many molecules exhibit phosphorescence when bimolecular reactions that can quench triplet molecules (e.g., quenching by molecular oxygen) are precluded. Bimolecular processes become negligible in very rigid systems, and so phosphorescence is usually observed when fluid samples are frozen at very low temperatures.

2.4.3. Quantum Yields, Lifetimes, and Rates of Relaxation Processes

The quantum yield Φ for some molecular photophysical or photochemical event is defined as

$$\Phi = \frac{\text{number of molecular events of interest}}{\text{number of photons absorbed}} \tag{2-16}$$

Assume that the photons are absorbed only by the molecules of interest and that excitation occurs to the S_1 state.

The quantum yield of fluorescence Φ_f is then

$$\Phi_f = \frac{\text{number of photons emitted as fluorescence}}{\text{number of photons absorbed}} \tag{2-17}$$

which, on the basis of the assumptions made above, is equivalent to the fraction of molecules excited to S_1 that emit fluorescence. The fraction that emits fluorescence is given by $k_f(k_f + {}^1k_{ic} + {}^1k_{isc})^{-1}$, where k_f is the rate of fluorescence, ${}^1k_{ic}$ is the rate of internal conversion from S_1, and ${}^1k_{isc}$ is the rate of intersystem crossing. That is, the fraction of S_1 molecules that fluoresce is given by the rate of fluorescence divided by the rate of relaxation of S_1 by all paths (photochemistry is excluded in these examples). The lifetime of the S_1 state, ${}^1\tau$, is the reciprocal of its decay rate or $(k_f + {}^1k_{ic} + {}^1k_{isc})^{-1}$. The radiative lifetime ${}^1\tau^0$ is k_f^{-1}, i.e., the lifetime of S_1 if it decays only by fluorescence. Thus, $\Phi_f = {}^1\tau/{}^1\tau^0$.

Φ_f can be determined experimentally from measurements of the intensity of the fluorescence and the intensity of the light absorbed by the molecules of interest. ${}^1\tau$ can be determined experimentally by measuring the time dependence of the concentration of molecules in the S_1 state after exciting the system with a very short pulse of light. One measure of the concentration of S_1 is the fluorescence intensity. Thus, $^1\tau$ is the measured fluorescence lifetime. An estimate of k_f and hence $^1\tau^0$ can be obtained from the absorption spectrum.

The quantum yield of phosphorescence Φ_p is

$$\Phi_p = \frac{\text{number of photons emitted as phosphorescence}}{\text{number of photons absorbed}} \qquad (2\text{-}18)$$

Φ_p is also given by the product of the fraction of molecules in S_1 that cross to the triplet state, and the fraction of the T_1 molecules that phosphoresce, i.e.,

$$\Phi_p = [^1k_{\text{isc}}(k_f + {}^1k_{\text{isc}} + {}^1k_{\text{ic}})^{-1}][k_p(k_p + {}^3k_{\text{isc}})^{-1}] \qquad (2\text{-}19)$$

where k_p is the rate of phosphorescence, and $^3k_{\text{isc}}$ is the rate of intersystem crossing from T_1 to S_0. Thus, Φ_p can be small because either $^1k_{\text{isc}}$ or k_p is negligible, and the yield of phosphorescence is not necessarily indicative of the yield of triplet states.

Excluding photochemical and bimolecular quenching processes, the lifetime of T_1, $^3\tau$, is $(k_p + {}^3k_{\text{isc}})^{-1}$. There are several experimental methods for measuring the concentration of triplet states.[11] Thus, it is possible to estimate the rate constants for all the relaxation processes that govern the path of relaxation of an electronically excited molecule.

2.5. INTERACTIONS OF EXCITED MOLECULES[16,20]

2.5.1. Effects of the Microenvironment

Because the electron distribution and geometry of a molecule in an electronically excited state are often very different from those of the ground state, the interactions of the excited molecule with the surrounding molecules (solvent, etc.) are often different than the interactions of the ground-state molecule. These differences cause the spectral properties and relaxation paths, rates, and yields to change when the microenvironment of the molecule is changed. Information about the molecule and/or the microenvironment can be obtained from the way the microenvironment affects the spectroscopic properties of the molecule.

Various solute–solvent interactions can alter spectra (energies) and luminescence yields. Both static and dynamic aspects of the solvation must be considered. Because of the delay between the creation and relaxation of an excited molecule, the solvent can reorganize around it, and the effect of the solvent on the emission spectrum may be different than the effect on the absorption spectrum. Solvent effects upon the spectroscopic properties of molecules is a subject that is large and complex, and cannot be treated here. Suffice it to say that there are many molecules whose absorption and/or fluorescence spectra shift (change in wavelength) significantly, or whose fluorescence yield changes dramatically in response to one or more solvent properties such as polarity, polarizability, hydrogen-bonding capacity, pH, or viscosity. Such fluorescent molecules can be powerful "probes" of the microenvironment especially because the fluorescence can be measured with very high sensitivity.[21]

2.5.2. Bimolecular Reactions of Excited Molecules

Excited molecules may undergo many kinds of reactions with other ground-state molecules. Most of these reaction types are discussed in Chapter 3 as photochemical primary processes. There are many bimolecular reactions of excited molecules that lead directly to the quenching (deactivation) of the excited molecule, or that result in the formation of transient excited bimolecular complexes called *exciplexes*.

One family of bimolecular quenching reactions of excited molecules is electronic energy transfer. In this process the electronic energy is transferred from the originally excited molecule to an acceptor molecule. This important reaction type is discussed separately in Section 2.7.

A simple scheme that describes the formation and reactions of exciplexes is

$$A^* + B \rightleftharpoons (AB)^* \longrightarrow A + B \text{ (quenching)}$$
$$\searrow A + B + \text{fluorescence} \qquad (2\text{-}20)$$
$$\downarrow \quad \text{products}$$
$$\left\{ \begin{array}{c} A^- + B^+ \\ \text{or} \\ A^+ + B^- \end{array} \right\}$$

Molecules A and B do not form a complex when both molecules are in the ground state. When one (A) is excited, complex formation takes place. The exciplex (complex) is stable only in the excited state, and may be reversibly formed. The exciplex may relax to the ground state, which is dissociative, giving the original unexcited starting molecules. Other decay paths may lead to chemical change. Electron transfer from one partner to the other, which is sometimes reversible, is one particularly interesting exciplex reaction. Many exciplexes fluoresce, and the spectrum is characteristically shifted to longer wavelengths than the fluorescence from the uncomplexed excited partner.

Complex formation between an excited molecule and an identical ground-state molecule may also occur:

$$A^* + A \rightleftharpoons (AA)^* \qquad (2\text{-}21)$$

The term *excimer* has been coined for this excited state dimer, which is a special kind of an exciplex. In the excimer, the electronic excitation energy is shared equally by the two partners.

The interactions that hold the partners together in an exciplex are essentially nonexistent in the ground state. These interactions include primarily: (1) The excitation–resonance interaction in which the excitation is shared between the partners leading to an overall reduction in energy ($A^*B \longleftrightarrow AB^*$) and (2) charge-transfer interactions ($A^+B^- \longleftrightarrow A^-B^+ \longleftrightarrow AB^*$). A molecule in an excited state is both a better electron donor and electron acceptor than it is in the ground state.

2.5.3. Bimolecular Kinetics. Stern–Volmer Quenching

A convenient method to determine the rate constants of bimolecular reactions of an excited state is to measure the competition between the unimolecular decay and the bimolecular quenching of the excited state. The general idea behind the method was formulated by Stern and Volmer in 1919. As an example, consider an excited molecule A^*, which undergoes a unimolecular decay to its ground state with rate constant k_d and which reacts with a ground-state molecule B with specific rate constant k_r, i.e.,

$$A^* \xrightarrow{k_d} A \tag{2-22}$$

$$A^* + B \xrightarrow{k_r} A + B \text{ or products} \tag{2-23}$$

the rate of the first step (which includes all unimolecular deactivation paths) is given by $k_d[A^*]$, and of the second step is given by $k_r[A^*][B]$. If the decay of A^* is directly measurable by some method, i.e., by the rate of decay of its emission, then the effect of adding B to a system containing A^* may be studied. As more B is added, more A^* will be quenched. As a result of this quenching, the rate of decay of A^* increases. The rate of decay of A^* in the presence of B is equal to $k_d[A^*] + k_r[A^*][B]$. Since measurement of the concentration of excited states is difficult, it is desirable to be able to evaluate k_r without directly measuring $[A^*]$. This can be done by measuring the efficiency of quenching of A^* by B as a function of $[B]$.

Let the quantum efficiency of a reaction of A^* in the absence of B equal Φ^0 and the efficiency of the same reaction in the presence of B be Φ. From the definition of quantum efficiency in terms of rate constants it can be shown that

$$\Phi^0/\Phi = 1 + k_r[A]/k_d \quad \text{(Stern–Volmer relationship)} \tag{2-24}$$

If a plot of Φ^0/Φ vs. [A] yields a straight line, then the Stern–Volmer relationship holds. The slope of this line equals k_r/k_d. Since k_d may be measured directly from the decay of A^*, k_r may be evaluated. If k_r can be estimated independently, then, of course, this approach can be used to obtain estimates of k_d. Thus, by employing the Stern–Volmer relationship, rate constants for bimolecular photochemical reactions may be measured.

2.6. DETECTION OF EXCITED MOLECULES

Excited molecules can be detected by their emission of light (for S_1 and T_1), by their absorption of light, or by some interaction with another molecular species that leads to some measurable event. An extensive literature exists concerning the various experimental techniques and the instrumentation, and many instruments are commercially available. A few principles are discussed here.

2.6.1. Luminescence Spectroscopy[13–16]

A luminescence spectrophotometer records the emission spectrum of a sample that is usually excited with monochromatic light. The light emitted from the sample, usually at 90° to the direction of the exciting beam, is focused onto the slit of a monochromator that has a photomultiplier tube detector at its exit slit (Fig. 2-9). The monochromator scans the wavelengths, and the spectrum is displayed on an $x–y$ recorder. If the emission monochromator is set at some wavelength in the emission spectrum, the excitation monochromator may scan the wavelengths, and an excitation spectrum is recorded. In the simplest case, the excitation spectrum, which is an action spectrum (see Sections 1.5.4. and 3.7.), is identical to the absorption spectrum of the emitting chromophore.

Luminescence lifetimes may be obtained in a number of ways. One may excite the sample with a pulse of light of duration that is short compared to the lifetime to be measured, and observe the decay of luminescence using a detector with a suitably fast response. Many photomultiplier tubes have nanosecond response times, and pulsed lasers may be used as sources of intense nanosecond light pulses, so that most fluorescence lifetimes are easily measured.

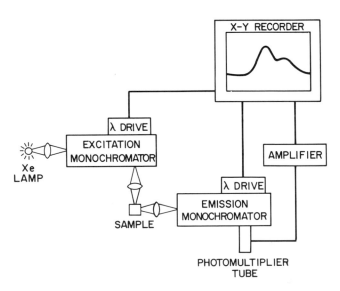

Fig. 2-9. A schematic representation of a luminescence spectrophotometer as described in Section 2.6.1.

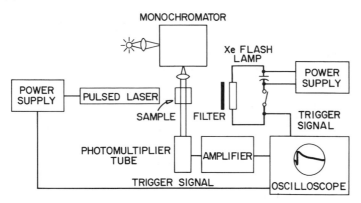

Fig. 2-10. A schematic representation of a kinetic flash photolysis apparatus. Filtered light from a xenon flash lamp or the light from a pulsed laser excites the sample. The time dependence of the absorption of a monochromatic monitoring light beam is recorded. The recording device, an oscilloscope here, is triggered in synchrony with the photolysis flash. If the photomultiplier tube is replaced by a spectrograph and the sample sensed with continuous-wavelength light, the full absorption spectrum of the transient species may be obtained in one experiment. Modern instruments employ electronic streak cameras that can record the transient absorption spectrum over a time course with subnanosecond resolution.

2.6.2. Flash Photolysis[22,23]

Excited molecules may absorb light and thus be excited to still higher-lying states (Fig. 2-8a). Since the separation between electronic states becomes smaller the higher the excitation energy, most excited molecules absorb longer-wavelength light than do the corresponding ground-state molecules. Many short-lived intermediates such as exciplexes, ions, etc., also have characteristic absorption spectra. To achieve a sufficiently large concentration of the excited state molecule or intermediate so that it may be observed by its light absorption, a powerful light pulse (flash) must be employed. The pulse must be of a duration that is short, relative to the lifetime of the species of interest if the lifetime of the latter is to be determined. This technique is called flash photolysis. A simple experimental scheme is shown in Fig. 2-10. Gas flash lamps may be employed for species that live microseconds or longer. Pulsed lasers must be employed for shorter time measurements. The advent of intense laser pulses of picosecond duration has allowed the direct measurement of such fast processes as vibrational relaxation and internal conversion between excited states.

2.7. ELECTRONIC ENERGY TRANSFER[24]

2.7.1. Theory

For our purposes, electronic energy transfer is described by $D^* + A \rightarrow D + A^*$. In the initial state, the donor D is electronically excited; in the final state, the donor has been demoted to a lower electronic state, and the acceptor A has been promoted to a higher electronically excited state. The excitation transferred is denoted by the asterisk. The transfer is a one-step radiationless process (no photon involved), which involves the simultaneous occurrence of the transitions $D^* \rightarrow D$ and $A \rightarrow A^*$, and requires some interaction between the donor and acceptor.

Radiative ("trivial") excitation transfer, a two-step process in which the donor emits a photon that is absorbed by the acceptor, is not considered in this discussion, not because it is unimportant (our lives depend on the radiative transfer of energy from the sun to photosynthetic organisms) but because it is easy to understand. The relative probability of radiative compared to nonradiative excitation transfer depends on many factors, but it is generally expected that nonradiative excitation transfer is the more probable process in biological systems. There are cases in which the opposite is true, of course; e.g., radiative excitation transfer is involved in the process by which male fireflies find female fireflies (bioluminescence; see Chapter 14).

Excitation transfer is a resonance process; the energies of the transitions in the donor ($D^* \rightarrow D$) and acceptor ($A \rightarrow A^*$) have the same total energy change. Thus, the transition in the acceptor must involve the same or smaller change in electronic energy than the donor transition. Any differences may be made up by including vibrational levels in the donor and acceptor (Fig. 2-11). The energy transferred corresponds to a frequency that is common to the appropriate emission spectrum of the donor, and appropriate absorption spectrum of the acceptor; i.e., a frequency within the region of overlap of the two spectra (Fig. 2-11). With increasing spectral overlap, the number of possible coupled transitions increases, and thus also the transfer probability.

The electronic interaction between the donor and acceptor, which couples the initial and final states, is usefully partitioned into two terms, called coulombic and exchange. It can be shown that the *coulombic interaction* is proportional [Eq. (2-23)] to the product of the appropriate transition dipole moments (Section 2.3.6.), of the donor ($\mathbf{M}_D$) and acceptor ($\mathbf{M}_A$), the squares of which are proportional to the strengths of the optical transitions in the isolated donor and acceptor. Thus, the coulombic interaction can be estimated very well from the appropriate spectra. The dipolar coulombic interaction falls off as the cube of the distance between the donor and acceptor [Eq. (2-25)]. This interaction predominates when the transitions in both the donor and acceptor are strong (allowed), and may lead to transfers over distances as large as 10 nm:

$$k_{et}\text{(dipolar coulombic)} \propto \left(\frac{\mathbf{M}_D \mathbf{M}_A}{R^3} \right)^2 \qquad (2\text{-}25)$$

Förster was able to relate the parameters in Eq. (2-25) to parameters which can be obtained experimentally. In Eq. (2-26), $\bar{\nu}$ is the wavenumber,

$$k_{et}\text{(Förster)} = \frac{8.8 : 10^{-25} \kappa^2 \Phi_D}{n^4 \tau_D R^6} {}_g F_D(\bar{\nu}) \, \epsilon_A \, (\bar{\nu}) \frac{d\bar{\nu}}{\bar{\nu}^4} \qquad (2\text{-}26)$$

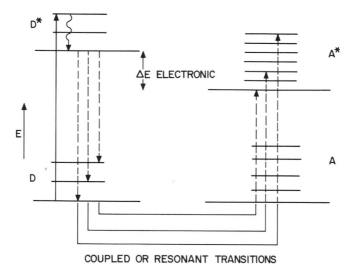

COUPLED OR RESONANT TRANSITIONS

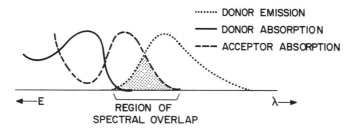

········ DONOR EMISSION
——— DONOR ABSORPTION
– – – ACCEPTOR ABSORPTION

←—E   λ—→

REGION OF
SPECTRAL OVERLAP

Fig. 2-11. Coupled transitions for a donor–acceptor pair with an electronic energy difference ΔE and their relation with the overlap of the donor emission and acceptor absorption spectra.

$F_D(\bar{\nu})$ is the spectrum of the donor emission, $\epsilon_A(\bar{\nu})$ is the molar extinction coefficient of the acceptor absorption, n is the refractive index of the solvent, κ^2 is an orientation factor that accounts for the geometric dependence of the dipolar interaction, Φ is the quantum yield of the donor emission in the absence of the acceptor, τ_D is the donor emission lifetime (s), and R is the distance between the donor and acceptor chromophores (cm). Thus, the rate of transfer is proportional to the strengths of the optical transitions in the donor and acceptor so that the selection rules for excitation transfer by this mechanism are the same as the selection rules for optical transitions. The transfer efficiency (number of transfers during the donor lifetime) may be expressed conveniently in terms of the critical separation R_0 at which transfer is half-efficient (transfer rate equals the isolated donor decay rate, $k_{et} = \tau_D^{-1}$). The efficiency of transfer is independent of the donor transition strength, and depends only upon the acceptor transition strength. This is because a weaker transition in the donor is balanced by a longer lifetime for the donor excited state (neglecting radiationless deactivation).

In terms of spin states, the transfer of type

$$D(S_1) + A(S_0) \rightarrow D(S_0) + A(S_1) \qquad (2\text{-}27)$$

or so-called singlet–singlet transfer is fully allowed by the coulombic or Förster mechanism.

The process called triplet–triplet transfer,

$$D(T_1) + A(S_0) \rightarrow D(S_0) + A(T_1) \qquad (2\text{-}28)$$

is doubly forbidden by the Förster mechanism. However, it may occur by virtue of the *exchange interaction*. The exchange interaction is dependent on the spatial overlap of the wavefunctions (electron orbits) of the donor and acceptor, so it is very short range (collisional distances). The magnitude of the intermolecular exchange interaction is not easily related to the optical properties of the donor and acceptor. If the transitions in the donor and acceptor are both spin-forbidden, the dipole–dipole term is very small, and the exchange term may predominate.

Triplet–triplet transfer and, of course, singlet–singlet transfer are allowed by the spin selection rules for exchange mitigated excitation transfer. The magnitude of the exchange interaction between molecules like the aromatic hydrocarbons is on the order of 10 cm^{-1} for nearest neighbors. This is sufficient for very fast transfer, 10^{-10} to 10^{-11} s, if spin-allowed. The transfer rate falls off very rapidly, about a factor of 100/0.1 nm as the donor–acceptor distance is increased. A transfer rate of 10^{10} to 10^{11} s^{-1} is sufficient to allow transfer within the time that two molecules in solution spend in a collision complex (encounter).

2.7.2. Phenomenological Types

It is useful to categorize radiationless excitation transfer into three phenomenological types:

1. Short-range, collisional transfer; one that requires the approach of donor and acceptor on the order of collisional diameters. This would be the case if only the exchange interaction mechanism were operating. If every encounter between donor and acceptor in solution leads to transfer, the transfer rate is the diffusion-controlled rate constant, which is about 5×10^9 liters mol^{-1} s^{-1}.

2. Long-range, single-step transfer; one that occurs over distances large compared to molecular diameters. Such a transfer necessitates a coulombic dipole–dipole (Förster) mechanism. For donor and acceptor molecules in solution, such long-range transfer can lead to transfer rates in excess of the diffusion-controlled rate.

3. Long-range, multistep transfer; one that may occur if the concentration of donor molecules is high. Efficient and fast transfer of excitation among donor molecules may occur if the donor molecules interact sufficiently, followed by transfer from a donor molecule to an acceptor that is sufficiently close.

If the donor and acceptor are immobilized and are far apart, only the long-range single step or long-range multistep transfer modes are possible. In fact, most of the cases of Förster-type singlet–singlet transfer have been those in which the donor and acceptor are relatively immobilized with respect to each other, e.g., singlet–singlet transfer between two chromophores that are part of the same molecule. The utility of singlet–singlet excitation transfer measurements in the determination of distances between particular chromophores, e.g., two groups on the same macromolecule, is frustrated by the uncertainty in κ^2, and in knowing how fast the relative motions of the donor and acceptor are, so that the proper average value of κ^2 can be computed.

When the coulombic interaction becomes so weak that the Förster critical distance becomes on the order of collisional diameters, the rate of singlet–singlet excitation transfer in solution cannot be faster than the diffusion-controlled rate. In this situation, the exchange mechanism is expected to contribute significantly to the transfer rate. Triplet–triplet transfer, for all intents and purposes, requires a collisional encounter between the donor and acceptor, and thus the rate of this process in solution cannot exceed the diffusion-controlled rate.

2.7.3. Bimolecular Excitation Transfer in Fluid Solution

In fluid solution, the singlet–singlet and triplet–triplet transfer processes are the most common. Triplet–triplet transfer rates, as expected, do not exceed the diffusion-controlled rate, and achieve the latter only when the triplet energy level of the donor is several kilocalories per mole higher than that of the acceptor. Endothermic transfers occur with an activation energy equal to the energy difference between the donor and acceptor. Anomalous rates ensue when special and infrequent geometrical factors obtain. The usually long life (10^{-6} to 10^{-3} s) of triplet states in de-aerated solution allows efficient transfer to an appropriate quencher when the latter is present at concentrations of 10^{-4} to 10^{-6} M.

Exothermic singlet–singlet transfer can proceed at a rate faster than the diffusion-controlled rate if the long-range coulombic mechanism obtains. If the latter is too slow, only collisional transfer occurs and the maximum rate is the diffusion-controlled rate. The short, 1–100 ns, lifetimes of singlet states require high concentrations, 10^{-1} to 10^{-3} M, of the acceptor for efficient transfer.

2.7.4. Electronic Energy Transfer in Biological Systems[25]

It is usual, in biological systems, to find the potential donors and acceptors in environments that restrict their translational and rotational motions. Restriction of translational motion leads to inefficiency of excitation transfer if the donor and acceptor cannot diffuse close enough to achieve transfer during the excited donor lifetime. The restricted motion of large molecules gives a great advantage to small freely diffusing molecules like oxygen.

In proteins and certainly in nucleic acids, the monomer units with low-lying excited states lie close enough together so that excitation transfers among them is expected to be fast. The aromatic amino acids tryptophan (Trp), tyrosine (Tyr), and phenylalanine (Phe) fluoresce and can exchange singlet energy. The singlet energies are in the order Phe > Tyr > Trp, and this is the expected direction of excitation transfer. The Förster distance (R_0) for 50% efficient downhill transfer of singlet excitation between pairs of these groups is on the order of 1.5 nm, or on the order of protein radii. Indeed such transfers are easily demonstrated in proteins, and on account of this transfer, many proteins that contain a Trp exhibit chiefly Trp fluorescence, even though the exciting light is absorbed mainly by Phe or Tyr. It is possible to estimate distances between these amino acids in protein from measured transfer efficiencies. The use of prosthetic groups as donors and acceptors greatly expands the possibilities for intramolecular distance determinations using excitation transfer.

Triplet–triplet transfer among aromatic amino acids in proteins should be more restricted than singlet–singlet transfer, because the former can proceed efficiently only if the chromophores are essentially nearest neighbors. Proteins containing Trp show mainly Trp phosphorescence, but this is probably because singlet–singlet transfer processes localize the singlet state on Trp. After intersystem crossing, the triplet excitation remains at Trp because it has the lowest triplet excitation energy of the amino acids.

In nucleic acids, the monomer units (bases) lie very close to each other, ~0.4 nm in DNA, and it would seem that excitation migration should be rampant. Indeed, the values for the electronic interactions between neighboring bases are large. However, it appears that both singlet and triplet excitation migration among the bases in nucleic acids are only of short range, on the order of 10 bases at most. Nearest-neighbor singlet transfer rates are fast ($\sim 10^9$ s^{-1} for Förster-type transfer), but relaxation of the electronic excitation energy by competitive pathways is usually even faster, making the actual transfer efficiencies low.

Depending upon which two bases are the donor and acceptor, triplet–triplet transfer rates between neighboring bases can vary over several orders of magnitude, 10^4 to 10^9 s^{-1}, at room temperature where the monomer triplet state lifetimes are on the order of 10^{-6} s. Thus, transfers between certain neighboring bases are precluded while certain others can occur many times during the triplet state lifetime. Since neighboring base combinations that preclude transfer occur very frequently, triplet excitation migration is highly restricted. In homopolynucleotides, such as polyadenylic acid, triplet migration over hundreds of monomer units occurs. Even the short-range triplet migration, ~ 10 bases, in polymers with essentially random base distribution, allows the excitation to localize on the bases of lowest triplet energy, e.g., thymine in DNA.

While there are many examples of electronic energy transfers involving biological molecules, there are few documented examples of excitation transfer connected with an *in vivo* photobiological process. The best-known example is that of excitation migration among antennae chlorophyll molecules in the chloroplast until the energy is finally transferred to special chlorophyll molecules in reaction centers (see Chapter 13). Many damage processes of the photodynamic type involve transfer of excitation energy from the sensitizing dye to oxygen (Chapter 4). Such processes undoubtedly occur *in vivo,* but there is no direct proof of this to date. Ketone sensitizers have been used to induce the formation of pyrimidine dimers in bacteria, phage, and cultured mammalian cells (Section 5.2.4.). This necessarily involves triplet excitation transfer from the ketones to the DNA.

The jellyfish *Aequorea aequorea* displays a blue-green bioluminescence when disturbed (Section 14.6.). The excited species first produced is associated with a protein called aequorin, which emits a blue chemiluminescence. A blue-green fluorescing protein can be extracted from the luminescent organ. The latter does not display chemiluminescence. Förster theory predicts that excitation transfer from the chromophore of the blue-green fluorescing protein can take place if these proteins are arranged as neighbors in the luminescent organ.

2.8. EXERCISES

1. How many photons impinge on a 1-cm² surface during 1 h of irradiation with light of wavelength 400 nm, if the intensity at the surface is 1000 J/m²?

2. Construct the three-level Jablonski diagram for the molecule whose absorption and emission spectra are shown in Fig. 2-12. What are its rates for

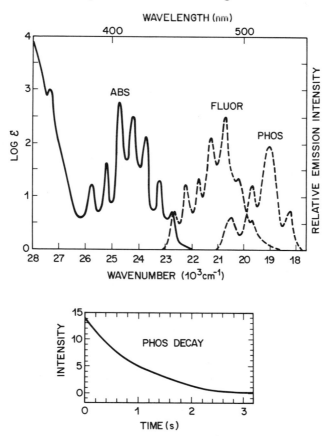

Fig. 2-12. Upper graph: The absorption (ABS), fluorescence (FLUOR), and phosphorescence (PHOS) spectra of an organic compound. The scale for the absorption spectrum (log ϵ) is to the left, and that for the emission spectra (relative intensity) is to the right. Lower graph: The decay of the phosphorescence of the compound (intensity vs. time after the exciting light is extinguished).

fluorescence, phosphorescence, intersystem crossing, and internal conversion? The $\Phi_p = 0{,}2$, and $\Phi_{isc} = 0.6$; the phosphorescence decay profile is also given in Fig. 2-12.

3. The unimolecular decay constant, k_d, for triphenylene triplets in hexane solution is $2 \times 10^4 \text{ s}^{-1}$. The concentration of naphthalene required to quench 50% of the triphenylene triplets is 2×10^{-5} M. What is the bimolecular quenching rate constant?

4. Would singlet–singlet electronic energy transfer occur from 1,2,5,6-dibenzanthracene (Fig. 2-1) to carbazole (Fig. 2-7)? Explain.

2.9. REFERENCES

1. H. H. Jaffé and Milton Orchin, *Theory and Applications of Ultraviolet Spectroscopy*, Wiley, New York (1962).
2. J. N. Murrell, *The Theory of the Electronic Spectra of Organic Molecules*, Wiley, New York (1963).
3. W. West (ed.), Chemical applications of spectroscopy, in: *Technique of Organic Chemistry* (A. Weissberger, ed.), Vol. 9, Wiley-Interscience, New York (1956).
4. A. E. Gillam and E. S. Stern, *Electronic Absorption Spectroscopy*, 2nd ed., Arnold, London (1957).
5. W. G. Herkstroeter, Special methods in absorption spectrophotometry, in: *Creation and Detection of the Excited State* (A. A. Lamola, ed.), Vol. 1, Part A, pp. 1–51, Marcel Dekker, New York (1971).
6. J. D. Roberts, *Notes on Molecular Orbital Calculations*, Benjamin, New York (1962).
7. L. Salem, *The Molecular Orbital Theory of Conjugated Systems*, Benjamin, New York (1966).
8. M. Kasha, Molecular photochemistry, in: *Comparative Effects of Radiation* (M. Burton, J. S. Kirby-Smith, and J. L. Magee, eds.), pp. 72–96, Wiley, New York (1960).
9. M. Kasha, The nature and significance of $n \rightarrow \pi^*$ transitions, in: *Light and Life* (W. D. McElroy and B. Glass, eds.), pp. 31–64, The Johns Hopkins University Press, Baltimore, Md. (1961).
10. R. B. Leighton, *Principles of Modern Physics*, pp. 233–251, McGraw–Hill, New York (1959).
11. S. P. McGlynn, T. Azumi, and M. Kinoshita, *Molecular Spectroscopy of the Triplet State*, Prentice–Hall, Englewood Cliffs, N.J. (1969).
12. H. Suzuki, *Electronic Absorption Spectra and Geometry of Organic Molecules*, Academic Press, New York (1967).
13. F. Dörr, Polarized Light in Spectroscopy and Photochemistry, in: *Creation and Detection of the Excited State* (A. A. Lamola, ed.), Vol. 1, Part A, pp. 53–122, Marcel Dekker, New York (1971).
14. Th. Förster, *Fluoreszenz Organischen Verbindungen*, Vanderhoech und Ruprecht, Göttingen (1951).
15. I. B. Berlman, *Handbook of Fluorescence Spectra of Aromatic Molecules*, 2nd ed., Academic Press, New York (1971).
16. R. S. Becker, *Theory and Interpretation of Fluorescence and Phosphorescence*, Wiley-Interscience, New York (1969).
17. M. Kasha, Paths of molecular excitation, *Radiat. Res. Suppl.* **2**, 243–275 (1960).
18. R. M. Hochstrasser, Some principles governing the luminescence of organic molecules, in: *Excited States of Proteins and Nucleic Acids* (R. F. Steiner and I. Weinryb, eds.), pp. 1–30, Plenum Press, New York (1971).
19. J. Jortner, S. A. Rice, and R. M. Hochstrasser, *Adv. Photochem.* **7**, 149–173 (1969).
20. N. Mataga and T. Kubota, *Molecular Interactions and Electronic Spectra*, Marcel Dekker, New York (1970).
21. J. W. Longworth, Luminescence spectroscopy, in: *Creation and Detection of the Excited State* (A. A. Lamola, ed.), Vol. 1, Part A, pp. 343–370, Marcel Dekker, New York (1971).

22. W. G. Herkstroeter, Flash photolysis, in: *Physical Methods of Chemistry* (A. Weissberger and B. W. Rossiter, eds.), Part 3B, pp. 521–576, Wiley-Interscience, New York (1972).

23. M. M. Malley, Lasers in spectroscopy and photochemistry, in: *Creation and Detection of the Excited State* (W. Ware, ed.), Vol. 2, pp. 99–148, Marcel Dekker, (1974).

24. A. A. Lamola, Electronic energy transfer in solution: Theory and applications, in: *Technique of Organic Chemistry* (P. A. Leermakers and A. Weissberger, eds.), Vol. 14, pp. 17–126, Wiley-Interscience, New York (1969).

25. R. F. Steiner and I. Weinryb (eds.), *Excited States of Proteins and Nucleic Acids,* Plenum Press, New York (1971).

3

Photochemistry

3.1. INTRODUCTION

A photobiological response is predicated upon some chemical modification of the biological system through the action of absorbed light energy. Chemical reactions that result from irradiation of chemical systems with light have traditionally been called photochemical reactions, and their study comprises that science called photochemistry. With the advent, during the last two decades, of many techniques that allow direct measurements on reactive intermediates, and on molecules in electronically excited states, the term photochemical reaction has taken on a more specific definition: a *photochemical reaction starts in one of the electronically excited states of a reactant and ends with the appearance of the first ground-state product(s)*. That is, photochemical reactions comprise that class of electronic relaxation processes that do not lead back to the starting molecule.[1–3] With this definition comes the insight provided by the notion that photochemical reactions differ from conventional thermal reactions in that photochemical reactions always involve a molecule possessing an excited electron.

In Chapter 2 it was shown that the absorption of a photon by a molecule promotes the molecule to an excited electronic state, thereby initiating a sequence of events that continuously lowers the energy of the excited molecule.

Nicholas J. Turro • Department of Chemistry, Columbia University, New York, New York   **Angelo A. Lamola** •   Bell Laboratories, Murray Hill, New Jersey

In addition to the radiative and radiationless (the so-called photophysical) processes, which return an excited molecule to its original ground state, a number of bond-breaking and bond-making (the so-called photochemical) processes are also available for the relaxation of electronically excited molecules. If, relative to the ground state, an old bond is completely broken or a new bond is completely made, then an excited molecule is said to have undergone a photochemical reaction (commonly called a primary photochemical process). Sometimes a stable and isolable product is the immediate result of a photochemical reaction. Such reactions are termed *concerted* photochemical reactions, i.e., no intermediate chemical species occur along the pathway from the excited state to the final product. Most photoreactions, however, produce *reactive intermediates* (i.e., radicals, biradicals, ions, or other unstable species), which are not isolable but which react further to produce the final product as a result of *secondary (thermal) reactions*. The flow diagram in Fig. 3-1 reviews these concepts and definitions. This view of photochemical reactions is not meant to deemphasize the final products, which may result from reactions of photochemically produced intermediates. On the contrary, these end products are what usually bring about the biological responses that almost always take place on comparatively longer time scales than do the photochemical reactions. It is the isolation and characterization of the end products that form the backbone of molecular photobiology. However, the separation of (primary) photochemical reaction from (secondary) thermal reaction(s) of photochemically produced species is very useful. For example, very often the factors that influence the primary photochemical events are different from the factors that influence the subsequent secondary reactions. This makes for many possibilities for controlling the chemistry. As will be evident below, the separation of primary and secondary reactions is useful for categorizing photochemical reactions.

For purposes of the treatment given here, the chemistry that starts with the absorption of light and ends with the formation of a stable product is referred to

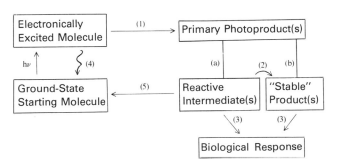

Fig. 3-1. Relations between excited states and photoproducts. The photoexcited molecule gives rise (1) to the primary photochemical product, which may itself be the final product (b) or, is a reactive intermediate (a) which undergoes transformation in the dark (2) to the final product. The biological response is effected (3) by the final product, although it is sometimes useful to consider that a reactive intermediate is the effector. Processes (4) and (5) are, respectively, degenerate electronic relaxation and degenerate chemical reaction, and lead back to the starting molecule.

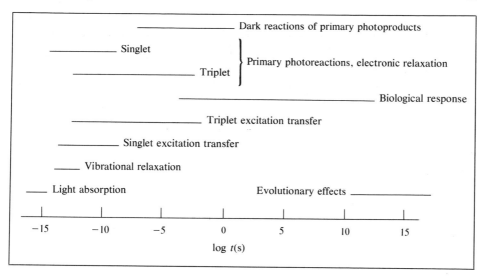

Fig. 3 2. Time scales in photophysics, photochemistry, and photobiology.

as the *overall* photochemical reaction. That part of the overall reaction that is associated with an excited state molecule, i.e., the photochemical reaction *per se,* is called the *primary* photoreaction or primary photochemical process.

Although the number of different overall photoreactions of organic molecules is large, the number of primary photochemical processes common to organic molecules is quite small. In the sections to follow, the important primary photoreactions are described, and by example it is shown how some overall reactions may be interpreted in terms of these initial, chemistry-controlling steps.

In Fig. 3-2, the time regimes for the processes of concern in photobiology are shown on a logarithmic scale from 10^{-15} to 10^{15} s. The phenomenological photobiologist records a biological response that takes place from milliseconds (ms) to perhaps years (e.g., sunlight-induced human skin cancers) after exposure to light, and relates the response to various experimental parameters, and notably to the wavelength and fluence of the light. The molecular photobiologist relates the biological response to biochemical events, and eventually to observable photoproducts. To complete the description of the mechanism requires an understanding of the photophysical and photochemical relaxation processes of the molecule(s) that absorb the light. Only after this complete picture is achieved, can the most effective means of assisting, modifying, or thwarting the photobiological response be decided upon.

3.2. COMMON OVERALL PHOTOREACTIONS[4–9]

From the phenomenological standpoint, the following classification of some of the most common organic photoreactions is convenient. Two examples of

each type of reaction are given; one involves a molecule of biological significance, and the other some simpler organic molecule. In this way the reader can
see that often there are simple analogs that may help in the study of the
photochemistry of biological molecules.

   1. Linear addition to unsaturated systems (also termed reductive addition):

$$Ph_2CO + (CH_3)_2CHOH +$$ $$\xrightarrow{hv}$$ $$+ (CH_3)_2CO \qquad (3\text{-}1)$$

uracil   cysteine   5-S-cysteine-6-hydrouracil

$$(3\text{-}2)$$

   2. Cycloadditions of unsaturated systems:

$$Ph_2CO +$$ $$\xrightarrow{hv}$$ $$(3\text{-}3)$$

$$2$$ $$\xrightarrow{hv}$$ $$(3\text{-}4)$$

thymine   thymine dimer
                                              (cyclobutane-type)

3. Substitution reactions:

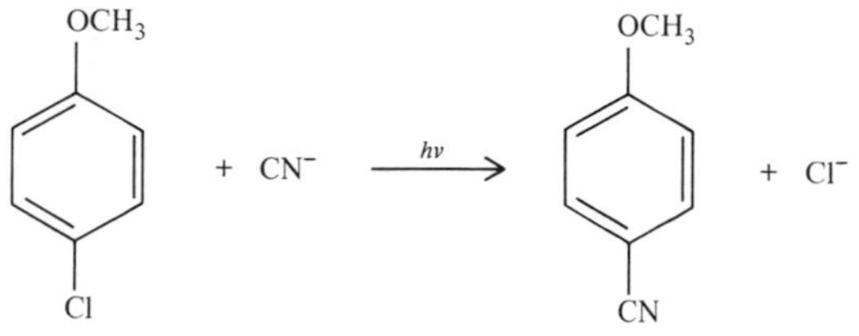

$$(3-5)$$

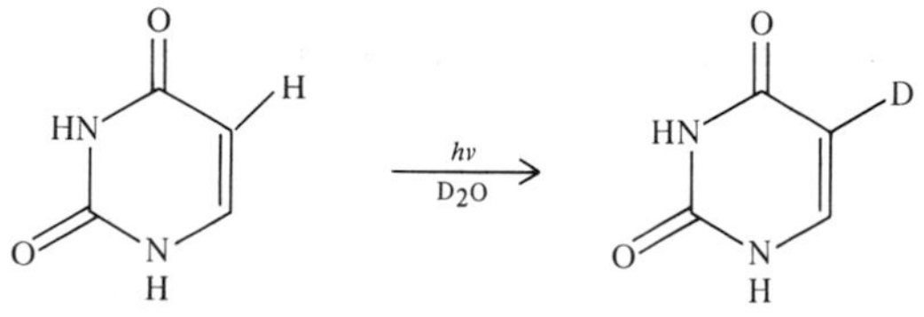

$$(3-6)$$

uracil   5-deuterouracil

4. *cis-trans* isomerizations:

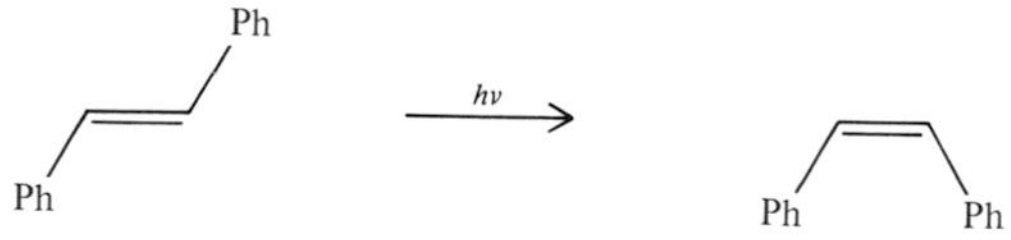

$$(3-7)$$

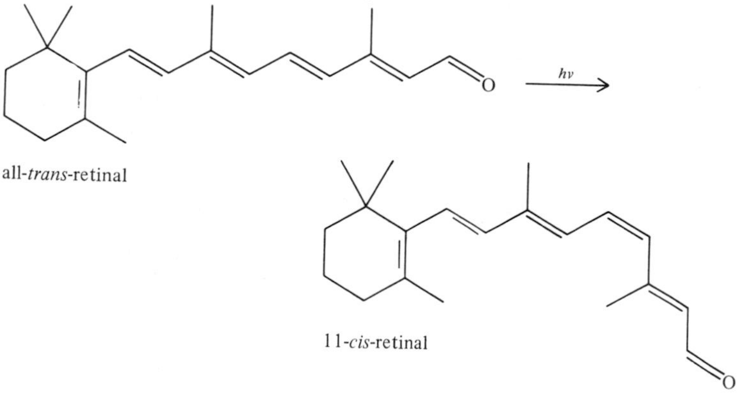

all-*trans*-retinal

$$(3-8)$$

11-*cis*-retinal

5. Structural rearrangements (skeletal):

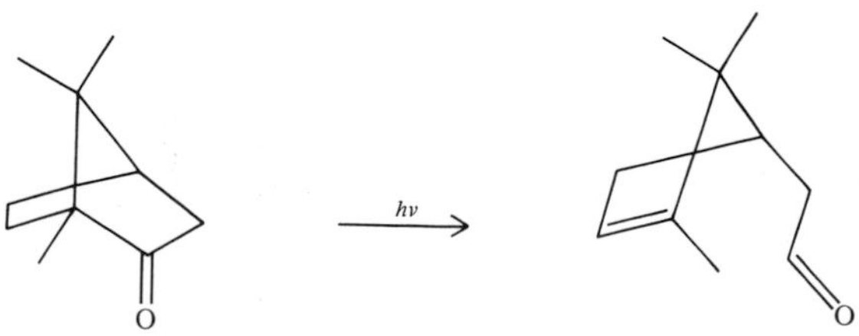

$$(3-9)$$

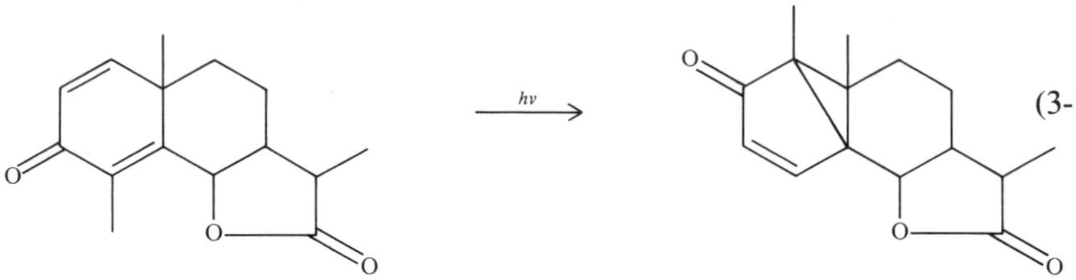
hν
O
O
O
O
O
O

santonin lumisantonin

(3-10)

6. Structure rearrangements (electrocyclic or valence):

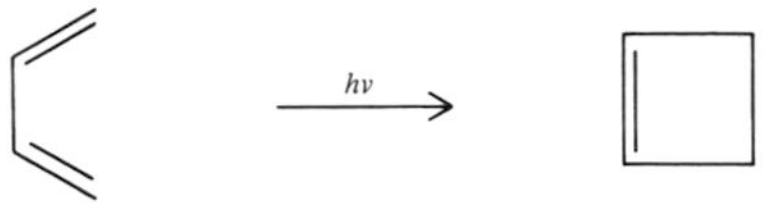
hν

(3-11)

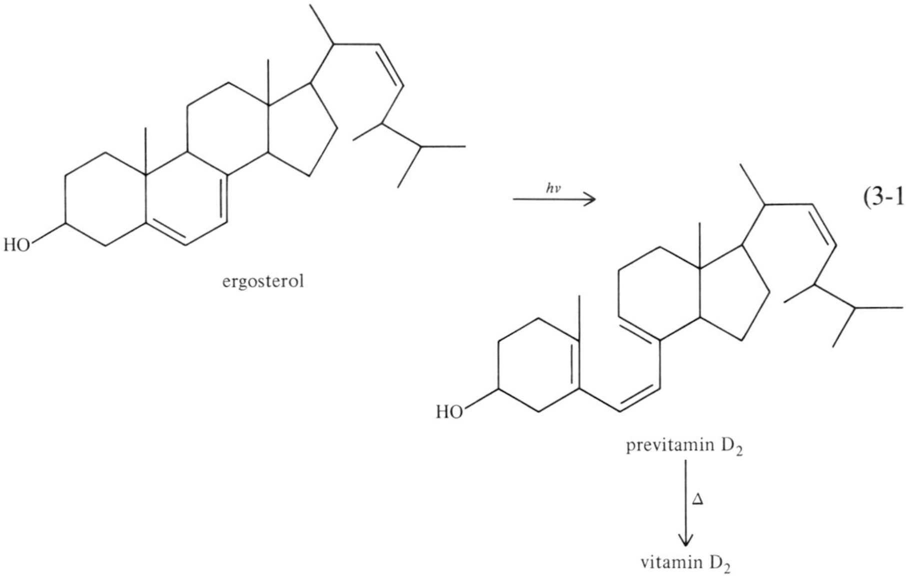
HO
ergosterol
hν
HO
previtamin D$_2$
Δ
vitamin D$_2$

(3-12)

7. Fragmentations:

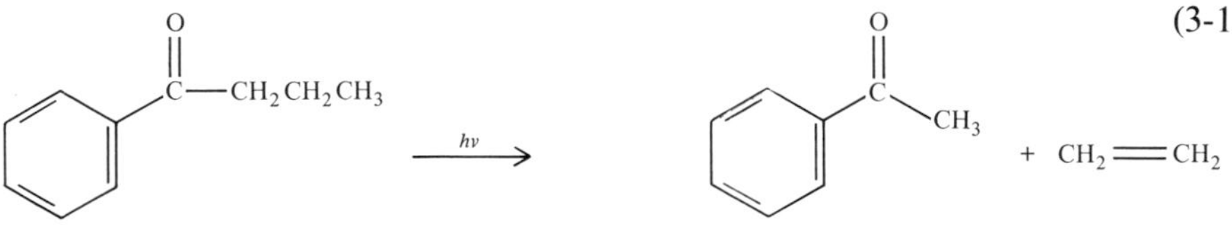
O
C—CH$_2$CH$_2$CH$_3$
hν
O
C
CH$_3$
+ CH$_2$=CH$_2$

(3-13)

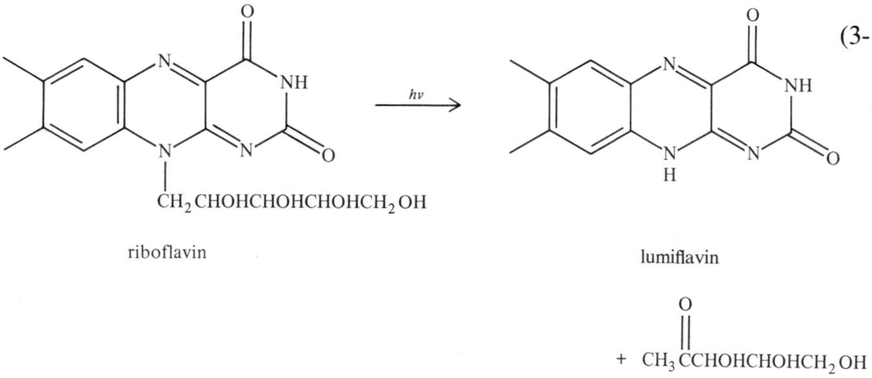

(3-14)

riboflavin   lumiflavin

$$+ \quad CH_3\overset{O}{\overset{\|}{C}}CHOHCHOHCH_2OH$$

8. Photooxidation:

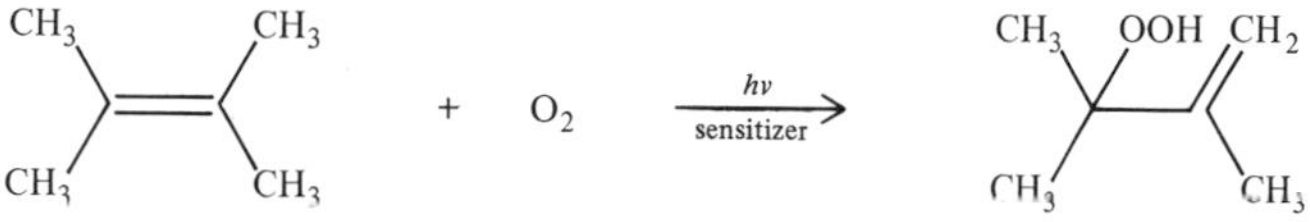

(3-15)

cholesterol

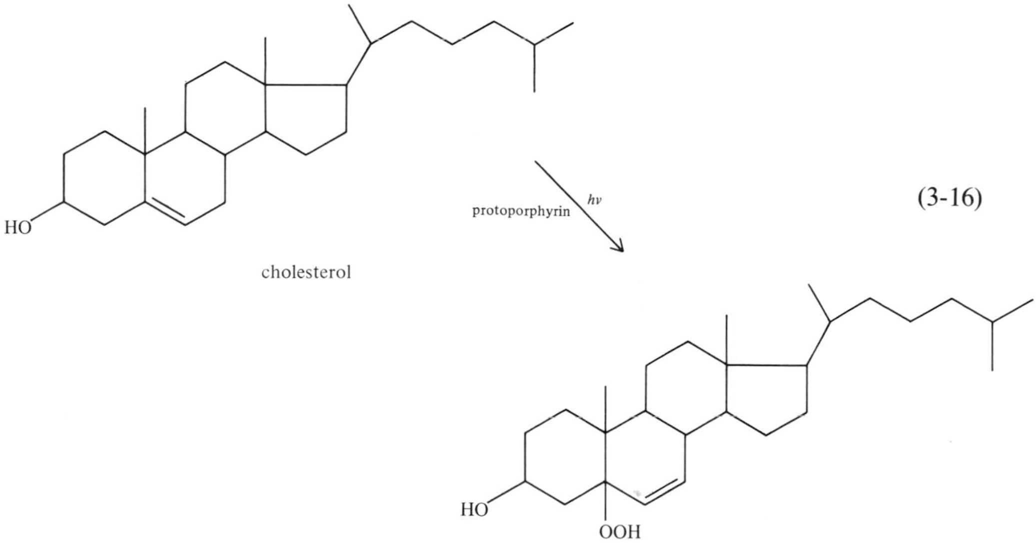

(3-16)

3β-hydroxy-5α-hydroperoxy-Δ^6-cholestene

3.3. PRIMARY PHOTOREACTIONS OF ORGANIC MOLECULES

The common primary photoreactions may be classified into two general types: (1) inherently unimolecular, and (2) inherently bimolecular. The distinction is based on whether the photoreaction may be perceived as taking place essentially within an electronically excited chromophore, or whether, in addition, ground-state species also take part in the process. The eight most common primary photochemical processes are listed below.

Inherently unimolecular and concerted
1. *cis-trans* isomerization [see Eqs. (3-7) and (3-8)].
2. Bond rearrangements of the electrocyclic type [see Eq. (3-11)]. This primary reaction type involves some concerted reorganization of single and double bonds (see below).
3. Bond rearrangements of the so-called sigmatropic type. This reaction type involves a concerted reorganization in which some atom or group in the molecule changes from one single-bonded partner to another. Other bonding arrangements in the molecule may also change concomitantly. For example,

$$(3-17)$$

Inherently unimolecular leading to the formation of a reactive intermediate
4. Simple homolytic bond cleavages, i.e., the cleavage of a single bond into two radicals:

$$(3-18)$$

Inherently bimolecular and concerted
5. Cycloadditions (and retrocycloadditions), i.e., the addition of two components to make a ring (or the cleavage of a ring into two components):

$$(3-19)$$

Inherently bimolecular leading to the formation of a reactive intermediate

6. Hydrogen abstraction:

$$Ph_2CO + (CH_3)_2CHOH \xrightarrow{h\nu} Ph_2\dot{C}OH + (CH_3)_2\dot{C}OH \qquad (3\text{-}20)$$

7. First step of a two-step addition to unsaturated systems:

$$(3\text{-}21)$$

8. Electron transfer:

$$(3\text{-}22)$$

The extensions and implications of each of these primary reactions will usually suffice to provide an adequate mechanism for most known overall photoreactions of organic compounds. Two general rules will assist in the application of the set of eight primary photoreactions to a mechanistic analysis:

Rule 1: With the exception of *cis-trans* isomerization, only molecules in singlet excited states undergo *concerted* or one-step overall photoreactions. *Comment:* Of the concerted photochemical processes only *cis-trans* isomerization may occur from triplet as well as singlet states.

Rule 2: The overall reaction products derived from reactive intermediates produced in primary photochemical processes are determined by the secondary reactions of the intermediates. *Comment:* In predicting or explaining an overall photochemical reaction product, one must consider the plausible reactions of the intermediates produced from primary photochemical processes.

Possible mechanisms for some of the overall photoreactions are given as examples in Section 3.2. The analysis of linear addition [Eq. (3-1)], an apparently complex reaction, is straightforward. The most likely primary process is *hydrogen abstraction* from isopropanol by photoexcited benzophenone to produce a diphenyl ketyl radical, which then adds to cyclopentenone:

$$Ph_2CO \xrightarrow{h\nu} Ph_2CO^* \xrightarrow{(CH_3)_2CHOH} Ph_2\dot{C}OH + (CH_3)_2\dot{C}OH \qquad (3\text{-}23)$$

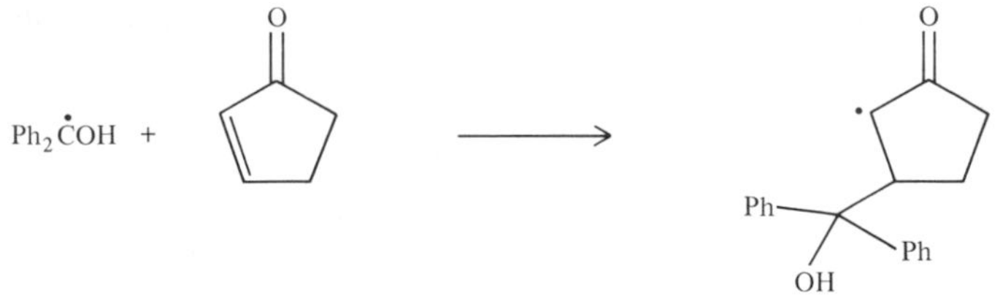

$$(3\text{-}24)$$

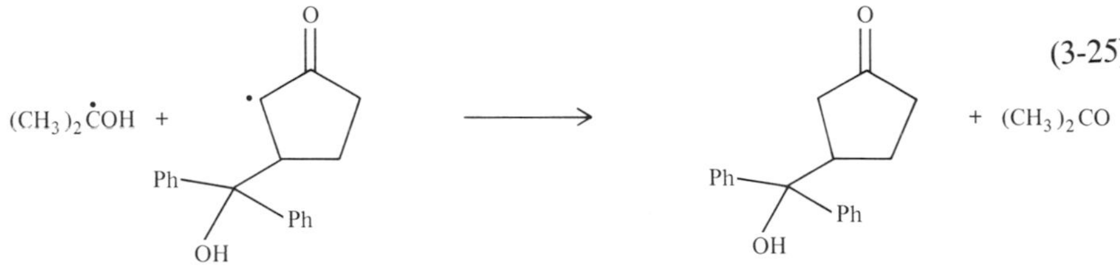

$$(3\text{-}25)$$

The cycloaddition reaction of Eq. (3-3) could proceed by a concerted (singlet) mechanism or by way of an intermediate formed by the attack of triplet benzophenone on thymine [Eq. (3-21)]. From experimental data it is known that singlet Ph_2CO is too short-lived to participate in bimolecular photoreactions. Therefore, the primary process shown in Eq. (3-21) is probably involved.

The reactions in Eqs. (3-7) and (3-8) may be either singlet or triplet *cis-trans* isomerizations.

The reaction in Eq. (3-9) involves the formation of a biradical intermediate [see Eq. (3-18)].

The reactions in Eqs. (3-10) and (3-12) are sigmatropic and electrocyclic rearrangements, respectively, and both probably are singlet reactions.

The reaction in Eq. (3-14) probably proceeds via the diradical intermediate shown below, which results from an intramolecular hydrogen atom abstraction.

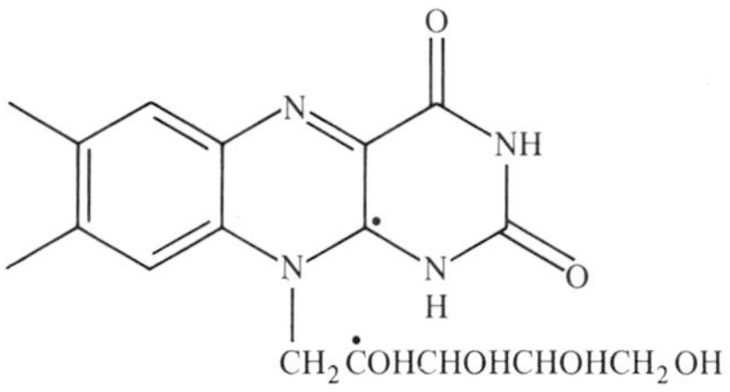

$$(3\text{-}26)$$

In summary, the scheme given in Fig. 3-3 indicates a general framework for the analysis of organic photoreactions in terms of rationalization and/or prediction.

3.4. STRUCTURE–REACTIVITY RELATIONSHIPS IN PHOTOCHEMISTRY

Since the number of primary photochemical processes is quite small, it might be asked whether general structure–reactivity relationships exist in photochemistry. The answer to the question is positive.

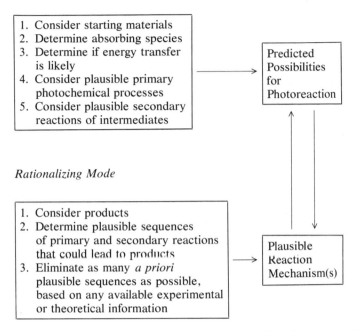

Predicting Mode

1. Consider starting materials
2. Determine absorbing species
3. Determine if energy transfer is likely
4. Consider plausible primary photochemical processes
5. Consider plausible secondary reactions of intermediates

Predicted Possibilities for Photoreaction

Rationalizing Mode

1. Consider products
2. Determine plausible sequences of primary and secondary reactions that could lead to products
3. Eliminate as many *a priori* plausible sequences as possible, based on any available experimental or theoretical information

Plausible Reaction Mechanism(s)

Fig. 3-3. A scheme for working out the mechanism of a photochemical reaction.

As discussed in Section 2.3.5., commonly encountered excited states of organic molecules may be classified as singlet or triplet states with respect to their overall electronic spin, and as n,π^* and π,π^* states with respect to electronic configuration. Some general statements can be made concerning the reactions of singlet and triplet states, and of n,π^* and π,π^* states.

The first general postulate concerning reactivity relates to differences between singlet and triplet states that correspond to the same electronic configuration.

Postulate 1: Singlet states tend to react as if the key valence electrons involved in the reaction move as pairs, whereas triplet states tend to react as if the key valence electrons involved in the reaction move as individual, radical-like species.

The next general postulate concerning reactivity relates to the correlation between the "best" single electronic configuration for a state, and the types of reaction of that state.

Postulate 2: The chemistry expected for a given excited state molecule may often be predicted by examination of a "best" single valence bond structure that conforms to the electronic configuration, and is consistent with Postulate 1.

To demonstrate the use of these postulates, the reader is referred to Fig. 3-4 where two prototype systems are shown, namely, ethylene and acetone. The

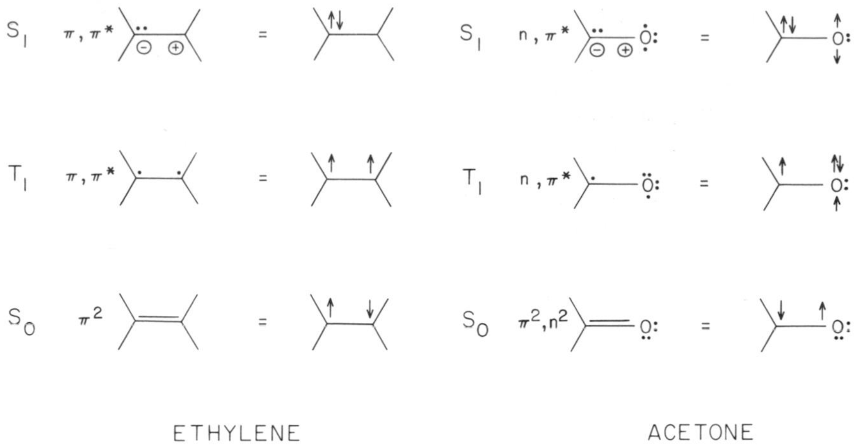

ETHYLENE   ACETONE

Fig. 3-4. The "best" one-structure description of S_0, S_1, and T_1 of ethylene and acetone. Only one of the two equivalent resonance structures for ethylene in S_1 is shown. For acetone, the electrons below the oxygen atom are viewed as the n electrons involved in the n,π^* state. The electrons above the oxygen atom represent π or π^* electrons. The energetic separations are not to scale.

best single electronic configuration for the lowest excited singlet and triplet states (S_1 and T_1) of ethylene is π,π^*. In accordance with Postulates 1 and 2, the ionic valence structure $\bar{C}H_2$—$\overset{+}{C}H_2 \longleftrightarrow \overset{+}{C}H_2$—$\bar{C}H_2$ (a resonance between two identical structures) corresponds to S_1, and the diradical structure $\dot{C}H_2$—$\dot{C}H_2$ corresponds to T_1.

The best single electronic configuration for the S_1 and T_1 states of acetone is n,π^*. In accordance with Postulates 1 and 2, the ionic structure $(CH_3)_2\bar{C}$—$\overset{+}{O}$ corresponds to S_1 and the diradical structure $(CH_3)_2\dot{C}$—$\dot{O}$ corresponds to T_1.

The extension of these postulates to other molecules involving conjugation is relatively straightforward.

The associated primary photochemical reactions discussed earlier may now be correlated with *valence structures,* i.e., structure–reactivity relations may be generated. As a first example, consider a ketone. Of the eight primary photochemical reactions, we expect acetone to participate only in *nonconcerted* reactions even in S_1 because the n,π^* configuration of S_1 restricts the possibilities for the simultaneous overlap of orbitals needed for continuous bond formation, which is a requirement for concerted reactions. Thus, of the inherently unimolecular processes, we expect only *bond cleavages,* e.g.,

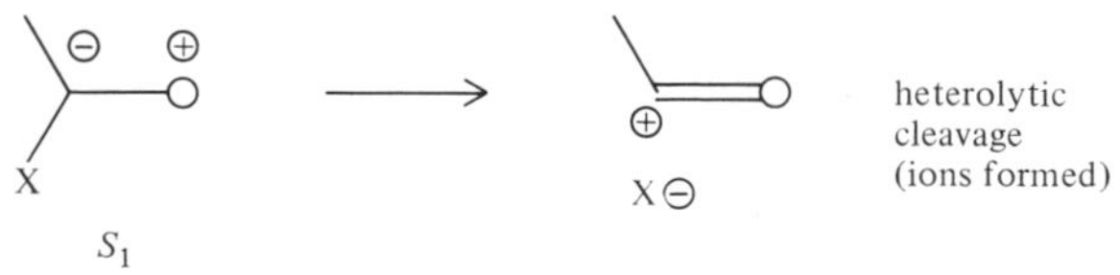

heterolytic   (3-27)
cleavage
(ions formed)

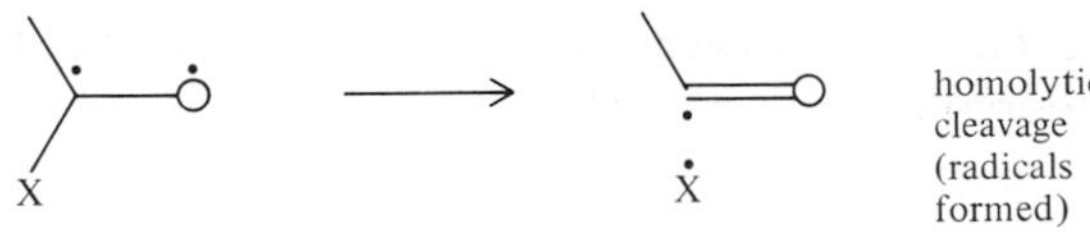

homolytic
cleavage
(radicals
formed)   (3-28)

As a second example, consider a conjugated diene. We expect 1,3-buta-
dienes to have the possibility of participating in most of the eight primary
photochemical reactions, depending on the reaction conditions. For illustration,
cis-trans isomerization (3-29) is an expected reaction of both S_1 and T_1.

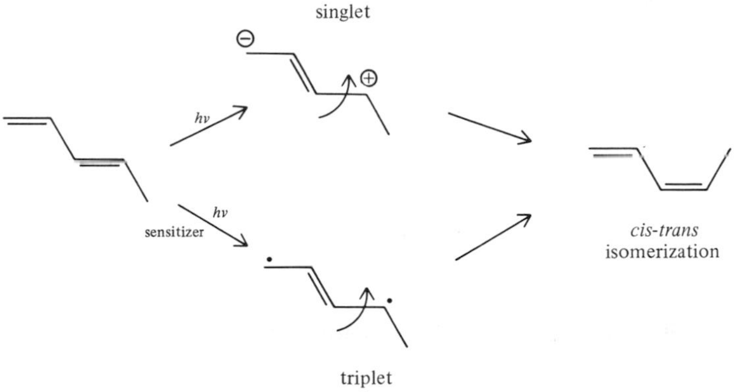

(3-29)

Since they are generally concerted reactions, electrocyclic reactions are
expected in S_1 but not T_1. Because spin must be conserved, cyclization of T_1
would produce a higher energy triplet [Eq. (3-31)]; cyclization of S_1 produces a
lower-energy ground state product [Eq. (3-30)].

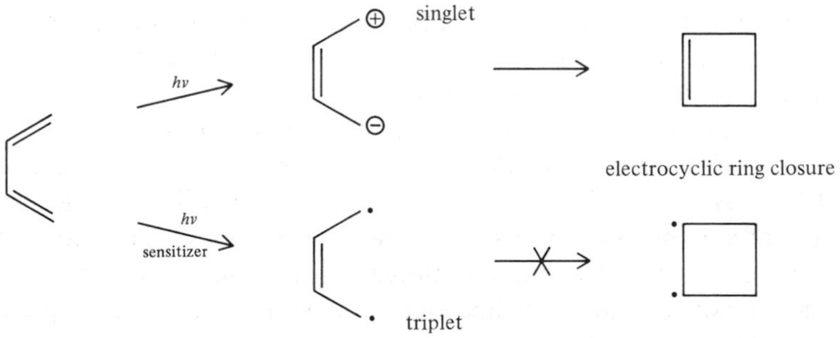

(3-30)

(3-31)

It is important to note that bimolecular cycloadditions are favored for T_1

relative to S_1, because the longer lifetime of T_1 provides it with a greater opportunity to undergo reaction [Eq. (3-32)].

triplet   +   diradical
                                                                                     addition

                                                                                          (3-32)

3.5. ANOTHER VIEW OF PHOTOCHEMISTRY[3,10]

A molecule in a particular electronic state can exist with various geometries (configurations of its nuclei). Each geometry corresponds to a particular potential energy content. The maps of potential energy vs. nuclear configuration (geometry) for each electronic state are called the potential energy surfaces of the molecule.

One can classify photochemical reactions on the basis of whether the process occurs on one potential surface (i.e., in one electronic state), or involves going from one surface or electronic state to another. The former is called an adiabatic process, and the latter a diabatic process.

For reactions in solution, adiabatic photochemical reactions are restricted to those in which the essential chemical change occurs on the same excited state surface. Photodissociation of diatomic molecules, e.g., I_2, is one example. In fact, all of the clear examples involve some association or dissociation reaction, e.g., reactions of excited molecules that involve protonation or deprotonation $(RH^* \rightarrow R^{\overset{*}{\pm}} + H^+)$. Usually the backreaction occurs rapidly in the ground state.

Most photochemical reactions are diabatic. The chemical change is part and parcel of the electronic deexcitation process, i.e., the chemistry occurs as a consequence of going from an excited electronic state to a lower-lying electronic state, usually the ground state.

Many photoreactions involve both adiabatic and diabatic changes. The adiabatic process involves movement over the excited state surface until a diabatic reaction (electronic deexcitation) occurs at a place (geometry) where the

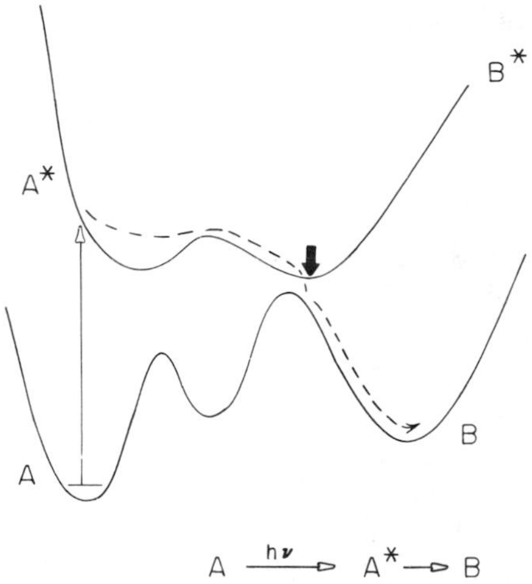

Fig. 3-5. A hypothetical pathway for a complicated photochemical primary process A*
→ B. Starting molecule A is promoted to an
excited state surface where rapid motion on
the surface across a low barrier brings the
molecule to a point (bold arrow) where rapid
electronic deexcitation occurs, followed by
further motion along the reaction pathway to
a thermally stable ground-state product B.

upper and lower surfaces lie close to each other. An example is shown in Fig. 3-
5, which represents a cross section through a set of hypothetical surfaces along a
reaction pathway.

There is much activity today in mapping the potential energy surfaces of
photochemically active compounds using spectroscopic data (absorption and
emission spectra, photoelectron spectra, etc.), molecular orbital calculations,
and symmetry arguments, in order to understand the features of the surfaces that
are important for controlling the photochemistry.[10]

3.6. QUANTUM YIELDS AND KINETICS

Consider the typical overall photochemical reaction where the product (R),
which can be assayed, results from dark reactions of a photochemically produced
intermediate (X):

$$A \xrightarrow{h\nu} A* \qquad \text{absorption} \qquad (3\text{-}33)$$

$$A* \xrightarrow{k_{dA}} A \qquad \text{deactivation } (d) \qquad (3\text{-}34)$$

$$A* \xrightarrow{K_{rA}} X \qquad \text{primary photochemical reaction } (r) \qquad (3\text{-}35)$$

$$X \xrightarrow{k_{dX}} A \qquad \text{degenerate reaction} \qquad (3\text{-}36)$$

$$X \xrightarrow{k_{rX}} R \qquad \text{secondary chemistry} \qquad (3\text{-}37)$$

$$(R \rightarrow \qquad \text{biological response}) \qquad (3\text{-}38)$$

The quantum yield (Section 2.4.3.) for production of R (Φ_R) is

$$\Phi_R = \frac{\text{number of molecules of R produced}}{\text{number of photons absorbed by A}} \qquad (3\text{-}39)$$

In a typical experiment, what can usually be measured is the number of molecules of R produced, and the number of photons (wavelength specified) that have been absorbed by the sample. From the latter data, and the absorption spectrum of A, its concentration and the time course of its disappearance, and the spectra and concentrations of other substances present that may absorb the exciting light, one can usually calculate the number of photons absorbed by molecules of A during the course of the experiment.[11]

The quantum yield is the most important single parameter that characterizes a photochemical reaction. Knowing its value allows one to control the amount of product by limiting the fluence of the light. The quantum yield is also extremely valuable in elucidating the mechanism of photoproduct formation. However, knowing the quantum yield for a single set of experimental variables rarely allows the mechanism to be defined in any detail.

In mechanism studies, the dependence of the quantum yield (or some quantity directly proportional to it) upon several experimental parameters, e.g., light intensity, concentration of reactants, solvent, and temperature, is determined. Only in this way can one begin to elucidate the various intermediate processes that occur between the light absorption, and the formation of the assayable product. Another approach is to vary the structure of one of the reactants, e.g., A in the scheme given above, in some logical fashion in order to find structure–reactivity relations that might give information about intermediates. To avoid wrong conclusions and predictions about photochemical reactivity, it is imperative that quantitative information about the yields of key intermediates in the overall process be obtained. This becomes clear once the expression for the quantum yield is expanded in terms of efficiencies of intermediate processes. Take the example given above for which

$$\Phi_R = (\text{efficiency of X} \rightarrow \text{R})\, \Phi_X \qquad (3\text{-}40)$$
$$= [k_{rX}/(k_{rX} + k_{dX})]\Phi_X \qquad (3\text{-}41)$$

and on further expansion:

$$\Phi_R = \Phi_{A*}[k_{rA}/(k_{rA} + k_{dA})][k_{rX}/(k_{rX} + k_{dX})] \qquad (3\text{-}42)$$

If A* is not the state achieved directly by photon absorption, e.g., if it is the triplet state, then Φ_{A*} itself can be expanded in terms of relative rates of photophysical processes.

An increase or decrease in the value of Φ_R could ensue from an increase or decrease of the rate of one of the several processes that the intermediates in the overall scheme can undergo. That is, *the overall quantum yield is related to ratios of rate constants of competing processes* so that a change in its value does not carry a unique interpretation. For example, say that an analog of A gives the analogous photoproduct under the same experimental conditions, but with a higher quantum yield; this could result from increased values for Φ_{A^*}, k_{rA}, or k_{rX} or from lower values of k_{dA} or k_{dX}, or from combinations of these changes. Of course, the same result would ensue if, e.g., there was a decrease in the value of k_{dA} such that the value of $k_{rA}/(k_{rA} + k_{dA})$ was increased. Thus, it can be seen that a correct interpretation of reactivity (in this case the rate constant k_{rA}) with respect to an overall photoprocess like the conversion of A to R, in the example under discussion, requires that the yields of the intermediates (A* and X) be determined.

Many approaches to the sorting out of intermediate processes have been developed in recent years. Fast spectroscopic techniques such as flash photolysis (Section 2.6.2.) allow even very short-lived intermediates to be observed directly, and their yields and lifetimes determined. Much can be learned by the introduction of specific sensitizing or quenching compounds into the system under study. Quenchers are compounds that intercept specific intermediates, and thus decrease the yield of the desired product. If the rate of the quenching action is known, then the rate of the reaction of the intermediate can be calculated from the reduction in quantum yield (see Section 2.5.3.). Sensitizers are compounds that can increase the yield of specific intermediates, usually triplet-state molecules (see Section 3.8.), and thus increase the yield of the product. Sensitizers may also totally change the overall chemistry of a photochemical system.

To recapitulate: The quantum yield of an overall photochemical reaction is a quantity that measures the efficiency with which absorbed light produces chemical changes that can usually be measured easily, and that serves as a functional parameter for characterizing the reaction. However, there is usually no one-to-one correspondence of overall quantum yield and photochemical reactivity (rate of the primary photochemical process). The product of the efficiencies of all the steps leading to the reactive excited state molecule, the efficiency of the primary process, and the efficiencies of all the steps that occur between primary reaction and formation of the overall product gives the overall quantum yield. To predict what effect some change in experimental conditions would have on the overall photochemical quantum yield, one would have to consider the effect of the change on each of the processes in the reaction sequence. In order to predict what effect some change in experimental conditions would have on the strength (with respect to light fluence) of a photobiological response, one would have to consider, in addition, the effect of the change on each of the sequential steps that takes place between the formation of the key photochemical product, and the occurrence of the biological response.

3.7. ACTION SPECTRA[12]

Consider the overall photochemical reaction of the last section, $A \overset{h\nu}{\rightarrow} R$, and further, that over the wavelength range examined only the starting material A can absorb the excitation light. Consider the irradiation of A with monochromatic (very small bandwidth $\delta\lambda$) light of wavelength λ_1, and the irradiation of an identical sample with light of wavelength λ_2. The amount N_1 of product R produced with light of wavelength λ_1 for a particular intensity I_1 and irradiation time t_1 is given by

$$N_1 = I_1 t_1 \sigma_1 [A] \Phi_1 \tag{3-43}$$

where σ_1 is the absorption cross section (a quantity proportional to ϵ, the molar extinction coefficient) of A at λ_1, and Φ_1 is the quantum yield for the reaction. It is assumed, of course, that the fraction of incident light absorbed by the sample is sufficiently small so that all the molecules of A "see" the same intensity, and further that during the course of the experiment only a negligible fraction of the starting material was converted to product. Similarly for irradiation at λ_2,

$$N_2 = I_2 t_2 \sigma_2 [A] \Phi_2 \tag{3-44}$$

If the irradiations are carried out such that $N_1 = N_2$ then

$$I_1 t_1 \sigma_1 \Phi_1 = I_2 t_2 \sigma_2 \Phi_2 \tag{3-45}$$

and

$$\frac{I_1 t_1}{I_2 t_2} = \frac{D_1}{D_2} = \frac{\sigma_2 \Phi_2}{\sigma_1 \Phi_1} \tag{3-46}$$

where $D = It$ (the total photon fluence). The ratio D_1/D_2 is a measure of the relative efficiencies of light of wavelengths λ_1 and λ_2 to cause the photochemical change for the conditions described.

One can determine relative efficiencies for the photoreaction at many wavelengths (at constant bandwidths). A plot of *relative efficiency vs. wavelength* is the *action spectrum* for the production of R. With the assumptions given, the action spectrum is a plot of relative $\sigma(\lambda)\Phi(\lambda)$ vs. λ. If Φ, the quantum yield, is independent of λ, then the action spectrum has the same shape as the absorption spectrum, $\sigma(\lambda)$ vs. λ. Thus, the action spectrum can be an important tool for determining the molecular species that is concerned with the light-absorption step in a photochemical or photobiological process.

The action spectrum, like the quantum yield, provides an important empirical characterization of a light-induced process. It tells, for example, what light source would lead to the most efficient results. But, like the quantum yield, the action spectrum may give useful information about molecular mechanisms in complicated systems only with difficulty; the action spectrum for some photobiological responses is a faithful representation of the absorption spectrum of the key chromophore only when several conditions are met. These conditions are reviewed in Section 1.5.4.

Despite all the possible pitfalls (see Section 1.5.4.), there are a number of photobiological responses whose action spectra follow the absorption spectra of the key component extraordinarily well, e.g., the action spectrum for the killing of several microorganisms follows the absorption spectrum of DNA. In many species, the action spectrum for the visual response follows the absorption spectrum of the visual pigment.

A counterexample is perhaps best provided by the action spectrum for the sunburn response (delayed erythema) in man. It has two peaks in the ultraviolet (UV) region, one at 295 nm and one at 255 nm, with a minimum near 280 nm (Fig. 3-6). It is thought by some workers that photodamage to DNA is the key step in the sunburn response. This leads to an interpretation of the action spectrum as the product of the absorption spectrum of DNA (a broad band centered near 260 nm) and the transmission spectrum of protein (a narrower band centered near 280 nm). The protein, which is acting as an inner filter or screen, is that located in the horny (uppermost) layer of the skin.

It is sometimes useful to consider the product of the action spectrum and the spectrum of the light source (usually the sun). A good term for this product spectrum is the *effectiveness spectrum* (coined by F. Urbach). The effectiveness spectrum for sunburn (Fig. 3-6) is peaked at about 308 nm (Section 7.3.1.), a very important fact for those who formulate sunscreen preparations.

3.8. PHOTOSENSITIZATION

An extensive treatment of photosensitization with special reference to biological systems is given in Chapter 4. The brief treatment given here is meant to serve as an introduction.[13]

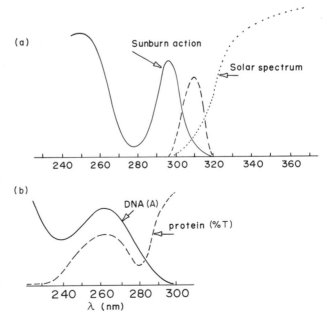

Fig. 3-6. (a) The action spectrum for the delayed cutaneous erythemal response called sunburn (solid line), a typical spectrum of sunlight at the earth's surface (dotted line), and the product of the two spectra or the effectiveness spectrum (dashed line). (b) The absorption spectrum of DNA (solid line), and the transmission spectrum of a protein that contains aromatic amino acids (dashed line). The product of these two spectra should be the action spectrum for DNA damage in the case where a layer of protein lies between the light source and the DNA. This product spectrum (not shown) is similar to the sunburn action spectrum.

The phenomena that are called "photosensitization" are manifold. One definition of photosensitization that encompasses all cases is: *The action of a component (photosensitizer) of a system that causes another component of the system to react to the radiation.* The reaction is chemical in the cases of interest here.

In general, one can study independently a system that does not contain the sensitizer, and does not exhibit the photochemical change of interest when irradiated with light of particular wavelengths. For example,

$$(X,Y,Z) \xrightarrow{h\nu} (X,Y,Z) \qquad \text{(no reaction)} \tag{3-47}$$

When the sensitizer (S) is present, photochemical reaction does occur. For example,

$$(X,Y,Z,S) \xrightarrow{h\nu} (X,Y,P,S) \qquad \text{(sensitized reaction } Z \to P) \tag{3-48}$$

To elucidate the mechanism of a photosensitization, one must determine: (1) whether or not the sensitizer absorbs the exciting light; (2) whether or not the sensitizer is consumed in the reaction; (3) whether or not electronic energy transfer could be involved in the sensitization; (4) whether or not electron transfer could be involved in the sensitization; (5) whether or not molecular oxygen is part of the reactive system. The answers to all these questions are relatively easy to obtain experimentally, assuming, of course, that all the relevant components of the system are known. The detailed molecular mechanism for a photosensitization reaction can range from being very simple to being extremely complex, and the answers to the five questions posed above serve only to form broad categories. What needs to be elucidated is the key interaction of the sensitizer with the system. A comparison of the action spectrum for the reaction with the absorption spectrum of the sensitizer (keep in mind the limitations discussed in Section 1.5.4.) usually answers question (1).

The action of a photosensitizer may be catalytic in nature, i.e., sooner or later it may be found in its original form, and may function again. In fact, a sensitizer is only referred to as such, by most photochemists, if it is not consumed in the reaction. This is not the case in photobiology, however.

Excitation transfer provides perhaps the simplest scheme by which a sensitizer may function without being consumed. The usual scheme for sensitization involving excitation transfer is

$$S \xrightarrow{h\nu} S^* \tag{3-49}$$
$$S^* + A \to S + A^* \tag{3-50}$$
$$A^* \to P \tag{3-51}$$

or

$$A^* + B \to P' \tag{3-52}$$

where S is the sensitizer that absorbs the light, A and B are reactants in a photochemical step that leads to products P or P', and A is a suitable acceptor of excitation from S*. In practice, the most useful schemes in solution photochemistry have involved the triplet state of the sensitizer. Without the sensitizer, there may be no reaction because either the light is not absorbed by the reactants, or, if it is, there is only inefficient crossing to the triplet state that gives the reaction of interest.

The production of singlet oxygen ($^1O_2$) is most efficiently achieved by excitation transfer to ground-state molecular oxygen ($^3O_2$) from a suitable triplet-state sensitizer (triplet energy > 23 kcal/mol, the energy required to excite ground-state oxygen to singlet oxygen):

$$^3S* + {}^3O_2 \rightarrow S + {}^1O_2 \tag{3-53}$$

The optical transition between $^3O_2$ and $^1O_2$ is so weak that practical concentrations of $^1O_2$ can be produced by direct excitation only by use of laser intensities.

Considerations of electronic energy levels obtained from spectral data are usually a sufficient basis on which to exclude excitation transfer as a possibility. When the energetics are appropriate, excitation transfer must be considered to be a possibility.

Electron transfer between the sensitizer and another component of the system is the primary process in many sensitized photoreactions:

$$S* + A \rightarrow S^+ + A^- \tag{3-54}$$

or

$$S* + A \rightarrow S^- + A^+ \tag{3-55}$$

The photoscission of thymine dimer (T <> T) in solution is sensitized by anthraquinone sulfonic acid (A) by way of an electron transfer mechanism:

$$A \xrightarrow{h\nu} {}^1A* \rightarrow {}^3A* \tag{3-56}$$

$$^3A* + T <> T \rightarrow A^- + T <> T^+ \rightarrow A^- + T + T^+ \rightarrow A + 2T \tag{3-57}$$

The energetics of electron transfer can be calculated from the ground-state redox potentials together with the excitation energy of the excited partner in the transfer (usually the sensitizer). If the energetics are unfavorable, electron transfer can usually be safely excluded from consideration. Favorable energetics, however, is not a sufficient condition for electron transfer to occur.

There are many examples of photosensitized reactions in which it appears that the formation of a complex (exciplex) occurs between the excited sensitizer and the substrate. These reactions are characterized by the following observations: The sensitizer is not consumed, excitation transfer is precluded on energetic grounds, the chemistry that the substrate undergoes is unimolecular, i.e., a rearrangement, and no discrete intermediates appear to be involved. The strength and nature of the substrate–sensitizer interaction appear to vary widely. For various cases the complex has been written $(SA)^*$, $(S^{\delta-}A^{\delta+})^*$, $(S^{\sigma+}A^{\delta-})^*$, $(\dot{S}\text{-}\dot{A})$, reflecting a range from excitation-resonance interactions to charge-transfer interactions, to actual covalent bond formation.

The photoisomerization of norbornadiene to quadricyclene, sensitized by aromatic hydrocarbons such as naphthalene, is thought to involve an exciplex (Section 2.5.2) intermediate in which charge transfer is important [Eq. (3-58)]. It has been suggested that the photoisomerization of olefins sensitized by ketones, which have insufficient excitation energy for excitation transfer to occur, may involve covalently bound ketone–olefin intermediates.

$$\text{norbornadiene} \xrightarrow[\text{naphthalene}]{h\nu} \text{quadricyclene} \qquad (3\text{-}58)$$

In photobiology and medicine, the definition of photosensitization includes cases in which the sensitizer is consumed in the reaction of interest. Usually the sensitizer is not naturally part of the system, but when it is introduced, purposely or inadvertently, the photoeffect is obtained. The simplest scheme for photosensitization in which the sensitizer is consumed is the one in which the sensitizer becomes part of the photochemical reaction product. For example, the skin pigmentation photosensitized by psoralens probably involves photoaddition of the psoralen to the DNA of epidermal cells (see Sections 4.4.1. and 7.6.2.1.).

It may not be apparent that the sensitizer is being consumed if it takes part in a primary process that produces a catalyst, such that only a small loss of sensitizer leads to a large yield of the reaction of interest. For example, the excited sensitizer may react in a variety of ways to form a relatively long-lived free radical, which can initiate a long radical chain process, such as an autooxidation.

This discussion ends with an example of photosensitization in a simple system in which the photosensitizer does not absorb the light. The direct irradiation of coumarin in ethanol yields the *cis-syn* photodimer with very low quantum yield. This dimer is formed from the singlet state of the coumarin. No dimer is formed in benzene solution. The addition of a small amount of benzophenone to coumarin solutions, in either benzene or ethanol, leads to the production of the *trans-syn* dimer of coumarin with high quantum yield, even when all of the exciting light is absorbed by the coumarin. The *trans-syn* dimer is formed from the triplet state of coumarin, and the sensitizing effect of benzophenone involves the production of coumarin triplets through two energy-transfer steps; coumarin singlets transfer energy to benzophenone where intersystem crossing takes place, and then triplet excitation is transferred back to coumarin. A lesson to be learned from this example is that an important controlling or sensitizing component may have been overlooked, even after the photochemical products and light-absorbing components of a complex system have been identified.

3.9. SUMMARY

Except for responses purely to the heating effects of light absorption, a photobiological response means that the absorbed light has effected some underlying chemical change. A quantum of light absorbed by a molecule produces an electronically excited molecule. The excited molecule relaxes (sheds its excess electronic energy) in a short time (10^{-12} to 10^{-3} s) by several possible paths among which are paths that lead to ground-state molecules chemically different from the starting molecule. Chemical reactions effected by light in this way are called photochemical reactions.

The chemical change that occurs most directly as a result of the relaxation of an electronically excited molecule is called the primary photochemical process. The product(s) of the primary photochemical process may be a stable chemical species. However, the products of most primary photochemical process are highly reactive intermediates (i.e., radicals, ions, etc.) that react further in secondary (thermal) processes. The microenvironment of the photochemical system very often affects the primary photochemistry and secondary chemistry in different ways. Because of this, it is crucial in modeling photobiological systems to keep in mind that the overall efficiency of production of some final product may depend in a complicated way on changes in experimental conditions. One must proceed with caution in relating photochemical reactivity to photobiological response. With a detailed understanding of the mechanisms of the primary photochemistry and secondary chemistry one may find means to assist, modify, or thwart the photobiological response at the level of the underlying photochemistry.

3.10. EXERCISES

1. The antibacterial agents, halogenated salicylanilides (such as tetrachlorosalicylanilide), cause photoallergic reactions of the delayed hypersensitivity

tetrachlorosalicylanilide

type. The allergen is presumably some photochemically produced compound consisting of the salicylanilide bound to a carrier protein. Salicylanilides with different numbers and types (chlorine or bromine) of halogen substituents lead to different severities of the photoallergic reaction in a test animal. Write down as many steps as you can imagine for the sequence that starts with the topical application of the salicylanilide, and the observation of the delayed photoallergic response. Consider which of these steps might be sensitive to a change in the salicylanilide. Suggest a general outline for a study aimed at explaining the different responses elicited by different salicylanilides. (It has been suggested, but not proven, that the key primary photochemical process is the homolytic cleavage of a halogen from the phenol ring.)

2. 2-Methoxynaphthalene undergoes photodimerization when irradiated with 313 nm light in de-aerated benzene solution. When 0.05 M benzophenone is added to the solution, *no* dimer is formed when exposed to either light at 360 nm (absorbed only by the benzophenone) or at 313 nm (the methoxynapthalene absorbs most of this light even in the presence of the benzophenone). Explain. (*Hint:* The lowest excited singlet and triplet energies of 2-methoxynaphthalene are about 95 and 60 kcal/mol, respectively; those of benzophenone are 85 and 70 kcal/mol, respectively.)

3.11. REFERENCES

1. G. Porter, Photochemistry of complex molecules, in: *An Introduction to Photobiology* (C. P. Swanson, ed.), pp. 1–22, Prentice–Hall, Englewood Cliffs, N.J. (1969).
2. Th. Förster, Adiabatic and non-adiabatic processes in photochemistry, *Pure Appl. Chem.* **24,** 443–450 (1970).
3. G. S. Hammond, Reflections on photochemical reactivity, in: *Advances in Photochemistry* (J. N. Pitts, Jr., G. S. Hammond, and W. A. Noyes, Jr., eds.), Vol. 7, pp. 373–391, Interscience, New York (1969).
4. N. J. Turro, Photochemical reactions of organic molecules, in: *Energy Transfer and Organic Photochemistry* (P. A. Leermakers, ed.), pp. 133–296, Interscience, New York (1969).
5. O. L. Chapman, *Organic Photochemistry,* Vol. 1, Marcel Dekker, New York (1967).
6. D. O. Cowan and R. L. E. Drisko, *Elements of Organic Photochemistry,* Plenum Press, New York (1976).
7. N. J. Turro, *Modern Molecular Photochemistry,* Benjamin, San Francisco, Calif. (1977).
8. J. G. Calvert and J. N. Pitts, Jr., *Photochemistry,* Wiley, New York (1966).
9. A. Schonberg, *Preparative Organic Photochemistry,* Verlag Chemie, New York (1968).
10. L. Salem, Theory of photochemical reactions, *Science* **191,** 822–830 (1976).
11. R. B. Setlow and E. C. Pollard, *Molecular Biophysics,* Addison-Wesley, Reading, Mass. (1962). [The theory and practice of action spectra are discussed on pp. 277–288.]
12. H. E. Johns, Quantum yields and kinetics of photochemical reactions in solution, in: *Creation and Detection of the Excited State* (A. A. Lamola, ed.), Vol. IA, pp. 123–172, Marcel Dekker, New York (1971). [An especially detailed and lucid description of the experimental aspects of quantum yield measurements.]
13. A. A. Lamola, Fundamental aspects of the spectroscopy and photochemistry of organic compounds; Electronic energy transfer in biologic systems; and Photosensitization, in: *Sunlight and Man* (M. A. Pathak, L. C. Harber, M. Seiji, and A. Kukita, eds.), pp. 17–55, University of Tokyo Press, Tokyo (1974).

4

Photosensitization

4.1. INTRODUCTION[1-9]

A generalized view of photosensitization is presented in Section 3.8. as an introduction to the subject; the treatment here emphasizes photosensitization in biological systems. Most cells are rather insensitive to direct effects of visible light, since their major organic constituents do not absorb appreciably in this wavelength range. In the presence of an appropriate photosensitizer, however, organisms, cells, and many types of biologically important molecules can be damaged and destroyed by visible light. Cells are probably killed by selective photochemical effects on certain cell organelles; damage to the organelle results

John D. Spikes • Department of Biology, University of Utah, Salt Lake City, Utah

from selective alteration of macromolecules in the organelle, and alteration of the macromolecule results from selective damage to certain of its subunits. Thus, the study of photosensitization in biology ranges from an examination of the photochemistry of excited sensitizer molecules to the photosensitized killing of mammals.

Although we now realize that observations of photosensitization phenomena must have been made quite early,[1] we usually regard this area of photobiology as having started with the work of Oscar Raab, a medical student in Munich. He showed that acridines sensitized paramecia to killing by visible light; this work was published in 1900. A large amount of research has been done in this area in the intervening years. In addition to its fundamental interest as a photobiological phenomenon, the study of photosensitization reactions has assumed greater importance in recent years because of the rapidly expanding increase in the use of an enormous variety of chemicals in medicine, industry, agriculture, and the home, many of which sensitize humans to injury by light.

In most photosensitized reactions, the light energy involved is absorbed by the photosensitizer molecule rather than by the biological system affected. Light of wavelengths longer than 320 nm is generally used in photosensitization studies, since many biologically important molecules, including proteins and nucleic acids, absorb at shorter wavelengths. Photosensitizers are often termed "dyes" in the biological literature even though many are not actually dyes. To function as a photosensitizer, a molecule must be capable of more than merely absorbing light of the wavelength involved; as we shall see later, it must, in general, be able to be excited by light into a long-lived energy-rich form, the triplet state (see Section 2.3.5.). Many kinds of natural and synthetic molecules can act as photosensitizers.

Most of the photosensitized reactions in biological systems that have been studied involve the participation of molecular oxygen, i.e., they are sensitized photooxidation processes. Such reactions in biological systems are often termed "photodynamic action" or "photodynamic" reactions.[1,6] The sensitizer is often not consumed or destroyed in reactions of this type, but is used over and over again in a somewhat catalytic fashion. In contrast, a few types of photosensitizers, such as the furocoumarins, do not require molecular oxygen for their action, and may be consumed in the reaction.

4.2. MECHANISMS OF PHOTOSENSITIZED REACTIONS INVOLVING MOLECULAR OXYGEN[6–15]

Most photosensitized reactions involve excited electronic states of the sensitizer molecule. The excited sensitizer can react directly with the substrate (the molecule being altered in the reaction), or, alternatively, with some other molecule in the reaction mixture, giving products that in turn can react with the substrate. Many photosensitized reactions are quite complex, and involve the competition of two or more reaction pathways.

4.2.1. Excited States of Sensitizer Molecules

Photodynamic reactions involve electronically excited states of the sensitizer molecule, as discussed in Section 3.8. In the dark, sensitizer molecules exist in the so-called ground state, S_0. On absorption of a photon of light, the ground state of the sensitizer molecule is raised to a short-lived (10^{-9} to 10^{-6} s), energy-rich excited state, the singlet excited state, ^1S. Most effective photodynamic sensitizers then "cross over" with high efficiency to a long-lived (10^{-3} to 10 s) excited state, the triplet state, ^3S. Because of its longer lifetime, the triplet sensitizer has a good chance of interacting with other molecules before decaying back to the ground state. Thus, in general, photosensitized oxidations proceed via the triplet state of the sensitizer. These reactions are shown schematically below:

$$S_0 \overset{h\nu}{\rightarrow} {}^1S \tag{4-1}$$
$$^1S \rightarrow S_0 \tag{4-2}$$
$$^1S \rightarrow {}^3S \tag{4-3}$$
$$^3S \rightarrow \text{photosensitized reactions} \tag{4-4}$$

As may be seen, S_0 absorbs a photon, $h\nu$, to give ^1S via Eq. (4-1); this can decay back to the ground state by Eq. (4-2), or cross over to give the triplet state, ^3S, by Eq. (4-3). The triplet sensitizer can undergo its primary reaction with molecules in its vicinity [Eq. (4-4)] by a hydrogen atom or electron transfer process (termed a type I = radical = redox reaction), or by an energy transfer process (termed a type II reaction). These processes will be described in the next two sections. A triplet sensitizer can decay to the ground state by Eq. (4-5); it can also be converted to the ground state by certain compounds (physical quenchers, Q) as shown in Eq. (4-6):

$$^3S \rightarrow S_0 \tag{4-5}$$
$$^3S + Q \rightarrow S_0 + Q \tag{4-6}$$

4.2.2. Free Radical Photooxidations

The triplet sensitizer molecule can react directly with some kinds of substrate molecules, A, by an electron transfer process to give a semireduced form of the sensitizer, $S^{\bar{\ }}$, and a semioxidized form of the substrate, $A^{\dot{+}}$ [Eq. (4-7)]; these last two species are free radicals. The semioxidized substrate can react with molecular oxygen via Eq. (4-8) to give a fully oxidized product, A_{ox}:

$$^3S + A(\text{substrate}) \rightarrow S^{\bar{\ }} + A^{\dot{+}} \tag{4-7}$$
$$A^{\dot{+}} + {}^3O_2 \rightarrow A_{ox} \tag{4-8}$$

Other reactions not shown, including chain reactions, can also be involved.[9] The ground-state sensitizer can be regenerated from the semireduced form by reaction with oxygen [Eq. (4-9)]; the oxygen superoxide radical, O_2^-, can be produced in this process, and in turn oxidize another molecule in the system via Eq. (4-10). Oxygen superoxide can also be produced by an electron transfer between triplet sensitizer and ground-state oxygen; the other product is a semioxidized form of the sensitizer, S^+ [Eq. (4-11)]; this process is typically quite inefficient. Ground-state sensitizer can be regenerated from the semioxidized form as shown in Eq. (4-12), while, as in Eq. (4-10), the oxygen superoxide radical can oxidize some substrates:

$$S^- + O_2 \rightarrow S_0 + O_2^- \tag{4-9}$$
$$O_2^- + A \rightarrow A_{ox} \tag{4-10}$$
$$^3S + {}^3O_2 \rightarrow S^+ + O_2^- \tag{4-11}$$
$$S^+ + A \rightarrow S_0 + A^+ \tag{4-12}$$

One well-known simple reaction proceeding by a type I process is the photooxidation of secondary propanol. In the primary photochemical step, hydrogen is abstracted from the —OH group of the alcohol by the triplet-state sensitizer. The propanol free radical then gives rise to acetone, while the sensitizer radical reacts with oxygen to give, ultimately, ground-state sensitizer and hydrogen peroxide. If methionine is illuminated in the presence of flavin, the triplet flavin abstracts an electron from the sulfur atom of the amino acid; the resulting methionine radical undergoes deamination and decarboxylation reactions to give methional (β-methylmercaptopropionaldehyde). The flavin radical then participates in a series of reactions that give ground-state flavin and hydrogen peroxide. In both examples, ground-state dye is regenerated, which can then participate in the photooxidation of another substrate molecule. Certain photosensitizers, including anthraquinone dyes and ketones, tend to sensitize by a type I process; compounds that readily undergo reduction, such as amines and phenols, are good substrates for type I reactions.

4.2.3. Singlet Oxygen Photooxidations

The most common type II (energy-transfer) process is the interaction of triplet sensitizer with ground-state oxygen (which is in the triplet state). The principal path is that shown in Eq. (4-13), which gives ground-state sensitizer and a highly reactive, singlet state of oxygen, $^1O_2$:

$$^3S + {}^3O_2 \rightarrow S_0 + {}^1O_2 \tag{4-13}$$
$$^1O_2 + A \rightarrow A_{ox} \tag{4-14}$$

Singlet oxygen can exist in two excited states; the one with the longer lifetime, $^1\Delta_g$, is apparently the main species involved in photodynamic reactions. Singlet oxygen is much more reactive than ground-state oxygen and can interact with a wide variety of substrates to give a fully oxidized form of the substrate [Eq. (4-14)]. Sensitizers such as acridines, porphyrins, xanthenes (eosin Y, rose bengal, etc.), and thiazines (methylene blue, thionine, etc.) tend to give good yields of singlet oxygen, and thus sensitize reactions by this pathway.

Singlet oxygen pathways are very common in photodynamic reactions. Many techniques are available to determine whether a particular sensitizer photooxidizes via the singlet oxygen mechanism. For example, the lifetime of singlet oxygen is ~20 μs in heavy water (D_2O) as compared to only 2 μs in regular water (H_2O). Thus, if the rate of a photodynamic reaction, such as the inactivation of an enzyme, is increased when it is carried out in D_2O as compared to H_2O, this is taken as reasonable evidence that the photooxidation is mediated by singlet oxygen. Another technique is to examine the effect on the reaction rate of an added compound, such as sodium azide, which quenches singlet oxygen with high efficiency. If the rate of the reaction under consideration is sharply decreased by added azide, this is good evidence for a singlet oxygen mechanism.[16]

It should be stressed that the particular reaction pathway(s) involved in a sensitized photooxidation depend on the sensitizer, the substrate and the reaction conditions. In many cases photooxidation can proceed simultaneously by more than one pathway, and the overall process can be more complicated than shown in the preceding equations.

4.2.4. Photodynamic Sensitizers

Many kinds of compounds have been reported to sensitize biological systems to light. Over 400 are listed in the most recent compilation (reference 7, p. xi) and there must be many, many more. The structural formulas for a number of typical photodynamic sensitizers are shown in Fig. 4-1; as may be seen, most are triheterocyclic compounds. Almost all photosensitizing dyes are fluorescent, but fluorescence as such is not involved in photosensitization. The important factors are the lifetime and the efficiency of populating the triplet state of the dye. As shown in Table 4-1, the photodynamic efficiencies of fluorescein derivatives for the photodynamic inactivation of the enzyme trypsin are directly proportional to the quantum yields of triplet-state formation.

Free dyes are efficient photosensitizers for many substrates, such as amino acids and proteins. For nucleic acids and viruses, binding of the dye to the substrate is often necessary for efficient photooxidation. In cells and tissues, most of the photosensitizing dye is probably bound to or closely associated with particular cell structures. Binding often changes the photodynamic efficiency of a photosensitizer. In extreme cases, a good sensitizer can lose all activity on being bound, and there are some cases where a compound is inactive in solution but becomes a photosensitizer on binding. Thus, dyes that sensitize in one system are not always effective in others.

Azo dyes, indophenols, ketone imine dyes, methine dyes, nitro dyes, nitroso dyes, oxazine dyes, thiazole dyes, and triarylmethane dyes (see reference 17 for structures) do not, in general, sensitize substrates such as proteins. Photodynamically active dyes for proteins include representatives from the

TABLE 4-1. Photodynamic Efficiencies of Substituted
Fluorescein Sensitizers

Sensitizer	Quantum yields[a]		
	Φ_F	Φ_T	Φ_{PDA}
Fluorescein (uranine)	0.92	0.05	0.00017
Tetrabromofluorescein (eosin Y)	0.19	0.71	0.0021
Tetraiodofluorescein (erythrosin B)	0.02	1.07	0.003

[a] Φ_F is the quantum yield of fluorescence emission, Φ_T the quantum yield of triplet-state production, and Φ_{PDA} the quantum yield for the photodynamic inactivation of the enzyme trypsin; modified from M. J. Wade and J. D. Spikes, *Photochem. Photobiol.* **14,** 221–274 (1971).

following classes of compounds: acridines (acridine orange, proflavin, etc.) anthraquinones, azines (such as the safranins), flavins (riboflavin, FMN, lumi-flavin; but not FAD), prophyrins (including chlorophyll and some of its deriva-tives, hematoporphyrin, protoporphyrin, etc.; but not porphyrins containing iron or other paramagnetic metal atoms), thiazines (thionin, methylene blue, toluidine blue, etc.), thiopyronin, and xanthenes (especially the fluoresceins, including eosin Y, rose bengal, etc.). Recently, coordination compounds of some metals such as ruthenium have been found to be good photosensitizers for amino acids and proteins. For nucleic acids, basic dyes such as methylene blue and thiopy-ronin are often the most efficient photosensitizers; flavins and acridines are also quite effective. In cells, the sensitizing dye must be able to penetrate into a susceptible region of the cell structure to be effective. Even correcting for this, different sensitizers vary enormously in photodynamic efficiency. For example, *E. coli* is rapidly killed on illumination in the presence of thiopyronin, more slowly with methylene blue, and very very slowly with pyronine Y (reference 7, p. 107), even though these three dyes have very similar structures (Fig. 4-1).

4.3. BIOLOGICAL PHOTOOXIDATIONS INVOLVING MOLECULAR OXYGEN (PHOTODYNAMIC ACTION)

4.3.1. Alcohols and Carbohydrates[9,18]

Alcohols have been studied as model compounds for the sensitized photoox-idation of carbohydrates. On illumination, anthraquinone and ketonic sensitizers abstract a hydrogen atom from the alcohol α-carbon, giving a radical form of the alcohol; this reacts with molecular oxygen to yield aldehydes or carboxylic acids for primary alcohols, and ketones for secondary alcohols. The photooxidation of hexitols with these sensitizers gives the corresponding hexoses, and more

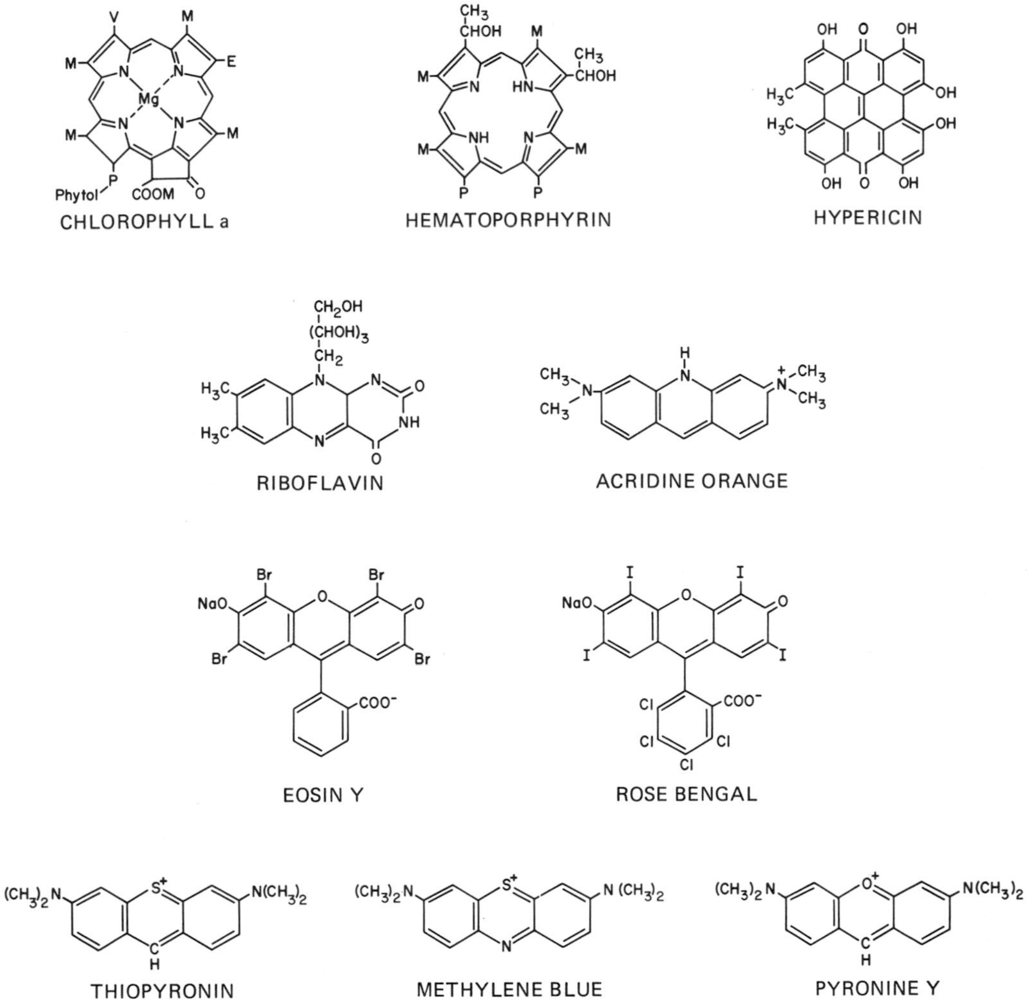

CHLOROPHYLL a

HEMATOPORPHYRIN

HYPERICIN

RIBOFLAVIN

ACRIDINE ORANGE

EOSIN Y

ROSE BENGAL

THIOPYRONIN

METHYLENE BLUE

PYRONINE Y

Fig. 4-1. Structures of selected photodynamic sensitizers. Chlorophyll *a* and hematoporphyrin are typical porphyrin sensitizers; E, M, P, and V in the structural formulas stand for ethyl, methyl, propionic acid, and vinyl groups, respectively. Hypericin is a polycyclic hydrocarbon. A number of effective sensitizers are based on fluorescein, including eosin Y (tetrabromofluorescein) and rose bengal (tetrachlorotetraiodofluorescein). Although the three sensitizers shown in the bottom row have very similar structures, they differ greatly in photosensitizing capability with thiopyronin being most effective, methylene blue less so, while pyronine Y is almost inactive.

slowly, the corresponding hexonic acids.[18] Much of the research on the sensitized photooxidation of carbohydrates has been carried out with the polysaccharide cellulose, because of its importance in the textile industry. Mechanistically, as for alcohols, the photooxidation of cellulose proceeds via a hydrogen abstraction pathway; cellulose is not oxidized appreciably by singlet oxygen.[18,19] The animal mucopolysaccharide hyaluronic acid is depolymerized on illumination in the presence of molecular oxygen and a number of different sensitizers, by what appears to be a free-radical mechanism.[18]

4.3.2. Fatty Acids and Lipids[9]

Unsaturated fatty acids are photooxidized with a variety of sensitizers. The methyl esters of stearic, oleic, linoleic, and arachidonic acids are photooxidized with protoporphyrin IX (a singlet oxygen sensitizer). The rate constants for photooxidation increase on going from the stearic to the arachidonic ester in a linear fashion corresponding to the number of double bonds in each molecule.[20] In some cases, photodynamic damage at the cellular level may involve the photooxidation of fatty acid chains in membrane lipids and phospholipids.

4.3.3. Amino Acids and Proteins[6,21,22]

Shortly after the discovery of photodynamic action, it was shown that crude preparations of the enzymes diastase, invertase, and papain were inactivated on illumination in the presence of eosin. A large amount of research on the mechanisms of photodynamic effects on amino acids, small peptides, and various kinds of proteins has been carried out in recent years. Over 100 purified proteins have been studied.[6]

Some 20 different amino acids occur in typical proteins. Only 5 of these, cysteine, histidine, methionine, tryptophan, and tyrosine are generally susceptible to sensitized photooxidation. The rate of photooxidation of free amino acids depends on the sensitizing dye, the solvent, the pH, and the concentrations of reactants, including oxygen. By varying the dye and reaction conditions, some selectivity in the photooxidation of amino acids can be gained. For example, histidine is photooxidized only above pH ~ 6, where the imidazole nitrogen is ionized; tyrosine is most rapidly photooxidized above pH 10, where the phenolic group is ionized.[21] The organic chemistry of amino acid photooxidation (primary products, intermediates, final products) is poorly understood in most cases. With methylene blue, proflavine, and rose bengal, the final product of methionine photooxidation is methionine sulfoxide; a singlet oxygen mechanism is involved. In contrast, with flavin sensitizers, the principal final product is methional; here an electron abstraction mechanism is involved.[21] With methylene blue, tryptophan gives two classes of photooxidation products, kynurenines and melanines. Cysteine is photooxidized to cysteic acid with crystal violet. Both histidine and tyrosine give rise to a large number of products on photooxidation.[21]

Essentially all proteins are susceptible to photooxidation. On illumination of proteins in the presence of a photosensitizer, oxygen is consumed and amino acid residues are destroyed. Disulfide and peptide bonds in proteins are not usually ruptured on photooxidation; chemical alteration involves the side chains of the five susceptible amino acids.[21] As with free amino acids, the rate of photooxidation of a protein depends on the sensitizer and the reaction conditions. Typically, certain residues of a given amino acid can be photooxidized rapidly in a protein suggesting that they are "exposed" at the surface of the molecule. Others are destroyed more slowly or not at all, indicating that they are protected to various degrees by being "buried" within the three-dimensional structure of the protein.[21] A number of physicochemical alterations occur in proteins during photooxidation, including changes in UV absorption, diffusion and sedimentation behavior, digestibility by proteases, electrophoretic mobility, heat sensitivity, light-scattering properties, optical rotation, solubility, surface tension, viscosity, mechanical properties, etc.[6]

Photodynamic reactions typically alter the biological properties of proteins. The most extensively studied change is the progressive loss of the catalytic activity of enzymes during photooxidation. The quantum yields for inactivation are in the order of 0.001, much smaller than the yields for free amino acids. Inactivation can occur by two mechanisms: directly, by the destruction of amino acid residues in the active site region of the molecule, or indirectly, by the destruction of residues necessary for the active conformation of the enzyme. Again, the mechanism depends on the sensitizer and the reaction conditions. Photooxidation can produce other kinds of biological alterations in proteins including loss of activity in protein hormones such as insulin, loss of ability to bind metal cofactors and coenzymes, changes in antigenic properties and antibody reactivity, loss of toxicity of snake venoms and bacterial toxins, etc.[6]

4.3.4. Purines, Pyrimidines, and Nucleic Acids[6,7,23]

With most sensitizing dyes, studies on the free bases, nucleosides and nucleotides show that guanine and its derivatives are photooxidized most rapidly at pH ~8, thymine derivatives much more slowly, while derivatives of adenine, cytosine, and uracil are essentially not affected. The sensitivity of guanine to photooxidation increases rapidly at higher pH values where a proton dissociates from the 1 - N of the molecule; this indicates that the guanine anion is much more sensitive than the neutral molecule.[7,23] The chemistry of the sensitized photooxidation of purines and pyrimidines is poorly understood. Both rings of guanine and its derivatives are broken on photooxidation; product studies show the formation of guanidine, parabanic acid, CO_2 and a variety of unidentified products.[6,7,24] Comparisons of the products resulting from sensitized photooxidation with those resulting from singlet oxygen oxidation (using singlet oxygen generated chemically and by radiofrequency discharge), show that both singlet oxygen and free-radical mechanisms are involved in the sensitized photooxidation of guanosine and related compounds.[24] In contrast to the results with most sensitizers, flavins sensitize the rapid photooxidation of adenine and its nucleotides.[25]

Guanine residues are selectively destroyed in DNA both *in vitro* and in cells on illumination in the presence of a wide variety of dyes. Although the binding of dye to substrate is not necessary for many photodynamic processes, it appears that binding of the sensitizer to nucleic acids often increases photodynamic efficiency.[23] Photooxidation brings about a variety of changes in the properties of DNA, including a decrease in the melting temperature, a decrease in viscosity (resulting from chain breaks), and an increase in susceptibility to enzymatic degradation.[6,7] Such chemical changes alter the biological activity of nucleic acids. For example, the infectivity of tobacco mosaic virus RNA is destroyed as well as the transforming capabilities of bacterial DNA transforming principle;[6] as will be discussed later, photodynamic treatment produces mutations in viruses and in cells. Finally, photodynamic treatment decreases the extractability of DNA from bacteria as a result of the photochemical cross-linking of DNA with proteins; such cross-linking also occurs *in vitro*.[26]

4.3.5. Viruses[6,23,27,28]

Bacteriophage were the first viruses to be used in photodynamic studies. Penetration of the sensitizing dye into the bacteriophage is an important factor in photodynamic sensitivity. For example, the T-odd *E. coli* phages are very sensitive to photodynamic treatment while the T-even phages are resistant; this difference results from the very slow rate of penetration of dyes through the protein head membrane of T-even phage (reference 7, p. 297). The photodynamic treatment of bacteriophage can damage nucleic acid and/or protein. Nucleic acid damage is the more important factor; as with free nucleic acids, guanine residues in the DNA of intact phage are preferentially destroyed on photodynamic treatment. Many breaks in the polynucleotide strands, as well as cross-linking of DNA fragments to the protein coat of bacteriophage are observed after extensive photodynamic treatment.[29] Photodynamically treated phage can be dark-reactivated in host cells possessing recombination and excision repair capabilities; however, the photoreactivation of photodynamically inactivated phage has not been observed.[30]

A number of animal viruses (adenoviruses, rabies virus, vaccinia virus, etc.) can be inactivated by photodynamic treatment *in vitro*.[27] Some viruses, such as polio virus, are not sensitive, apparently due to lack of penetration of dye into the virus particle; if polio virus is grown in cells in the dark in the presence of dyes, dye is incorporated into the viral structure. The resulting virus is rapidly inactivated upon illumination.[27] Many resistant animal viruses become sensitive to light if incubated with the dye under conditions of high pH; such treatment probably increases the permeability of the protein coat of the virus to the sensitizing dye.[27] Photodynamically inactivated viruses often retain their specific antigenic properties and can therefore be used for the preparation of antibodies for the treatment of viral infections.[6] The infectivity of some plant viruses (tobacco mosaic virus, alfalfa mosaic virus, etc.) can be abolished by photodynamic treatment.[6]

In 1931, Herzberg reported that photodynamic treatment with methylene blue suppressed vaccinia virus eruptions on the skin of the rabbit.[4] More recently, photodynamic treatment has been used on herpes simplex infections of the skin and genitals in humans.[31] Solutions of dyes such as neutral red are applied to the viral lesions and the area is then illuminated with light sources such as ordinary fluorescent lamps. The efficacy of these treatments is still being debated. Further, some concern has been expressed over possible side effects of the treatment, e.g., the possible photodynamic induction of tumors (see Section 4.3.11.).

As might be expected from the effects on nucleic acids described above, photodynamic treatment produces mutations in viruses (reference 7, p. 280). Mutations of various types have been observed in several different kinds of bacteriophage on illumination in the presence of a variety of photosensitizers.[6] Mutants of tobacco mosaic virus have also been produced by photodynamic treatment of intact virus, and of isolated viral RNA; some mutant viruses show amino acid changes in the protein coat.[6] Photodynamic treatment with proflavin also induces mutants in an animal virus, polio virus (reference 7, p. 280). It is presumed that mutation results from the photooxidation of guanine residues in the viral nucleic acid; it should be remembered, however, that photodynamic treatment can produce free radicals in nucleic acids, cross-linking of nucleic acids with protein, and a variety of reactive species, such as peroxides, which might chemically alter nucleic acids, thus producing mutations.

4.3.6. Subcellular Level[3,4]

The mechanisms by which cells are injured and killed by photodynamic treatment are poorly understood. Selective effects might be expected, since different dyes may localize in certain cell organelles or in particular parts of organelles, depending on their molecular properties. For example, it has been shown, using fluorescence microscopy, that a variety of mammalian cell types, cultured in the presence of porphyrins, concentrate the sensitizer in the lysosomes. Subsequent illumination increases the permeability of the lysosomal membranes, and leads to rupture of the lysosomes. Cell death follows, presumably due to the release of lysosomal hydrolytic enzymes into the cytoplasm.[32] Some sensitizers, such as rose bengal, appear to accumulate selectively in the plasma membranes of cells; on illumination, sensitized cells show changes in permeability with the loss of potassium ions, and subsequent extrusion of protoplasm.[32] Acridine orange accumulates selectively in chromosomes; illumination of acridine orange-treated plant cells in the presence of oxygen produces chromosome breakage.[4]

Effects of photodynamic treatment at the subcellular level can also be examined using isolated parts of cells. For example, illumination of isolated liver lysosomes in the presence of oxygen and eosin liberates significant amounts of β-glycerophosphatase, β-glucuronidase, and β-galactosidase.[33] Isolated *E. coli* ribosomes lose the ability to incorporate amino acids into polypeptides following photodynamic treatment.[34] Photodynamic studies on isolated mammalian mitochondria indicate that some step in oxidative phosphorylation is more sensitive than the respiratory pathway. The site of photodynamic interference with electron flow has not been established, although mitochondrial succinic dehydrogenase is much more sensitive than is cytochrome oxidase (reference 7, p. 464). Photodynamic studies at the subcellular level will be discussed further in Section 4.5.

4.3.7. Cells and Unicellular Organisms[4,8,23,35,36]

Photodynamic effects on cells have been examined with many different sensitizers, and with a wide variety of cell types, including microorganisms (both prokaryotic and eukaryotic), and cells from multicellular organisms. Many kinds of cell responses have been described, depending on the cell type and condition, the sensitizer, and the reaction conditions. In some cases, sensitizers are toxic in the dark; thus appropriate control experiments must always be carried out to check this possibility. Most kinds of bacteria are rapidly inactivated (in terms of subsequent colony formation) on illumination with photosensitizers in the presence of molecular oxygen. Photodynamic treatment can inhibit bacterial glycolysis, respiration and protein synthesis; membrane properties are also altered.[37] Wild-type cells of *Sarcina lutea,* which have membrane-bound carotenoids, are less sensitive to photodynamic killing with the sensitizer toluidine blue than are carotenoidless mutants.[36] In kinetic studies of the photodynamic inactivation of bacteria, both "single hit" and "multiple hit" inactivation curves have been observed[4,5] (see Fig. 4-2). Some strains of bacteria show dark recovery after photodynamic treatment with acridine orange or acriflavine, while others do not.[38]

Unicellular green algae are susceptible to photodynamic treatment, although they have not been used much in photodynamic studies.[4] Protozoa, however, have been used extensively for photodynamic studies, especially paramecium. The pattern of effects depends on the protozoan used, the sensitizing dye and the reaction conditions. Responses include delay of cell division, loss of motility, sensitization to subsequent heat treatment, inactivation of the contractile vacuole, and lysis.[4]

Many studies have been carried out with "free" cells from multicellular organisms. For example, mammalian erythrocytes have been used extensively

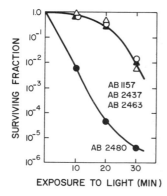

Fig. 4-2. Photodynamic killing of four different strains of *E. coli* K-12 as a function of time of illumination. The photosensitizer was acriflavine at a concentration of 10 μg/ml. Strain AB1157 is a "wild-type," strains AB2437 (*uvr⁻*) and AB2463 (*rec⁻*) are single mutants, and strain AB2480 (*uvr⁻rec⁻*) carries a double mutation affecting sensitivity to UV radiation. (Redrawn from reference 31.)

for photodynamic experiments ever since it was shown in 1908 that illumination of rabbit red blood cells in the presence of hematoporphyrin or chlorophyll preparations caused their hemolysis.[8] Photodynamic damage to erythrocytes appears to be a pure membrane phenomenon. The efficiency of photodynamic hemolysis is low; approximately 10^{10} photons must be absorbed by a sensitized cell to bring about lysis.[1] Erythrocytes from humans with certain kinds of porphyrias (conditions in which abnormal porphyrins accumulate in the body) are light-sensitive. For example, cells from patients with the condition erythropoietic protoporphyria are rapidly hemolyzed on illumination *in vitro* in the presence of oxygen; such cells contain free protoporphyrin, a known photodynamic sensitizer.[8] Amino acids in the cell membrane are destroyed, and potassium is released from the cells during illumination, suggesting membrane damage; also, membrane enzymes, such as acetylcholinesterase, are inactivated.[8] Recent studies suggest that a photooxidation product of cholesterol, which occurs in the red blood cell membrane, increases the osmotic fragility of the membrane leading to hemolysis.[39] Reducing agents and certain other compounds decrease the light sensitivity of cells from protoporphyric patients; β-carotene (a quencher of sensitizer triplet states, and singlet oxygen) is especially effective.[8] Cells isolated from plant and animal tissues, as well as cells maintained in tissue culture, have also been used extensively in photodynamic studies.[4,8] For example, skeletal muscle fibers and neurons are depolarized on illumination in the presence of sensitizing dyes.[4,40] Photodynamic treatment of human and animal cells in tissue culture causes killing, blocking of protein synthesis, interference with DNA replication and mitosis, destruction of protein and nucleic acids, loss of the ability to support the growth of viruses, etc.[4,8]

Finally, photodynamic treatment produces mutations in microorganisms; the efficiency of photodynamic mutagenesis depends on the sensitizer and the strain of microorganism used (see reference 7, p. 280). Mutations have been produced in *Neurospora,* yeast and a variety of bacteria.[4] Recent studies demonstrate the photodynamic production of both base-pair substitution and frameshift mutations in *Salmonella typhimurium* with methylene blue and the tranquilizer chlorpromazine as sensitizers.[41]

4.3.8. Multicellular Plants[1,4,36]

Relatively few studies have been carried out on photodynamic effects on multicellular plants. In general, however, root, stem, and leaf tissues of plants are damaged and killed by illumination in the presence of sensitizers; oxygen is necessary for the reaction.[1,4] Even without added sensitizers, leaves are bleached and killed when subjected to very high-intensity illumination in the presence of oxygen and the absence of carbon dioxide; this results from photooxidation processes sensitized by endogenous chlorophyll.[4,36] Under ordinary conditions, chloroplast components are protected from photooxidation by the high concentration of carotenoids present.[4,36] If the roots of light-insensitive plants, such as wheat, are permitted to take up sensitizing dyes and then illuminated unilaterally, they bend toward the light source as a result of growth inhibition on the illuminated side.[1]

4.3.9. Multicellular Animals[1,2,4,8]

4.3.9.1. Mammals

Two mechanisms of response to light occur in sensitized mammals, phototoxicity and photoallergy, depending on the photosensitizer.[42] The phototoxic reaction is immediate, and occurs in all mammals if sufficient sensitizer and light are used. The photoallergic response is delayed, and appears to result from the photosensitized formation of new compounds that interact with protein to give "photoantigens"; these then produce an immunological type of reaction.[42] The phototoxic response involves a characteristic array of symptoms on illumination including hyperactivity, skin itching and skin injury (including reddening, edema, necrosis, and ulceration).[1] In addition to localized responses, intestinal hemorrhage, decreased blood pressure, and circulatory collapse occur in severe photosensitization; these responses presumably result from generalized transport via the blood of toxic compounds formed photodynamically in the skin.[1]

Light sensitivity in humans, accompanied by the excretion of abnormal porphyrins in the urine, was observed in the last century.[8] Experimentally, photosensitization in humans was first demonstrated in 1913 by Meyer-Betz, a physician who injected himself with hematoporphyrin, and then exposed parts of his body to light in controlled ways; each exposure to light resulted in erythema, swelling, and subsequent pigmentation.[8] Photosensitivity persisted for two months. Photosensitization in man can result from ingestion of the sensitizer, from skin contact with the sensitizer, from an injection of a photosensitizing drug, or from certain diseased conditions including drug-induced and hereditary porphyrias (conditions involving abnormalities of porphyrin metabolism),[43] lupus erythematosus,[42] and pellagra.[42] Patients with the drug-induced condition porphyria cutanea tarda excrete quantities of uroporphyrin (a photosensitizer); they show marked photosensitivity. In the hereditary disease erythropoietic protoporphyria, abnormal amounts of another photosensitizer, protoporphyrin, accumulate in the body; some persons with this condition show a moderate photosensitivity (see reference 7, p. 464). The use of oral β-carotene as a photoprotective agent markedly decreases the photosensitivity of many patients with erythropoietic protoporphyria.[44] Photosensitivity in man induced by chemicals is often referred to as "drug photosensitivity," since many drugs used in medical practice do photosensitize, including certain anesthetics, antibiotics, antihistamines, antiseptics, barbiturates, diuretics, oral antidiabetics, psoralens, sulfonamides, sunscreens, and tranquilizers (especially phenothiazine derivatives such as chlorpromazine).[42] In addition, a large variety of chemicals commonly used in agriculture, industry, the home, etc., are also phototoxic. Drugs to be used in medical practice are typically screened for photosensitizing action using laboratory animals such as guinea pigs and hairless mice. Cosmetics should also be carefully screened. More detailed information on sensitized photoresponses in humans will be found in Chapter 7.

In many parts of the world, plants occur in quantity that contain potent photodynamic sensitizers; grazing animals can become seriously light-sensitive by consuming such plants.[1,2] The animals show responses similar to those described above for man; in severe cases they die. Sensitivity is greatest in unpigmented areas of the skin; historically white horses and white sheep could not be raised in certain parts of the world for this reason. Members of the plant genus *Hypericum,* which occurs worldwide, are among the worst offenders. Many species of this genus contain a potent photodynamic sensitizer, hypericin (or related compounds)[1,2]; the structure of hypericin is shown in Fig. 4-1. Plants of another genus, *Fagopyrum* (buckwheat), also produce pigments that photosensitize grazing animals, swine, and chickens. Because of their photosensitizing capabilities, buckwheat leaves are no longer used as fodder (the grain contains little photosensitizer, so eating buckwheat cakes is not hazardous!).[1]

Ruminants sometimes show a type of photosensitivity resulting from disturbances in liver function.[1,2] Photosensitivity results from an accumulation of phylloerythrin, a chlorophyll derivative, which is ordinarily excreted in the bile.[1,2] Guinea pigs become light-sensitive if fed small, repeated amounts of carbon tetrachloride (which results in liver damage), or if the common bile duct is ligated.[45] In both cases, photosensitization appears to result from the accumulation of phylloerythrin; presumably this results from the inability of the liver to properly excrete this material as a result of liver damage in the first case, and from mechanical obstruction of the bile duct in the second. Ligation of the bile duct in sheep can also result in photodynamic death if the animal is fed green plants; photosensitization does not occur on a chlorophyll-free diet. A hereditary type of porphyria is sometimes observed in cattle.[1,2] Cattle with this condition, which is inherited in a simple Mendelian way, show a reddish pigmentation of the urine, the teeth, and the bones; lightly pigmented skin areas are highly photosensitive. A large amount of work has been done on photosensitization using the usual laboratory animals such as rabbits, rats, and mice.[8] Porphyrin-sensitized phototoxicity in small mammals is a true photodynamic type of sensitization, since oxygen is necessary (in those cases that have been examined).[8]

4.3.9.2. Other Animals[1,4]

Relatively little has been done on photosensitization reactions in lower vertebrates and the invertebrates. Blum[1] describes earlier work on amphibians, fish, rotifers, and hydra. More recently, some work has been published concerning photodynamic studies with newts, amphibian larvae, and insects.[4] Many dyes sensitize mosquito larvae to killing by sunlight[4]; some dyes are effective at concentrations that have no effect on small fish, suggesting this as a possible technique for the control of mosquitos. Adult houseflies are killed on exposure to light after ingesting dyes such as rose bengal, rhodamine, and erythrosin B.[46] The mechanism of killing has not been established; however, the behavior of the sensitized insects on exposure to light indicated nervous system involvement. This work suggests that photodynamic dyes incorporated into appropriate baits could be used as insecticides for the control of positively phototactic insects. Although only a few animal species have been examined, it is reasonable to assume that any animal could be sensitized to light with dyes introduced in an appropriate way.

4.3.10. "Natural" Photosensitivity[3,4,47-49]

Certain multicellular animals are injured and/or killed by visible light; this results from photosensitization by endogenous pigments. In some cases, an oxygen requirement has been demonstrated. For example, the small aquatic annelid *Tubifex* is killed in a few minutes if exposed to bright light in the presence of oxygen; this organism normally lives in areas of water where both the oxygen content and the light intensity are quite low.[50] Furthermore, certain kinds of cells in some multicellular animals are light-sensitive. In particular, the mammalian retina is severely damaged in the presence of oxygen by high-intensity light of wavelengths shorter than 500 nm.[3] The metabolism of Yoshida hepatoma cells is sharply reduced on illumination in the presence of oxygen, but not in nitrogen.[3] Under some conditions, multicellular green plants are also injured by light in chlorophyll-mediated reactions.[49]

Natural photosensitivity is observed in a wide variety of microorganisms. For example, visible light selectively inactivates certain membrane active transport systems in *E. coli*;[51] this phenomenon may be a useful tool in identifying component molecules involved in membrane energy coupling and transport processes. In general, the chemical nature of the endogenous sensitizer involved in natural photosensitivity is not known. One exception is the case of the protozoan *Blepharisma*, where the cells contain a red fluorescent pigment termed zoopurpurin; this pigment is chemically similar to hypericin (Fig. 4-1), and has been shown to be a photodynamic sensitizer.[47] Mutant strains of photosynthetic bacteria that lack colored carotenoids are rapidly killed on illumination in the presence of oxygen; the action spectrum for this response is essentially the same as the absorption spectrum for bacteriochlorophyll, suggesting that chlorophyll is the photosensitizer.[47] Light inhibits respiration and growth in the "colorless" alga *Prototheca zopfii*; oxygen is necessary for the inhibitory effect.[49] The action spectrum for this phenomenon suggested the involvement of a heme protein, and further studies showed that blue light rapidly inactivated cytochrome a_3, a component of the respiratory electron transport chain.[49] Mutagenesis is produced in bacteria and some other microorganisms by visible and by 365 nm near-UV light.[48] Presumably a natural, internal chromophore, which acts as a sensitizer, is involved; possible chromophores, most of which are known to be present in bacteria, include cytochromes, cytochrome oxidase, flavins, hemoproteins, NAD, NADH, porphyrins, and quinones. As has been pointed out,[48] the accumulating array of examples of natural photosensitivity to near-UV radiation and to visible light should result in concern over the increasing use of near UV-emitting sources and high-intensity visible light sources in industry and in home lighting.

4.3.11. Light, Photosensitizers, and Cancer[1,3,8,9,14]

It is well known that UV radiation in the 280–320 nm range can induce skin cancer in experimental animals (and presumably in humans); it is not so well known that long-wavelength UV radiation and visible light can also induce skin cancer when used with an appropriate photosensitizer. This was first demonstrated by Büngeler in 1937, who found that extended illumination of mice injected subcutaneously with solutions of eosin or hematoporphyrin produced skin tumors; neither illumination nor sensitizer alone gave this effect.[3] Presumably this same phenomenon would occur in humans; in fact, it has been suggested that the known accumulation of porphyrins in human skin with aging may result in photodynamic carcinogenesis.[3] Many carcinogenic polycyclic hydrocarbons, such as 3,4-benzpyrene, are photodynamic when illuminated with long-wavelength UV radiation. Studies with a large number of such compounds show that there is a rather good correlation between chemical carcinogenicity and photosensitizing efficiencies.[3] The appearance of skin tumors in mice painted with 3,4-benzpyrene solution is accelerated by illumination with long-wavelength UV radiation (reference 7, p. 671). The mechanism of photodynamic carcinogenesis with sensitizers of this type is not known, although photodynamic treatment with 3,4-benzpyrene as a sensitizer increases the frequency of chromosome breaks in plant cells (reference 7, p. 671). More recent speculations on photodynamic carcinogenesis suggest the possible involvement of photochemically generated singlet oxygen.[9,52]

In 1904, Jesionek and von Tappeiner reported the successful treatment of skin tumors in humans by a photodynamic technique, which involved the repeated painting and injecting of the tumorous area with eosin solution, followed by illumination with sunlight or an arc lamp.[1] Recently, more selective photochemotherapeutic techniques have been used for the treatment of tumors in experimental animals. Hematoporphyrin and certain of its derivatives are taken up and concentrated in neoplastic tissues; such an accumulation would be expected to sensitize selectively the photodynamic destruction of tumor cells on illumination. Using this technique, a variety of tumor types growing in mice and rats can be killed with high efficiency.[53,54]

4.3.12. Photodegradation of Food

Photodegradation of food is a well-known phenomenon and accounts for the care taken in the storage and packaging of many kinds of susceptible items to protect them from light. Although the mechanisms of the photodegradation of foods are not known in general, some cases are clearly photodynamic. For example, olive oil is often rapidly degraded on exposure to light in the presence of air. This results from the photooxidation of unsaturated fatty acid side chains in the oil, perhaps sensitized by traces of chlorophyll in the oil; a similar behavior is shown by soybean oil, where photooxidation appears to occur by a singlet oxygen mechanism.[55] Exposure of milk to light results in a riboflavin-sensitized photooxidative degradation of milk proteins[56]; in addition, methionine is photooxidized to methional giving the milk a characteristic off-flavor. A similar ''sunlight flavor'' develops in beer exposed to light; this is one reason that bottles for beer are usually made from a dark brown glass. Potato chips and other snack foods stored in transparent containers also develop off-flavors on illumination; this has been shown in some cases to result from the photooxidation of the cooking oils used.[57] Attempts are being made to prevent such photochemical deterioration by incorporating carotenoids in such foods. It has been usually considered that naturally occurring photosensitizers were responsible for the photodegradation of foods; recent work shows, however, that certain artificial food colors may also be involved.[58] Fatty acids in detached leaves undergo photooxidation on illumination, presumably sensitized by chlorophyll; linoleic acid is degraded most rapidly.[59] During the drying process in the sun, sensitized photochemical processes go on in hay that decrease its nutritional value; storing of fresh green plant material as silage prevents this decrease.

4.4. PHOTOSENSITIZATION REACTIONS NOT INVOLVING MOLECULAR OXYGEN[47,60]

Although this chapter has stressed oxygen-requiring photosensitization reactions, several important photosensitized processes in biological systems do not require molecular oxygen, and in fact are inhibited by its presence.

4.4.1. Furocoumarins

A number of furocoumarins (derived from the condensation of a coumarin nucleus with a furanoic ring) are potent photosensitizers for certain biological systems, when irradiated with long-wavelength UV radiation. Many furocoumarins (based on psoralen and isopsoralen) occur naturally in certain plant species; the structures of some typical compounds are shown in Fig. 4-3. Unlike the photodynamic sensitizers described earlier in this chapter, photosensitization by furocoumarins does not require the participation of molecular oxygen; furthermore, temperature has little effect on the rate of furocoumarin-sensitized reactions. Carotenoids, which are potent protective agents against photodynamic action by virtue of their ability to quench singlet oxygen, have no effect with furocoumarins. A number of biological systems that are very sensitive to photodynamic treatment are not appreciably affected by furocoumarins plus light. For example, furocoumarins do not sensitize the photohemolysis of red blood cells.[60]

Furocoumarins do sensitize a number of systems, however. For example, some furocoumarins applied topically or ingested sensitize an erythema of mammalian skin (including humans) on subsequent UV treatment; typically dark

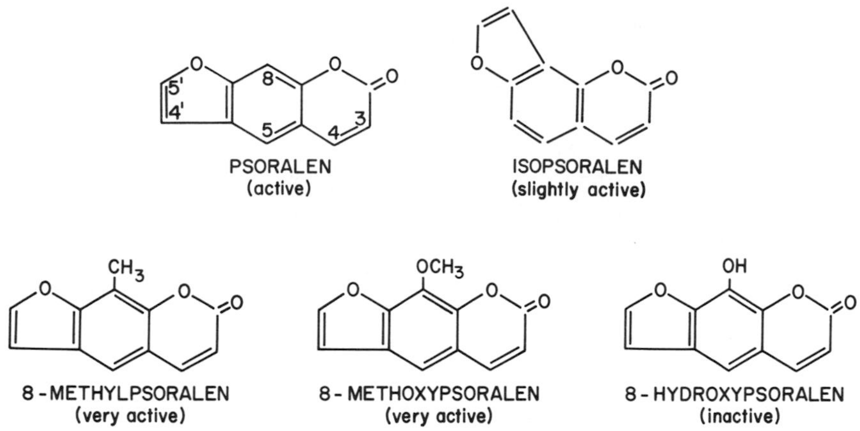

Fig. 4-3. Structures of selected photochemically active and inactive furocoumarins. It will be noted that substitution of a methyl or methoxy group in the 8-position of psoralen increases photochemical activity, while substitution of a hydroxyl group eliminates activity. A similar pattern is observed with substituents in the 5-position. Isopsoralen is also known as angelicin, while 8-methoxypsoralen is also termed methoxsalen or xanthotoxin.

tanning results. Such a treatment has been used cosmetically in humans to darken the white areas of skin that occur in the condition termed vitiligo (see Chapter 7); this treatment was used as early as 1400 B.C. in India.[47] Very recently the skin condition termed psoriasis in humans has been successfully treated with long-wavelength UV radiation following the ingestion of 8-methoxy-psoralen; presumably the ameliorating effect of this "photochemotherapy" results from the inhibition of DNA synthesis and mitosis in the affected areas of epidermis.[61] These studies are being carried out with some caution because of reports of skin cancer production in animals subjected to psoralen photosensitization.[47,62]

Furocoumarins are effective photosensitizers at the cellular level. Bacteria are rapidly killed by illumination with 365 nm radiation in the presence of furocoumarins; gram positive strains are most susceptible. No photoreactivation is observed.[47,60] Animal cells in culture are also killed. Furocoumarins markedly increase the frequency of mutations in bacteria and yeast on illumination at 365 nm[60] (also see reference 7, p. 280). Irradiation of certain bacteriophage and animal viruses in the presence of psoralen gives complete inactivation; in general, RNA viruses are more resistant.[47,60]

A variety of studies have attempted to identify those biomolecules susceptible to photosensitized alteration by furocoumarins. Furocoumarins active in the sensitization of skin, bacteria, etc. are found to undergo photoreactions with flavin mononucleotide (FMN) giving products that lack coenzyme activity. More importantly, it was found that the sensitizing furocoumarins react with DNA when irradiated at 365 nm; studies with labeled furocoumarins show a photobinding reaction, i.e., a UV-induced linking of the furocoumarin with DNA.[60]

Studies with the nucleic acid bases show that only the pyrimidine bases and their nucleosides and nucleotides react with furocoumarins; no photobinding with purines and their derivatives occurs.[60] The photobinding reaction is temperature-independent, and does not require oxygen; in fact, oxygen and other paramagnetic species decrease the reaction rate. Two types of photoadducts are formed with free pyrimidines, with the 5,6 double bond of the base forming a cyclobutane ring with either the 3,4 or the 4',5' double bond of the furocoumarin. On illumination of photoactive furocoumarins in the presence of DNA, the nucleic acid is cross-linked; presumably this results from an interaction of the 3,4 double bond of the furocoumarin with a pyrimidine base in one strand while the 4',5' double bond reacts with a pyrimidine base in another strand. Such reactions occur in cells as well as in solution. Photobinding of furocoumarins to DNA decreases the template activity of the nucleic acid for RNA synthesis.[60]

4.4.2. Other Photosensitization Reactions

Only a few studies have been made of other non-oxygen-requiring photosensitization reactions in biological systems. Uranium compounds act as efficient sensitizers under some conditions. For example, many organic acids, alcohols and carbohydrates are photooxidized under anaerobic conditions in the presence of uranyl ions.[63] The uranyl oxalate system has been studied extensively and has been used for many years as a chemical actinometer (a chemical system used to measure light intensities) (Section 1.4.4.); it is very useful for measuring radiation in the UV and blue range, where uranyl ions absorb.[63] Ferric compounds also sensitize. For example, on illumination under anaerobic conditions with light of wavelengths greater than 310 nm, uridine, cytidine, thymidine, and guanosine are destroyed in the presence of ferric iron as a photosensitizer[64]; tobacco mosaic virus RNA is also inactivated on illumination in the presence of iron salts.[65] Actinometers are also based on the ferric oxalate system.

UV irradiation of DNA leads to the production of a variety of products as a result of the direct absorption of the radiation; in particular, thymine–thymine dimers are produced in the nucleic acid strand (see Chapter 5). The *sensitized* photochemical formation of thymine–thymine dimers in the DNA of T4 bacteriophage has also been demonstrated using sensitizers such as acetophenone and its derivatives; the photosensitizer must have a higher triplet energy than the thymine triplet, and oxygen is not required for the reaction.[66] The reaction is mediated by energy transfer from triplet sensitizer to thymine; the phage is inactivated, and mutant phage are also produced. In addition, a variety of compounds photosensitize dimer splitting, including serotonin, lysyl–tryptophyl–lysine, and potassium ferricyanide.[67]

4.5. CONCLUSIONS

Considerable insight has been gained in recent years on the initial processes involved in photodynamic reactions, including the production of radiation-induced excited states of sensitizers, and the immediate interactions of excited sensitizer with substrate via electron and/or hydrogen transfer processes, and with oxygen by an energy transfer process to give singlet oxygen. Our knowledge is much less satisfactory concerning the next steps, i.e., the detailed mechanistic organic chemistry of the oxidation of substrate molecules. For example, we do not even know most of the intermediate and final products involved in the sensitized photooxidation of amino acids. The situation is even more hazy with macromolecules, and detail is almost completely lacking on photosensitization at the subcellular, cellular, and organismal levels.

In the non-oxygen-requiring reactions, we have a reasonable knowledge of the molecular level interactions of radiation-excited furocoumarin molecules with free pyrimidines and with pyrimidines in DNA. A little information is also available on the reactions in which photosensitizers transfer excitation energy to thymine residues in nucleic acids with the subsequent formation of thymine dimers. However, much remains to be learned about the overall effects of these reactions on chromosome structure, and on the mechanisms of mutation and killing. Thus, it is clear that a large amount of work still remains to be done on photosensitization reactions in biology, not only at the organismal and cellular levels, but also at the molecular level.

Studies of photosensitization phenomena are, of course, important in their own right in terms of increasing our understanding of this important biological phenomenon. In addition, photosensitized reactions are becoming an increasingly important *tool* in biological research. For example, sensitized photooxidation reactions can be used to study structure–function relationships of proteins as well as the conformation of proteins in solution. In a number of cases it has been possible to obtain the selective photooxidation of certain kinds of amino acid residues, or even of specific individual residues in a protein by controlling the reaction conditions, by the use of dyes selective for the photooxidation of certain residues, by controlling the degree of "exposure" of certain residues, and by the use of sensitizers linked or associated with specific sites in the molecule[18,68] (see reference 7, p. 137). Also, by the use of specifically bound sensitizers, and by the use of photodynamic "protective" agents with different protective radii bound to specific sites in the protein, photodynamic action has been used as a tool to determine the spacings between certain amino acid residues of proteins in solution.[68,69]

Photodynamic techniques can be used to selectively inhibit or inactivate specific parts of cells. Several different strategies have been used for studying photodynamic effects at the subcellular level. For example, cells are treated with sensitizing dyes that accumulate in localized regions. The location of the dye is determined by absorption or fluorescence microscopy, and then, during subsequent illumination, those regions can be observed for damage.[32] Techniques like this for the photochemical generation of reactive species at specific subcellular sites may have great future applications in cell biology, developmental biology, etc. Another approach is to stain cells generally, and then to illuminate a localized area with a microbeam of light.[35,70] A third approach is to isolate particular cell organelles, and then study their biochemical and structural responses to photodynamic treatment *in vitro*[33,34] (also see reference 7, p. 464).

Finally, sensitized photochemical reactions will probably be used more and more in applied ways in the future. For example, such reactions are beginning to be used for analytical measurements. In effect, light can be used to generate a chemical reagent conveniently and rapidly by using an appropriate photosensitized reaction. Photochemical assays for riboflavin,[71] and for compounds such as ascorbic acid, epinephrine, nicotine, and caffeine[72] have been described. Many herbicides and pesticides in the environment are degraded by UV radiation from the sun;[73] it may become feasible to incorporate photosensitizing or photoprotective groupings into herbicide and pesticide molecules such that their persistence in the environment can be controlled more precisely. Photochemical techniques have been suggested for use in cleaning up chemical pollutants in industrial waste waters; for example, photodynamic sensitizers have been used in preliminary studies to photooxidize phenols, a major type of toxic water pollutant. Much concern has developed over the accumulation of cellulose-containing wastes from agricultural and industrial sources. Preliminary studies have shown that photodynamic sensitizers increase the rate of degradation of cattle feed lot wastes.[74] Historically, a great deal of research was carried out to develop plastics that would not deteriorate in sunlight. More recently, photosensitizing compounds have been incorporated into plastics used for food packaging, cold drink cups, etc., so that they will undergo a rapid photodegradation when littered by the roadside, and in other outdoor areas. Although some beverages and foods are traditionally packed in ways that prevented light-induced deterioration, transparent containers are often used that permit rapid, sensitized deterioration of many types of food items under the high-intensity levels of fluorescent light found in supermarkets. Much of this waste could be prevented by using appropriate light-absorbing materials for packaging, and by developing acceptable additives that would block the sensitized photooxidation of sensitive food components. As described in Section 4.3.11., some success has been obtained in treating superficial tumors[53,54] as well as the skin condition psoriasis[61] by photochemotherapy. Hopefully, such techniques will be improved and expanded in the future (see Chapter 7). Finally, with an increase in our understanding of the fundamental mechanisms involved in photosensitization reactions, it may be possible to develop better ways of preventing photosensitization in humans.

4.6. EXPERIMENTS

One of the simplest demonstrations of photosensitization in biology is the photodynamic killing of paramecium (the same organism used by Raab in his discovery of photodynamic action). This experiment can be done in many ways; the following directions should be regarded only as a general outline. Prepare a 1×10^{-4} M stock solution of the dye rose bengal (MW = 974; ~10 mg/100 ml distilled water). Add 10 drops of a culture of paramecium (can be obtained at little cost from any biological supply house) to each of 3 small dishes or shallow vials numbered 1–3. Then, in dim light, add 10 drops of distilled water to dish No. 1 and 10 drops of rose bengal solution to dishes No. 2 and No. 3, and mix well. Immediately cover dish No. 3 and place in the dark to serve as the dark control. Place dishes No. 1 and No. 2 under a light source, and examine at intervals with a low-power microscope. If the dishes are placed ~5 cm from a 15-W cool white fluorescent bulb, a swelling of the cells, and a decrease in the swimming rate will be observed in ~1 min with *Paramecium caudatum;* by 2 min most of the organisms will be immobile or swimming erratically, and by 5 min the majority will be immobile, and some will have broken open. The reciprocal of the time of illumination (in minutes or seconds) necessary for the immobilization of 90 or 100% of the paramecia can be used as the rate of photodynamic immobilization. You should observe not only motility, but the morphology of the cell, the behavior of the contractile vacuole, etc. Dish No. 1 serves as a light control (without dye); when most of the animals are dead in dish No. 2, examine those in the dark control (dish No. 3). The dye solution may be too concentrated for some cultures of paramecia. If the reaction goes too fast, or if appreciable effects are observed in the dark control, try 3×10^{-5} M or 1×10^{-5} M dye.

This experiment is quite open-ended in that the rate of photodynamic immobilization, and the development of morphological changes can be measured as a function of dye concentration, dye type (methylene blue, neutral red, eosin Y, etc., can be used), light intensity, light color, temperature, etc. Also, the sensitivity of other organisms can be compared (other protozoans, *Euglena, Daphnia,* brine shrimp, nematodes, small tadpoles, small fish, etc.); bacteria could also be used, but samples would have to be plated out after successively increasing periods of illumination to determine inactivation. A very sensitive assay for carcinogenic hydrocarbon pollutants in air and water has been developed, based on the sensitization of the photodynamic immobilization of paramecia by such compounds.[75]

Another very simple, but spectacular experiment is the photodynamic hemolysis of red blood cells; after an appropriate period of illumination in the presence of a sensitizing dye the cells lyse, and the somewhat opaque tube of highly scattering erythrocytes becomes quite transparent. The experiment differs from that described above in that all solutions must be made up in isotonic saline to prevent osmotic effects on the cells. A detailed protocol for a student type of experiment on photodynamic hemolysis has been published.[76]

ACKNOWLEDGMENT

The preparation of this chapter and the original work of the author included were supported in part by the Energy Research and Development Agency under Contract No. EY-76-S-02-0875.

4.7. REFERENCES

1. H. F. Blum, *Photodynamic Action and Diseases Caused by Light,* Reinhold, New York (1941). (Reprinted in 1964 with an updated appendix by Hafner Publ., New York).
2. N. T. Clare, Photodynamic action and its pathological effects, in: *Radiation Biology* (A. Hollaender, ed.), pp. 693–723, McGraw–Hill, New York (1956).
3. L. Santamaria and G. Prino, The photodynamic substances and their mechanism of action, in:

Research Progress in Organic, Biological and Medicinal Chemistry (U. Gallo and L. Santamaria, eds.), Vol. 1, pp. 260–336, Società Editoriale Farmaceutica, Milan (1964).

4. J. D. Spikes, Photodynamic action, *Photophysiology* **3**, 33–64 (1968).

5. K. C. Smith and P. C. Hanawalt, *Molecular Photobiology* (Inactivation and Recovery), Chapter 9, Photodynamic action, pp. 179–191, Academic Press, New York (1969).

6. J. D. Spikes and R. Livingston, The molecular biology of photodynamic action: Sensitized photoautoxidations in biological systems, *Adv. Radiat. Biol.* **3**, 29–121 (1969).

7. U. Gallo and L. Santamaria (eds.), *Research Progress in Organic, Biological and Medicinal Chemistry,* Vol. III, Parts 1 and 2, North-Holland, Amsterdam (1972). [This volume reports the proceedings of a NATO-sponsored conference on photosensitization phenomena.]

8. J. D. Spikes, Porphyrins and related compounds as photodynamic sensitizers, *Ann. N.Y. Acad. Sci.* **244**, 496–508 (1975).

9. C. S. Foote, Photosensitized oxidation and singlet oxygen: consequences in biological systems, in: *Free Radicals in Biology* (W. A. Pryor, ed.), Vol. II, pp. 85–133, Academic Press, New York (1976).

10. J. Bourdon and B. Schnuriger, Photosensitization of organic solids, in: *Physics and Chemistry of the Organic Solid State* (D. Fox, M. M. Labes and A. Weissberger, eds.) Vol. 3, pp. 59–131, Interscience, New York (1967).

11. C. S. Foote, Mechanisms of photosensitized oxidation, *Science* **162**, 963–970 (1968).

12. L. I. Grossweiner, Molecular mechanisms in photodynamic action, *Photochem. Photobiol.* **10**, 183–191 (1969).

13. T. Wilson and J. W. Hastings, Chemical and biological aspects of singlet excited molecular oxygen, *Photophysiology* **5**, 50–95 (1970).

14. I. R. Politzer, G. W. Griffin, and J. L. Laseter, Singlet oxygen and biological systems, *Chem. Biol. Interact.* **3**, 73–93 (1971).

15. J. D. Spikes and F. Rizzuto, Photodynamic oxidation not involving singlet oxygen, in: *Progress in Photobiology, Proceedings of the Sixth International Congress on Photobiology,* Bochum, Germany, 1972 (G. O. Schenck, ed.), 009, Deutsche Gesellschaft für Lichtforschung e.v., Frankfurt (1974).

16. R. Nilsson and D. R. Kearns, A remarkable deuterium effect on the rate of photosensitized oxidation of alcohol dehydrogenase and trypsin, *Photochem. Photobiol.* **17**, 65–68 (1973).

17. H. J. Conn, *Biological Stains,* 7th ed., Williams and Wilkins, Baltimore, Md. (1961).

18. J. D. Spikes and M. L. MacKnight, The dye-sensitized photooxidation of biological macromolecules, in: *Photochemistry of Macromolecules* (R. F. Reinisch, ed.), pp. 67–83, Plenum Press, New York (1970).

19. K. Davies, G. A. Gee, J. McKellar, and G. O. Phillips, Primary photochemical processes of two phototendering dyes on cellulose substrates, *Chem. Ind. 1973,* 431–432.

20. F. H. Doleiden, S. R. Fahrenholtz, A. A. Lamola, and A. M. Trozzolo, Reactivity of cholesterol and some fatty acids toward singlet oxygen, *Photochem. Photobiol.* **20**, 519–521 (1974).

21. J. D. Spikes and M. L. MacKnight, Dye-sensitized photooxidation of proteins, *Ann. NY Acad. Sci.* **171**, 149–162 (1970).

22. G. Jori, Photosensitized reactions of amino acids and proteins (yearly review), *Photochem. Photobiol.* **21**, 463–467 (1975).

23. E. -R. Lochmann and A. Micheler, Binding of organic dyes to nucleic acids and the photodynamic effect, in: *Physico-chemical Properties of Nucleic Acids,* (J. Duchesne, ed.), Vol. I, pp. 223–267, Academic Press, London (1973).

24. A. Kornhauser, N. I. Krinsky, P. -K. C. Huang, and D. C. Clagett, A comparison of photodynamic oxidation and radiofrequency-discharge generated $^1O_2$ oxidation of guanosine, *Photochem. Photobiol.* **18**, 63–69 (1973).

25. K. Uehara and T. Hayakawa, Photooxidation of adenine and its nucleotides in the presence of riboflavin. IV. Photochemical reaction and products of NAD, *J. Biochem.* **71**, 401–415 (1972).

26. K. C. Smith, The radiation-induced addition of protein and other molecules to nucleic acids, in: *Photochemistry and Photobiology of Nucleic Acids* (S. Y. Wang, ed.), Vol. 2, pp. 187–218, Academic Press, New York (1976).

27. C. Wallis and J. L. Melnick, Photodynamic inactivation of animal viruses: A review, *Photochem. Photobiol.* **4,** 159–170.

28. C. W. Hiatt, Methods for photoinactivation of viruses, in *Concepts in Radiation Cell Biology* (G. L. Whitson, ed.), pp. 57–89, Academic Press, New York (1972).

29. A. Jaffe-Brachet, N. Henry and M. Errera, The photodynamic inactivation of λ bacteriophage particles in the presence of methylated proflavine, *Mutat. Res. 12,* 9–14 (1971).

30. W. Harm, Dark repair of acridine dye-sensitized photoeffects in *E. coli* cells and bacteriophages, *Biochem. Biophys. Res. Commun.* **32,** 350–358 (1968).

31. T. D. Felber, E. B. Smith, J. M. Knox, C. Wallis, and J. L. Melnick, Photodynamic inactivation of Herpes simplex, *JAMA* **223,** 289–292 (1973).

32. A. C. Allison, I. A. Magnus, and M. R. Young, Role of lysosomes and of cell membranes in photosensitization, *Nature* **209,** 874–878 (1966).

33. D. S. Williams and T. F. Slater, Photosensitization of isolated lysosomes, *Biochem. Soc, Trans.* **1,** 200–202 (1973).

34. R. T. Garvin, G. R. Julian, and S. J. Rogers, Dye-sensitized photooxidation of the *Escherichia coli* ribosome, *Science* **164,** 583–584 (1969).

35. I. L. Cameron, A. L. Burton, and C. W. Hiatt, Photodynamic action of laser light on cells, in: *Concepts in Radiation Cell Biology* (G. L. Whitson, ed.), pp. 245–258, Academic Press, New York (1972).

36. N. I. Krinsky, The protective function of carotenoid pigments, *Photophysiology* **3,** 123–195 (1968).

37. N. I. Krinsky, Membrane photochemistry and photobiology, *Photochem. Photobiol.* **20,** 532–535 (1974).

38. J. Das, B. Bagchi, and U. Chaudhuri, Liquid holding recovery of photodynamic damage in *E. coli, Photochem. Photobiol.* **19,** 317–319 (1974).

39. A. A. Lamola, T. Yamane, and A. M. Trozzolo, Cholesterol hydroperoxide formation in red cell membranes and photohemolysis in erythropoietic protoporphyria, *Science* **179,** 1131–1133 (1973).

40. J. Pooler, Photodynamic alteration of lobster giant axons in calcium-free and calcium-rich media, *J. Membrane Biol.* **12,** 339–348 (1973).

41. F. P. Imray and D. G. MacPhee, Induction of base-pair substitution and frameshift mutations in wild-type and repair-deficient strains of *Salmonella typhimurium* by the photodynamic action of methylene blue, *Mutat. Res.* **27,** 299–306 (1975).

42. L. C. Harber and R. L. Baer, Pathogenic mechanisms of drug-induced photosensitivity, *J. Invest. Dermatol.* **58,** 327–342 (1972).

43. W. J. Runge, Photosensitivity in porphyria, *Photophysiology* **7,** 149–162 (1972).

44. M. M. Mathews-Roth, M. A. Pathak, T. B. Fitzpatrick, L. C. Harber, and E. H. Kass, Beta-carotene as a photoprotective agent in erythropoietic protoporphyria, *N. Engl. J. Med.* **282,** 1231–1234 (1970).

45. D. P. Bremner, Hepatogenous photosensitization. Induction and study in guinea pigs, *J. Comp. Pathol. 84,* 555–568 (1974).

46. T. P. Yoho, J. E. Weaver, and L. Butler, Photodynamic action in insects. I. Levels of mortality in dye-fed light-exposed house flies, *Environ. Entomol.* **2,** 1092–1096 (1973).

47. A. C. Giese, Photosensitization by natural pigments, *Photophysiology* **6,** 77–129 (1971).

48. A. Eisenstark, Mutagenic and lethal effects of visible and near-ultraviolet light on bacterial cells, *Adv. Genet.* **16,** 167–198 (1971).

49. B. L. Epel, Inhibition of growth and respiration by visible and near-visible light, *Photophysiology* **8,** 209–229 (1973).

50. H. B. Lamberts, Natural photodynamic sensitivity in *Tubifex,* in: *Progess in Photobiology* (B. Chr. Christensen and B. Buchman, eds.), pp. 431–432, Elsevier, Amsterdam (1961).

51. L. R. Barran, J. Y. Daoust, J. L. Labelle, W. G. Martin, and H. Schneider, Differential effects of visible light on active transport in *E. coli, Biochem. Biophys. Res. Commun.* **56,** 522–528 (1974).

52. A. U. Khan and M. Kasha, An optical-residue singlet-oxygen theory of photocarcinogenicity, *Ann. NY Acad. Sci.* **171,** 24–33 (1970).

53. S. Granelli, I. Diamond, A. McDonough, C. Wilson, and S. Nielsen, Photochemotherapy of glioma cells by visible light and hematoporphyrin, *Cancer Res.* **35**, 2567–2570 (1975).

54. T. Dougherty, G. Grindey, R. Fiel, K. Weishaupt, and D. Boyle, Photoradiation therapy II: Cure of animal tumors with hematoporphyrin and light, *J. Natl. Cancer Inst.* **55**, 115–121 (1975).

55. A. H. Clements, R. H. Van Den Engh, D. J. Frost, K. Hoogenhout, and J. R. Nooi, Participation of singlet oxygen in photosensitized oxidation of 1,4-dienoic systems and photooxidation of soybean oil, *J. Am. Oil Chem. Soc.* **50**, 325–330 (1973).

56. A. W. M. Sweetsur and J. C. D. White, Studies on the heat stability of milk protein: II. Effects of exposing milk to light, *J. Dairy Res.* **42**, 57–71 (1975).

57. R. Radtke, Storage behavior of potato chips exposed to light and in dark. I. Analysis of alterations of frying oil caused by light (in German), *Fette Seifen Anstrichmittel* **76**, 540–546 (1974).

58. H. W. -S. Chan, Artificial food colours and the photooxidation of unsaturated fatty acid methyl esters: the role of erythrosine, *Chem. Ind.* **1975** 612–614.

59. Ph. P. Van Hasselt, Photooxidation of unsaturated lipids in *Cucumis* leaf discs during chilling, *Acta Bot. Neer.* **23**, 159–169 (1974).

60. L. Musajo and G. Rodighiero, Mode of photosensitizing action of furocoumarins, *Photophysiology* **7**, 115–147 (1972).

61. J. A. Parrish, Methoxalem—UV-A therapy of psoriasis, *J. Invest. Dermatol.* **67**, 669–671 (1976).

62. R. E. Hakim, A. C. Griffin, and J. M. Knox, Erythema and tumor formation in methoxysalen-treated mice exposed to fluorescent light, *Arch. Dermatol.* **82**, 572–577 (1960).

63. E. Rabinowitch and R. L. Bedford, *Spectroscopy and Photochemistry of Uranyl Compounds,* Macmillan, New York (1964).

64. I. J. Černohorský and G. M. Blackburn, Photodynamic effects of Fe^{3+} upon bases of nucleic acids, in: *Radiation Biophysics, Free Radicals, Proc. First Eur. Biophys. Congr.* (E. Broda, A. Locker, and H. Springer-Lederer, eds.), Vol. 2, pp. 29–31, Verlag der Wiener Medizinischen Adkademie, Vienna (1971).

65. B. Singer and H. Fraenkel Conrat, Effects of illumination in the presence of iron salts on ribonucleic acid and model compounds, *Biochemistry* **4**, 226–233 (1965).

66. M. L. Meistrich and A. A. Lamola, Triplet-state sensitization of thymine photodimerization in bacteriophage T4, *J. Mol. Biol.* **66**, 83–95 (1972).

67. M. Charlier and C. Hélène, Photosensitized splitting of pyrimidine dimers in DNA by indole derivatives and tryptophan-containing peptides, *Photochem. Photobiol.* **21**, 31–37 (1975).

68. G. Jori, Photosensitized oxidation of biomolecules as a tool for elucidating three-dimensional structure, *Anais Acad. Brasil. Cien.* **45**, 33–44 (1973).

69. G. Jori, M. Folin, G. Gennari, G. Galiazzo, and O. Buso, Photooxidation of lanthanide ion-lysozyme complexes. A new approach to the evaluation of intramolecular distances in proteins. *Photochem. Photobiol.* **19**, 419–433 (1974).

70. M. W. Berns, *Biological Microirradiation: Classical and Laser Sources,* Prentice–Hall, Englewood Cliffs, N.J. (1974).

71. E. Clausen, Simple and fast assay method for riboflavine, *Lab. Pract.* **24**, 161–162 (1975).

72. V. R. White and J. M. Fitzgerald, Dye-sensitized continuous photochemical analysis: Identification and relative importance of key experimental parameters, *Anal. Chem.* **47**, 903–908 (1975).

73. R. Rabson and J. R. Plimmer, Photoalteration of pesticides: Summary of workshop, *Science* **180**, 1204–1205 (1973).

74. K. Eskins, B. L. Bucher and J. H. Sloneker, Sensitized photodegradation of cellulose and cellulosic wastes, *Photochem. Photobiol.* **18**, 195–200 (1973).

75. S. S. Epstein, M. Small, J. Koplan, N. Mantel, and S. H. Hutner, Photodynamic bioassay of benzo[α]pyrene with *Paramecium caudatum, J. Natl. Cancer Inst.* **31**, 163–168 (1963).

76. W. S. Hoar, Some effects of radiation on cells. I. Photodynamic action, in: *Experiments in Physiology and Biochemistry* (G. A. Kerkut, ed.), Vol. 1, pp. 132–135, Academic Press, London and New York (1968).

5

Ultraviolet Radiation Effects on Molecules and Cells

5.1. INTRODUCTION[1-3]

The detrimental effects of ultraviolet (UV) radiation are well documented, and appear to be mediated largely through its action upon deoxyribonucleic acid

Kendric C. Smith • Department of Radiology, Stanford University School of Medicine, Stanford, California

(DNA). The most detrimental effect, of course, is cell killing. Other effects include mutagenesis, carcinogenesis, interference with the synthesis of DNA, ribonucleic acid (RNA), and protein, delay of cell division, and changes in permeability and motility. Except for the production of vitamin D in the skin of man, the beneficial effects of UV radiation are not as well documented (see Section 7.6.).

UV radiation was discovered in 1801 by J. W. Ritter. While studying the blackening of silver chloride by different wavelengths of light, he observed that the dark region just beyond the violet end (380 nm) of the visible spectrum was more efficient in this regard than was visible light. The first observation of an effect of UV radiation on living systems occurred in 1877 when Downes and Blunt reported the inactivation of bacteria. The next landmark came in 1928 when Gates found that the relative effectiveness of different wavelengths of radiation for killing bacteria (an action spectrum for killing) paralleled the absorption spectrum of the nucleic acid bases.

Early photochemical studies on the nucleic acids dealt largely with the destructive cleavage of the pyrimidine ring; but in 1949, Sinsheimer and Hastings showed that the pyrimidines could undergo reversible photochemistry. This photochemical reaction was later established as the addition of a molecule of water across the 5,6 double bond of the pyrimidines. In the 1960s, Beukers and Berends observed that UV radiation caused the stable linkage of two adjacent thymine residues in a strand of DNA. This discovery of thymine dimer formation stimulated a resurgence of interest in UV photobiology that continues today. As the importance of thymine dimers to biological inactivation became apparent, there developed a tendency to give them credit for too much of UV photobiology. There are many other types of photoproducts produced in the nucleic acids of cells, and certain of these have been isolated and characterized. In some cases, their relative biological importance has also been determined.

Different strains of bacteria show tremendous differences in their sensitivity to killing by UV radiation. *Escherichia coli* K-12 *recA uvrB*, a strain that is deficient in two types of DNA repair systems (discussed in Section 5.4.2.), is killed to the 10% survival level by a fluence of UV radiation at 254 nm of 0.1 J m^{-2}, the single mutants *uvrB* by 9.5 J m^{-2}, and *recA* by 5.8 J m^{-2}, and the wild-type parent by 68.0 J m^{-2}. Thus, in *E. coli* K-12 there is a range of almost 700 in the sensitivity of the different strains that depends upon their ability to repair photochemical damage to their DNA.

Currently the major emphasis of research in the field of UV photobiology is on the mechanisms by which cells repair their DNA, and the possible role of inaccurate repair in the production of mutations. This emphasis will be maintained in this chapter.

5.2. PHOTOCHEMISTRY OF THE NUCLEIC ACIDS[4-8]

Many of the biological effects of UV radiation can now be explained in terms of specific chemical and physical changes in DNA. DNA does not exhibit the same sensitivity to UV radiation under all experimental conditions, however. The intrinsic sensitivity of DNA to photochemical alteration can be changed by a variety of biological (e.g., growth state of cells), chemical (e.g., base analog substitution), and physical (e.g., denaturation, freezing, drying) techniques. To give an example: one photoproduct that is produced in high yield, and appears to be the major cause of death in UV-irradiated bacterial cells, is not produced to a significant extent in bacterial spores. Different types of photoproducts appear to inactivate irradiated vegetative cells and spores, respectively. Simple generalizations, therefore, cannot be made as to which photoproduct in DNA is the most important to all irradiated cells under all experimental situations.

In addition to the intrinsic sensitivity of the DNA, we must also consider the ability of the cell to repair the damage. If one type of lesion is repaired accurately and quantitatively by a cell, it cannot be of biological importance to that cell, while lesions that are not repairable (or are not repaired) may be of biological importance.

5.2.1. Effects of UV Radiation on Deoxyribose

Although carbohydrates make up about 41% by weight of the nucleic acids, they show essentially no UV absorption at wavelengths above 230 nm, and therefore would not be expected to undergo photochemical reactions when irradiated with light of wavelengths greater than 230 nm.

Indirect effects of UV radiation on deoxyribose have been reported. A cell is much more sensitive to killing by UV radiation if 5-bromouracil (BrUra) replaces the thymine in its DNA. One effect of UV radiation on BrUra in DNA is debromination with the consequent production of a uracil radical. In the absence of another hydrogen donor, a hydrogen atom will be abstracted from the adjacent deoxyribose. This leads to the production of uracil, to the destruction of the deoxyribose (by mechanisms presently unknown), and ultimately to a chain break in the DNA.[9] Chain breaks in DNA are also produced in the presence of ketone sensitizers such as benzophenone. When these agents are excited by light at 313 nm they can abstract hydrogen from water to produce hydroxyl radicals, which then attack the DNA and produce chain breaks.[10] Thus, under certain conditions, chemical alterations in the carbohydrates of the nucleic acids can be brought about when the nucleic acids are exposed to UV and near-UV radiation, even though the primary absorption of photons does not occur in the carbohydrates.

5.2.2. Effects of UV Radiation on Purines

In pure solution, purines are approximately 10-fold more resistant to photochemical alteration than are the pyrimidines. Because of this difference in sensitivity to photochemical alteration, it has been implied that the photochemistry of the purines is not important biologically, since by the time a significant amount of purine damage has occurred the cells would have been inactivated by pyrimidine damage. Although statistically this hypothesis has much in its favor, the biological importance of purine photochemistry should not be so quickly dismissed. Although the absorption of UV radiation by the purines does not result in the photochemical alteration of the purine ring with a high efficiency, some of the absorbed energy may well be transferred to the pyrimidines or to the sugar-phosphate backbone of DNA, and thus result in chemistry. The transfer of energy from adenine has been implicated in the formation of a thymine radical, and the phosphorescence of thymine in UV-irradiated poly-dAT.

Inside cells, however, the nucleic acids are not in pure solution, and Elad and co-workers[11] have demonstrated that purines readily react photochemically with other organic compounds. In fact, the purines are much more photochemically reactive than the pyrimidines in undergoing these heteroaddition reactions. The biological importance of these heteroaddition reactions has only recently become appreciated.[7]

5.2.3. Hydration Products of the Pyrimidines

When solutions of uracil and its derivatives are UV-irradiated, they lose their characteristic UV absorbance, but this can be largely regenerated by heat, alkali, or acid treatment. It was postulated that the hydration of the 5,6 double bond of uracil could account both for the loss of the absorption spectrum, and the reversibility by subsequent treatment with acid or heat. The ultimate proof of this postulate came when 6-hydroxy, 5-hydrouracil (Fig. 5-1) was synthesized, and was shown to be identical with the reversible radiation product of uracil (note that 5-hydroxy, 6-hydrouracil is stable to heat). Photochemical hydrates of cytosine and thymine are also formed, but these hydrates (especially that of thymine) are much more labile than the uracil derivative.

The formation of the water addition photoproduct of cytosine in irradiated denatured DNA has been inferred from the appearance of a heat-reversible absorption peak around 240 nm. Dihydrocytosine derivatives (i.e., lacking the 5,6 double bond) exhibit a characteristic absorption peak at this wavelength. Irradiated native DNA, however, showed no such heat-reversible absorption peak. These and other data suggest that hydrates of cytosine are probably not formed in measurable yield in irradiated double-stranded DNA, but they are formed in single-stranded DNA. Other photoproducts (e.g., pyrimidine dimers), however, can distort the DNA and produce *local denaturation* (i.e., single-stranded regions), and thus may permit the formation of hydrates in otherwise double-stranded DNA.

During replication and/or transcription of the DNA, there may be short regions of single-strandedness, and in these regions the formation of pyrimidine

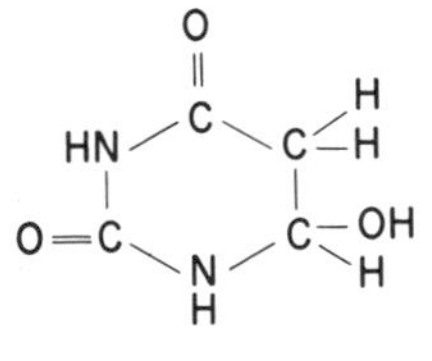

Fig. 5-1. 6-Hydroxy, 5-hydrouracil.

hydrates may well be of importance. The possible role of pyrimidine hydrates in causing mutations has been demonstrated in an *in vitro* model system.[12] When polycytidylic acid was UV-irradiated, its coding properties in an RNA polymerase system were altered. The irradiated polymer lost its ability to code for the incorporation of guanylic acid unless adenylic acid was also added to the medium. The increase in adenylic acid incorporation in the polymer, as a function of UV fluence, was heat-reversible under conditions known to reverse pyrimidine hydrates, and for this reason it was suggested that the code change might be the result of the formation of cytosine hydrates (i.e., the cytosine hydrate appears to base pair with adenine rather than guanine). The formation of hydrates in single-stranded regions of the DNA, therefore, may well be of significance in the production of mutations.

5.2.4. Cyclobutane-Type Dimers of Thymine, Cytosine, and Uracil

When an aqueous solution of thymine is UV irradiated (254 nm), it loses its characteristic absorption properties at a rate about one-tenth that of uracil ($\Phi = 0.4 \times 10^{-3}$ for thymine). If a solution of thymine is frozen and then UV irradiated, it shows a greatly increased photochemical reactivity ($\Phi = {\sim}0.2$), and the major product formed is a dimer. The probable effect of freezing is to bring the thymine molecules into an oriented juxtaposition favorable for a bimolecular photochemical reaction.

To form the thymine dimer (Fig. 5-2), two thymine molecules are linked to each other between their respective 5 and 6 carbon atoms, thus forming a cyclobutane ring (four-carbon ring) between the two thymines. There are six possible isomers of the thymine dimer and these have been isolated from irradiated thymine oligomers. The type I *(cis-syn)* thymine dimer (Fig. 5-2) is the one formed between adjacent thymines in the same strand of DNA. Certain of these isomers are stable to acid hydrolysis, while others are not. Since acid hydrolysis is the usual method for liberating photoproducts from irradiated DNA, such labile photoproducts would be destroyed.

There is a wavelength dependency for the formation and monomerization of the cyclobutane-type thymine dimer. After a sufficient fluence of UV radiation, a photosteady state between monomer and dimer is reached that is characteristic for the wavelength used. At the longer wavelengths (around 280 nm) the formation of the dimer is favored, while at the shorter wavelengths (around 240 nm) monomer formation is favored. This response is due to differences in the absorption spectra of thymine and its dimer (Fig. 5-3), and in the quantum yields for the formation and splitting of the dimer.

Five other cyclobutane-type dimers of the natural pyrimidines are also known. These are the dimers of uracil, cytosine, uracil–thymine, cytosine–thymine, and uracil–cytosine. The isolation of cytosine dimers is complicated not only by the competition of the hydrate reaction, but also by the fact that cytosine deaminates readily when its 5,6 double bond is saturated. Cytosine dimers are therefore readily converted to uracil dimers. It is apparent that if a cytosine dimer in the DNA of a cell were to deaminate to form a uracil dimer, and if this dimer were then split *in situ* by the photoreactivating enzyme (see below), a mutation could result since the uracil residues would base pair with adenine rather than with guanine.

Up to now we have talked about photochemical reactions that occur as the result of the direct absorption of photons by the reacting species. Thymine dimers can be formed, however, by wavelengths of light that are not absorbed by thymine, providing that the thymine is in the presence of suitable molecules that do absorb these wavelengths. This process is called triplet-state *photosensitization* (see Section 2.7.). It requires that the triplet state of the absorbing species (the photosensitizer) be higher in energy than the triplet state of the thymine. Upon collision, the triplet energy of the photosensitizer is transferred to the thymine, yielding thymine in its triplet state with the possibility for the subsequent formation of thymine dimers. Examples of this situation are the formation of thymine dimers by light of wavelengths above 300 nm when DNA is irradiated in the presence of 10^{-2} M acetophenone, or when bacteria are irradiated while suspended in 10% acetone. One advantage of the use of triplet state photosensiti-

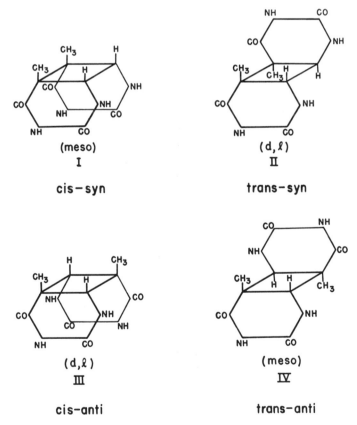

Fig. 5-2. Isomeric forms of the cyclobutane-type thymine dimers. Optical isomers are possible for types II and III.

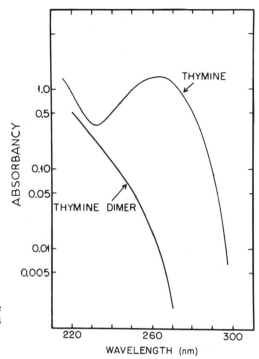

Fig. 5-3. The absorption spectra of thymine (25 μg/ml) and of purified thymine dimer (34 μg/ml). [Modified from R. Setlow, *Biochim. Biophys. Acta* **49**, 237 (1961).]

zation to drive a reaction is that it can be performed at wavelengths where the reverse reaction (dimer splitting) does not occur. With this technique one should achieve essentially a quantitative conversion of adjacent thymine residues to dimers, rather than achieving an equilibrium yield as obtained by the direct excitation of thymine residues.

5.2.5. Other Photochemical Reactions of the Pyrimidines

Many pyrimidine photoproducts other than the hydrates or the cyclobutane-type dimers are produced both *in vivo* and *in vitro,* but generally at much lower yields. Some of these photoproducts appear to be monomeric, while others appear to be bimolecular in nature. The chemical structure and relative biological importance of most of these photoproducts remain to be determined.[1,4-7]

The so-called "spore photoproduct" deserves special consideration (Fig. 5-4). After high fluences of UV radiation (2×10^4 J m^{-2} at 254 nm) about 30% of the thymine in bacterial spores can be converted to this product. The spore photoproduct does not exhibit the short-wavelength reversal properties of the cyclobutane-type thymine dimers. Therefore, in spores the yield of this product can approach the maximum determined by the number of thymines that are nearest neighbors in the DNA. The maximum yield of cyclobutane-type dimers, however, is dependent upon the equilibrium between the formation and splitting of dimers. Thus, in irradiated *E. coli,* where 30–40% of the thymines could theoreti-

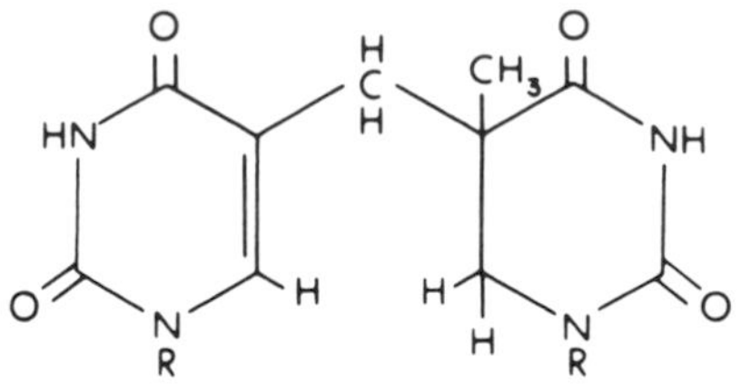

Fig. 5-4. The "spore photoproduct" (5-thyminyl-5,6-dihydrothymine).

cally dimerize, only about 15% are dimerized even after high fluences of UV radiation at 254 nm.

During the germination of spores, the yield of spore photoproduct decreases, and the yield of cyclobutane-type pyrimidine dimers increases, and the sensitivity to killing by UV radiation increases. The cyclobutane-type pyrimidine dimers appear to be about 11 times more effective in killing vegetative cells than are the spore photoproducts in killing spores. This could imply that the spore photoproduct is more efficiently repaired than are the cyclobutane-type dimers.

The pyrimidines (and purines) will also react photochemically with other compounds. For example, alcohols (R—OH), and the amino acid cysteine (R'—SH) photochemically add across the 5,6 double bond of the pyrimidines. Since the addition products of methanol and ethanol can be reversed by acid they appear analogous to the water (H—OH) addition product described above, and therefore are expected to be attached to position 6. Cysteine is known to add at the 5 position and forms a stable product (Fig. 5-5).

5.2.6. Macromolecular Changes in DNA

While X-irradiation is very effective in producing single- and double-strand breaks in DNA, UV radiation is very inefficient in this respect except in the presence of sensitizers such as benzophenone (Section 5.2.1.). In UV-irradiated cells, however, DNA chain breaks are rapidly produced enzymatically as necessary steps in the repair of damaged DNA.

If DNA is UV irradiated while dry or when it is very tightly packed, as in sperm heads, DNA–DNA cross-links leading to gel formation have been observed. This type of reaction appears to be of little importance to normal wet cells, however.

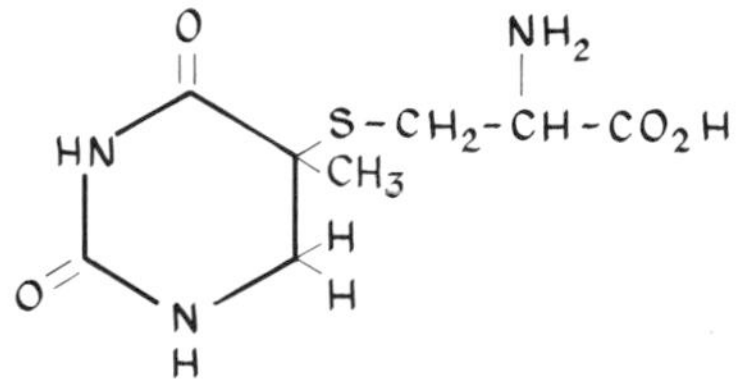

Fig. 5-5. 5-S-cysteine, 6-hydrothymine.

Another type of DNA–DNA cross-linking causes the two strands of a molecule of DNA to be connected so that they can no longer be separated when the DNA is denatured with heat or formamide. Since no interchain cross-links were detected in normal phage T7 irradiated to a survival of 1%, the biological importance of this lesion seems in doubt at low fluences of UV radiation. However, this lesion may achieve a position of greater biological importance at higher fluences of UV radiation in those cells that are relatively resistant to radiation.

5.2.7. Cross-Linking of DNA to Protein[1,4,7]

There is a progressive decrease in the amount of DNA that can be extracted free of protein from bacteria and mammalian cells following increasing fluences of UV radiation. The DNA that becomes nonextractable due to UV irradiation can be quantitatively accounted for in the precipitate containing the denatured proteins. Treatment of this precipitate with trypsin, however, yields free DNA. These results suggested that UV irradiation had caused the DNA to become cross-linked to protein. More direct proof came from experiments showing that DNA and protein could be cross-linked *in vitro* by UV irradiation.

The precise chemical mechanisms by which DNA and protein are cross-linked *in vivo* are not yet known; however, the isolation of a mixed photoproduct of thymine and cysteine (5-S-cysteine, 6-hydrothymine) (Fig. 5-5) from the *in vitro* UV irradiation of a solution of thymine and cysteine may serve as a possible model for the cross-linking phenomenon. Cysteine also adds photochemically to poly-rU, poly-dC, poly-dT, and to RNA and DNA.

In additin to cysteine, the following amino acids add photochemically to thymine: arginine, lysine, tyrosine, tryptophan, and cystine. The following amino acids add photochemically to uracil: glycine, serine, cysteine, cystine, methionine, lysine, arginine, histidine, tryptophan, phenylalanine, and tyrosine. The other common amino acids were unreactive under the conditions tested.

The experiments that best demonstrate the biological importance of DNA–protein cross-linking are those using bacterial cells irradiated while frozen. *E. coli* B/r cells showed marked differences in survival after UV irradiation as a function of the temperature at which they were irradiated (Fig. 5-6). When the temperature was reduced from +21 to −79°C, an increase in sensitivity to UV radiation was shown both by a change in shoulder and a change in slope in the survival curves. At −196°C the cells were not as sensitive as at −79°C, but were more sensitive than at +21°C due to the absence of a shoulder on the survival curve.

The rate of formation of cyclobutane-type thymine dimers *decreased progressively* as the temperature of the cells during irradiation was reduced from +21 to −79°C and to −196°C. Therefore, there was no correlation between the production of thymine dimers and the increased killing of *E. coli* by irradiation at −79 and −196°C. This suggests that cyclobutane-type thymine dimers do not play as significant a role in the events leading to the death of irradiated frozen

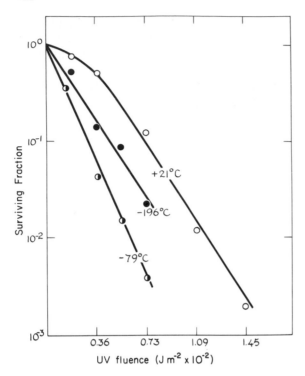

Fig. 5-6. Survival of *E. coli* B/r *thy* as a function of the fluence of UV radiation (254 nm) at different temperatures. [Modified from K. C. Smith and M. E. O'Leary, *Science* **155**, 1024 (1967).]

cells as they appear to play at room temperature. These results provide further evidence that the relative biological importance of a given photoproduct can change markedly, depending upon growth or irradiation conditions.

In contrast, however, the amount of DNA cross-linked to protein as a function of temperature during irradiation followed the same relative pattern as the changes in survival. Thus, the photochemical event that appears to correlate with viability when cells are irradiated while frozen is the cross-linking of DNA with protein. Freezing may alter the configuration or the proximity of the protein and the DNA within the cells so that the probability of forming DNA-protein cross-links by irradiation is greatly enhanced, thus leading to the greater lethality observed under these conditions.

The action spectrum for the killing of *Micrococcus radiodurans,* one of the most radiation-resistant organisms known, differs markedly from that for the more sensitive organism *E. coli* in that it shows a high component of sensitivity to irradiation at 280 nm as well as at 260 nm. Classically, a response at 280 nm has indicated an involvement of protein, while a response at 260 nm has implicated nucleic acid. It has been suggested that the resistance of this organism to UV radiation is due to its extraordinary ability to repair pyrimidine dimers, but what ultimately kills the organism is damage that involves both DNA and protein. The cross-linking of DNA and protein may constitute one type of such damage.

It is reasonable to assume that a different type of photochemistry might arise when protein and DNA are irradiated together as compared to when they are irradiated separately. Since DNA and protein do not exist in cells as pure solutions of the separate molecules, but are in intimate contact with each other, it might be expected that the photochemical interaction of DNA and protein (and other molecules) would play a significant role in the inactivation of UV-irradiated cells under certain conditions.[1,4,7]

5.2.8. Photochemistry of RNA

RNA differs from DNA by virtue of having ribose instead of deoxyribose, by having uracil instead of thymine, and by having more single-stranded regions. These differences do affect the photochemistry of RNA (e.g., more pyrimidine hydrates are formed), but in general it is quite similar to that of DNA.

The photochemistry of RNA *in vivo* has not been studied as extensively as has the photochemistry of DNA. This is perhaps understandable in view of the central importance of DNA to the survival of cells. Since RNA species are generally present within a cell in multiple copies, the inactivation of a few percent of these molecules would not be expected to have as adverse an effect upon a cell as the photochemical alteration of a small portion of its DNA. Nevertheless, under special circumstances the photochemical alteration of messenger RNA and transfer RNA might be expected to have serious consequences for a cell (Section 15.7.2.1.).

Some viruses contain RNA as their genetic determinant (e.g., tobacco mosaic virus). For such viruses the photochemical alteration of their RNA leads to rapid inactivation. The biological and chemical effects of UV radiation on RNA have been most extensively studied with RNA viruses.[4]

5.2.9. Conclusions

Our understanding of the chemistry and biological consequences of the formation of pyrimidine dimers in cells is quite sophisticated, but interest in the photochemistry of the nucleic acids has waned somewhat. We know very little about the chemistry of many of the non-cyclobutane-type pyrimidine photoproducts, and of their biological importance. A similar statement can be made about purine photochemistry.

While the absolute number of the possible types of monomeric and dimeric photochemical alterations in the purines and pyrimidines is finite, the number of possible photochemical adducts to DNA is almost infinite. There is currently renewed interest in the field of heteroaddition reactions with DNA (e.g., the cross-linking of protein to DNA, of carcinogens to DNA, etc.) because of their relevance to aging, carcinogenesis, and photobiology.[7] Thus, many exciting discoveries have yet to be made in the field of the photochemistry of the nucleic acids.

5.3. PHOTOCHEMISTRY OF AMINO ACIDS AND PROTEINS[1,8,13,14]

5.3.1. Relative Photochemical Sensitivity of the Amino Acids

Since light must be absorbed before photochemistry can occur, it is reasonable to assume that those amino acids that are transparent to the wavelengths normally used in photobiology (i.e., >240 nm) will undergo little or no photochemistry at these wavelengths. This then largely eliminates the aliphatic amino acids from our consideration. The strong absorption of radiation of wavelengths above 240 nm by the aromatic amino acids would certainly suggest their photochemical importance, although it should be recalled that absorbed light need not necessarily result in photochemistry (i.e., absorbed energy can be dissipated harmlessly as heat, or by fluorescence or phosphorescence).

There are two chromophores that are not present in monomeric amino acids but are present in proteins: peptide linkages and disulfide bridges. The absorption spectrum for peptide bonds shows a peak in the region of 180–190 nm that decreases rapidly at longer wavelengths. However, even at 254 nm, the quantum yield for peptide bond breakage is a significant value (Table 5-1).

The amino acids cysteine and cystine were considered at one time to be unimportant photochemically, since, according to their absorption spectra in acid solution, they were essentially optically transparent above 230 nm. At neutral and alkaline pH, however, the absorption by these compounds becomes appreciable even at wavelengths above 250 nm. This fact coupled with the high quantum efficiency for its photochemical alteration makes cystine the most sensitive target affecting protein function.

If the molar absorptivity (molar extinction coefficient) (ϵ) of a compound is a measure of the probability that light of a particular wavelength will be absorbed by that compound, and the quantum yield (Φ) is the probability that the absorbed light will cause a chemical change, then the product of these two values ($\epsilon \times \Phi$) is a measure of the photochemical sensitivity of that compound (Section 1.5.2.).

TABLE 5-1. The Relative Photochemical Liability of Amino Acids at 254 nm[a]

Compound	ϵ	Φ	($\epsilon \times \Phi$)
Cystine (—S—S— bond)	270	0.13	35.1
Tryptophan	2870	0.004	11.5
Phenylalanine	140	0.013	1.8
Tyrosine	320	0.002	0.6
Peptide bonds (acetylalanine)	0.2	0.05	0.01
Histidine	0.24	<0.03	<0.0072

[a] Data from reference 8, p. 97. ϵ is the molar absorptivity at 254 nm (the probability that radiation at 254 nm will be absorbed by the given compound; Φ is the quantum yield at 254 nm (the probability that the absorbed radiation at 254 nm will cause a chemical change); ($\epsilon \times \Phi$) is a measure of the photochemical sensitivity of a particular amino acid.

Values for these parameters at 254 nm for the most important chromophores in proteins are given in Table 5-1. When cystine is present in a protein, it is the most labile target. Following in decreasing lability are tryptophan, phenylalanine, tyrosine, peptide bonds, and histidine. It should be stressed, however, that at other wavelengths the UV sensitivity (due to changes in ϵ and Φ), and therefore the relative importance of these targets in the photochemical inactivation of a particular protein, will change.

Most of the early photochemical studies on the amino acids did not use monochromatic light, and the analyses for photochemical alteration relied upon easily measurable end products of destruction such as ammonia, carbon dioxide, sulfur, and loss of spectral properties. These data, however, provide very little useful information about the reactions that occur at biologically significant fluences of UV radiation. In a more recent study on the photochemistry of phenylalanine (selectively labeled at different positions with radioactive ^{14}C), it was shown that tyrosine, dihydroxyphenylalanine (Dopa), aspartic acid, benzoic acid, phenyllactic acid, and a high-molecular-weight melanic polymer were formed when this amino acid was irradiated at the wavelength of maximum absorption of the aromatic residue (258 nm). The reaction that occurred with the greatest quantum efficiency was the liberation of carbon dioxide by decarboxylation. These results indicate that there was cleavage of side-chain carbon bonds, and since all of the energy was initially absorbed by the benzene ring, it is therefore reasonable to conclude that energy migration must have occurred.

Excitation (i.e., electronic energy) migration in proteins is thought to occur in part through a *resonance-transfer mechanism*. A donor, with a fluorescence spectrum overlapping the absorption spectrum of the energy acceptor, can transfer electronic energy to the acceptor if the distance between the two is sufficiently small (~ 10 nm), and if the relative orientation between the two is suitable. Since it is possible to distinguish between the emission spectra of phenylalanine, tyrosine, and tryptophan, the fluorescence spectra of proteins can yield information on the direction of transfer of energy between these amino acids. Phenylalanine fluorescence is not observed from most proteins, although a tyrosine component is almost always observed. Tyrosine fluorescence, however, makes only a small contribution to the total fluorescence from those proteins containing tryptophan. Thus, the transfer of energy appears to proceed from phenylalanine to tyrosine to tryptophan (Section 2.7.4.).

In addition to descriptive chemical studies of the type mentioned above for phenylalanine, where the early products of the photochemical alteration of the amino acid were isolated and identified, more sophisticated studies, using flash photolysis, electron paramagnetic resonance techniques, and emission spectroscopy, have elucidated some of the photochemical intermediates that lead to photoproducts. The aromatic amino acids (RH) have been shown to undergo photoionization (RH^+) as a result of the ejection of a hydrated electron (e_{aq}^-). The cation radical (RH^+) can dissociate to yield a neutral radical ($R\cdot$) and a hydrogen ion (H^+). The neutral amino acid radicals and hydrated electrons can then undergo reactions with each other, and also with other molecules to yield a multiplicity of products.[13] The yields of cystine destruction are much higher in proteins that contain tryptophan. There is evidence to suggest that the mechanism by which tryptophan sensitizes the photochemical alteration of cystine is via an ejected electron. Consistent with this observation is the fact that the amino acid that reacts most readily with hydrated electrons is cystine.[13]

It has long been known that the major action of X-radiation is to produce radicals, and, therefore, radiation chemistry is really radical reaction chemistry. It has not been generally appreciated that certain types of photochemical reactions (as described above) also lead to the production of radicals; and radicals, whether produced photochemically or otherwise, undergo their own characteristic types of reactions. Thus, photochemistry and radiation chemistry have more in common than would seem to be the case at first.

5.3.2. General Photochemical Reactions of Proteins[14−16]

When proteins are UV-irradiated, both lower- and higher-molecular-weight products are formed, solubility changes occur, sensitivity to heat denaturation increases, digestibility by trypsin increases, and changes in optical and physical properties occur. In general, however, these changes occur at much higher fluences of UV radiation than does enzyme inactivation.[1,8]

5.3.3. Photochemical Inactivation of Enzymes

Two different theories of the mechanisms involved in the UV inactivation of enzymes have developed over a period of years. McLaren and Hidalgo-Salvatierra[15] have assumed, as a first approximation, that the alteration of any amino acid residue causes inactivation of an enzyme. If the quantity ($\epsilon \times \Phi$) for each amino acid (which is a measure of the photochemical sensitivity of each amino acid; see Table 5-1) is multiplied by the number of such amino acids (n) in the protein and the sum of these is divided by the molar absorptivity of the enzyme, one arrives at a calculated quantum yield for the inactivation of the enzyme that in some cases is very close to the observed quantum yield. Of the 10 proteins listed in Table 5-2, the calculated quantum yields for seven of these are within a factor of 2 of the experimentally determined values. The values for carboxypeptidase, pepsin, and insulin do not adequately fit the hypothesis.

The authors themselves have pointed up certain criticisms of these calculations. There is no assurance that either the molar absorptivity or the quantum yield for a particular amino acid (at a given pH and wavelength) is the same in a protein as it is in free solution; in fact, there is some evidence that these values change when the amino acid is incorporated into a protein. Also, it is known that all residues of a particular amino acid in a given protein are *not* of equal importance to enzymatic function.

For those few proteins studied, their absorbance at 254 nm does not differ markedly from that for an equivalent mixture of the free amino acids, although greater differences exist at other wavelengths. This fact, plus the rather close

TABLE 5-2. Calculated and Experimental Quantum Yields for Enzyme Inactivation by UV Radiation at 254 nm[a]

Protein	Chromophores					Quantum yield	
	Cys	His	Phe	Trp	Tyr	Found	Calculated[b]
1. Carboxypeptidase	2	8	15	6	20	0.001[c]	0.01
2. Chymotrypsin	5	2	6	7	4	0.005	0.01 (0.007)[d]
3. Lysozyme	5	1	3	8	2	0.024	0.014 (0.0097)[d]
4. Pepsin	3	2	9	4	16	0.002	0.01 (0.006)[d]
5. Ribonuclease	4	4	3	0	16	0.027	0.03 (0.026)[d]
6. Subtilisin A	0	5	4	1	13	0.007	0.0051
7. Subtilisin B	0	6	3	4	10	0.006	0.011
8. Japanese nagarse	0	6	3	4	10	0.007	0.01
9. Trypsin	6	1	3	4	4	0.015	0.02 (0.014)[d]
10. Insulin	18	12	18	0	24	0.015	0.06

[a] Data from reference 15.

[b] $\Phi_{ENZ} = \Sigma^1 n_1 \epsilon_1 \Phi_1/\epsilon_{ENZ}$ (see Section 5.3.3.).

[c] The quantum yield for carboxypeptidase has been redetermined by R. Piras and B. L. Vallee [*Biochemistry* **6**, 2269 (1967)] to be 0.0049 which now brings the calculated value (0.01) within about a factor of 2 of the observed value. Newer data (cited by Piras and Vallee) indicate that carboxypeptidase contains no S—S bridges.

[d] Often a much closer agreement is obtained between the observed and calculated quantum yields if only the cystine residues are used in the calculations instead of using all of the amino acids that absorb UV radiation. This is permissible as a special condition of the formulation but suggests that the statement by McLaren and Hidalgo-Salvatierra,[15] "the site of quantum absorption and the site of photochemical reaction in proteins are one and the same and that it is not necessary to invoke a mechanism of migration of absorbed energy from one kind of chromophore to another within an absorbing molecule" is not valid for all proteins.

agreement between the calculated and determined quantum yields cited in Table 5-2 would suggest that, under certain conditions, the amino acid residues are, in fact, independent absorbers unaffected by their neighbors. McLaren and co-workers concluded, therefore, that energy transfer among chromophores within molecules of enzymes need not be invoked in order to account for photochemical inactivation.

Augenstein and Riley,[16] however, have taken particular exception to this conclusion. They developed a theory that enzymes are inactivated by UV radiation as a consequence of the disruption of specific cystine residues, and of hydrogen bonds responsible for the spatial integrity of the active center of the enzyme. Their theory emphasizes the importance of energy migration in this process. Consistent with these postulates are the observations that the photochemical damage in the enzyme ribonuclease is nonrandom; at least two and perhaps three of the four constituent cystines must be disrupted before activity is lost, i.e., the most photosensitive cystines are not critical for enzymatic activity. Similarly, in both trypsin and lysozyme the integrity of the most photosensitive cystines does not appear to be critical for the retention of enzymatic activity. In insulin, however, all three cystines appear to be crucial for activity and have approximately equal photosensitivities. These differences in sensitivity of cystines in different proteins must depend specifically upon energy transfer and/or chemical interactions between the chromophoric groups. Cystine destruction is much greater in those proteins that contain tryptophan. There is evidence that the most photosensitive cystines lie close to tryptophan residues as a consequence of the tertiary structure of the enzyme, rather than of its primary structure.

While we seem to be faced with two opposing theories concerning the mechanism(s) by which enzymes are inactivated by UV irradiation, it is perhaps more reasonable to conclude that both theories are correct but for different enzymes. There must be unique structural features or unique amino acid sequences in some proteins that allow the amino acids to be independent absorbers, and for photochemical inactivation events to occur apparently at random, while different structural features in other proteins give rise to the opposite results, i.e., nonrandom photochemical events, and the involvement of energy migration in the inactivation of enzymes.[14]

5.3.4. Conclusions

Interest in the photochemistry of the amino acids and proteins has been overshadowed in recent years by the phenomenal surge in research activity on the photochemistry of the nucleic acids. Knowledge of the photochemistry of both nucleic acids and proteins is necessary, however, before we can properly understand the mechanisms of action of UV radiation on cells. The finding that proteins and nucleic acids interact photochemically within cells and viruses, necessitates a renaissance in studies on the photochemical reactions of amino acids and proteins, especially in the presence of other biological molecules.

5.4. REPAIR OF PHOTOCHEMICAL DAMAGE[17]

5.4.1. Introduction

The first indication that cells might have the capacity to recover from radiation damage was the observation that minor modifications in the handling of the cells (e.g., growth media, temperature, etc.) had a marked effect upon the ultimate viability of irradiated cells. Thus in 1937, Hollaender and Claus found that higher survival levels of UV-irradiated fungal spores could be obtained if they were allowed to remain in water or salt solution for a period of time before plating on nutrient agar. Roberts and Aldous[18] extended these observations by showing that the shapes of the UV survival curves for *E. coli* B could be changed quite drastically simply by holding the irradiated cells in media devoid of an energy source for various times before plating on nutrient agar (Fig. 5-7). This phenomenon, known as *liquid holding recovery* (LHR), has been shown to require intact *uvr* genes, the genes that control the first step in excision repair

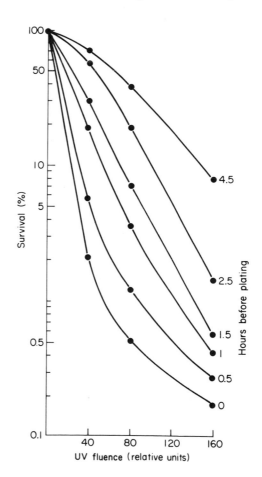

Fig. 5-7. UV radiation survival curves of *E. coli* B. After irradiation the cells were suspended in a liquid medium devoid of an energy source for the times indicated before being plated on nutrient agar. It is evident that both the slopes and the shapes of the survival curves can be altered by the postirradiation treatment of the cells (Modified from reference 18).

(see below). Thus, holding the cells in nonnutrient media appears to improve the efficiency of the excision repair process.[19] Other recovery phenomena whose molecular bases are not as well understood as liquid-holding recovery have been reviewed.[1,20]

A reinterpretation of the meaning of the diverse shapes of radiation survival curves has also led to the hypothesis that certain cells can repair radiation damage. According to classical target theory,[21] a shoulder on a survival curve should indicate multiple targets (or multiple hits on targets). However, closely related mutants of *E. coli* are not expected to have markedly different numbers or types of targets, yet their survival characteristics are markedly different (Fig. 5-8). The shoulders on the survival curves of the more resistant strains have been reinterpreted as implicating the capacity of these cells to repair radiation damage.[22] The shoulder of a survival curve thus represents the dose range within which the cells can cope with the damage produced. At higher doses where the survival curve becomes steep, the repair systems themselves may either have become inactivated by the radiation, or the number of lesions in the DNA may exceed the capacity of the repair systms to cope with this damage.

With the mapping of several genes that affect the radiation sensitivity of cells (Fig. 5-9), and the determination of the biochemical deficiencies of several of these mutants, both the presence and importance of enzymatic mechanisms for the repair of radiation damage became firmly established.

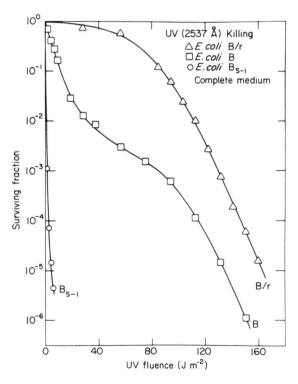

Fig. 5-8. UV radiation survival curves of different mutants of *E. coli* B. [Modified from R. H. Haynes, *Photochem. Photobiol.* **3**, 429 (1964).]

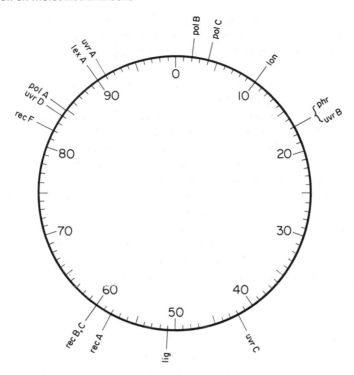

Fig. 5-9. Mutations in *E. coli* (see following tabulation) affecting radiation sensitivity. [Modified from B. J. Bachmann, K. B. Low, and A. L. Taylor, *Bacteriol. Rev.* **40**, 116 (1976).]

Mutant designation	Deficiency
lex (exr)	Postreplication repair (PRR); excision repair (ER)
lon	Septate formation (i.e., forms filaments)
phr	Photoreactivation enzyme
polA (res)	DNA polymerase I; ER
polB	DNA polymerase II; ER
polC (dnaE)	DNA polymerase III; ER
recA	Recombination; PRR; ER
recB, C	Recombination; exonuclease V; PRR; ER
recF	Recombination; PRR
uvrA, B	Incision nuclease; ER
uvrC	ER
uvrD	PRR; ER
lig	Polynucleotide ligase; ER, PRR

5.4.2. Mechanisms for the Repair of Photochemical Damage

Three modes of repair have been documented:

1. *The damaged part of the molecule is restored to its functional state in situ*. This may result from the spontaneous "decay" of the damage to an

innocuous form, e.g., dehydration of pyrimidine photohydrates or the recombination of radicals to yield a restored molecule (e.g., $R\cdot + H\cdot \rightarrow RH$); or it may be accomplished by some enzymatic mechanism, e.g., photoreactivation. *Photoreactivation,* the enzymatic splitting of cyclobutane-type pyrimidine dimers *in situ* mediated by exposure to visible light, was the first repair system to be elucidated at the molecular level.[23]

In 1949, Kelner observed that the survival of UV-irradiated bacteria could be greatly enhanced if the cells were subsequently exposed to an intense source of blue light. In 1960, Rupert demonstrated the existence of a photoreactivating enzyme, and established its basic properties. The enzyme combines with UV-irradiated DNA in the dark to form an enzyme–substrate complex. The absorption of light between 320 and 410 nm activates the complex, the substrate is converted from a cyclobutane-type pyrimidine dimer to monomeric pyrimidines, and the enzyme is released. *Under certain conditions,* as much as 80% of the lethal damage induced in bacteria by low fluences of UV radiation at 254 nm can be photoreactivated, thus indicating the importance of pyrimidine dimers as lethal lesions. Photoreactivating enzymes have been found in a wide range of species from the simplest living cells, the mycoplasmas, to the skin and white blood cells of man.[24]

2. *The damaged section of a DNA strand is removed and replaced with undamaged nucleotides to restore the normal function of the DNA.* This mechanism is the basis of the *excision repair* system that is known colloquially as ''cut and patch.''[1,25,26] This system constituted the first discovery of a mechanism of *dark repair,* to distinguish it from photoreactivation that is mediated by visible light. The excision repair system has been shown to repair a variety of radiation and chemically induced structural defects in DNA, but was originally observed as a mechanism for the repair of UV-induced cyclobutane-type pyrimidine dimers. A schematic representation of the probable steps involved in excision repair is shown in Fig. 5-10.

The first steps in this repair system are the recognition of damage (perhaps in part as a distortion of the DNA helix), and the introduction of a break in the DNA chain near the lesion (incision step). Resynthesis is then initiated by the action of DNA polymerase (repair replication) using the opposite strand of DNA as the template. The lesion is cut out to complete the excision process, and finally, when the excised region is filled with undamaged nucleotides, the single-strand interruption is closed enzymatically by polynucleotide ligase yielding repaired DNA. The enzymology of excision repair has been reviewed recently.[17,24,27]

Excision repair can be divided into two pathways (Fig. 5-11). In the major pathway, repair can proceed in buffer and requires DNA polymerase I (*polA*$^+$). In the absence of this enzyme (i.e., in a *polA*$^-$ mutant), DNA polymerase III (*polC*$^+$) can partially substitute. A minor pathway (in terms of the number of lesions repaired) requires complete growth medium, and a number of gene products. This pathway apparently gives rise to mutations, presumably by inaccurate repair. While these pathways have been delineated on the basis of the requirements of a few gene products, it is clear that more gene products (enzymes?) than are shown in Fig. 5-11 will be required to complete the repair in each pathway. For example, if a chain break in DNA is produced as a step in

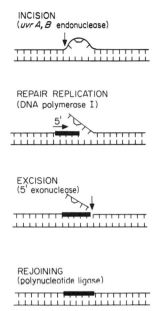

INCISION
(*uvr A, B* endonuclease)

REPAIR REPLICATION
(DNA polymerase I)

5'

EXCISION
(5' exonuclease)

REJOINING
(polynucleotide ligase)

Fig. 5-10. A general model for the major pathway of excision repair. An enzyme recognizes the lesion, shown here as a cyclobutane-type pyrimidine dimer, and makes an incision cut in the DNA strand. Repair replication (heavy line) commences using the opposite strand of DNA as the template. Finally, the damaged section of the DNA is excised, and the break in the DNA strand is sealed. The vertical arrows indicate the locations of nuclease cuts in the damaged DNA strand and the horizontal arrow indicates the direction of repair replication, beginning at the 3'OH end of the DNA strand. [Modified from P. C. Hanawalt, *Endeavour* **31,** 83–87 (1972).]

repair, the action of polynucleotide ligase will be required to reseal the chain (Fig. 5-10). These same pathways have been shown to repair the DNA single-strand breaks produced by X irradiation.[28]

A study of excision repair in mammalian cells has led to the observation of an apparent correlation between carcinogenesis and the defective repair of

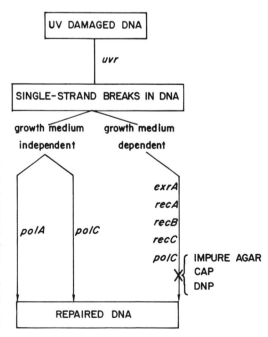

UV DAMAGED DNA

uvr

SINGLE-STRAND BREAKS IN DNA

growth medium independent

growth medium dependent

polA   *polC*

exrA
recA
recB
recC
polC   IMPURE AGAR
CAP
DNP

REPAIRED DNA

Fig. 5-11. The genetic and physiological control of the different pathways of excision repair in *Escherichia coli*. While these pathways have been delineated on the basis of the requirements of only a few gene products, it is clear that many more gene products than are shown will be required to complete the repair in each pathway. The mutations are defined in Fig. 5-9. Chloramphenicol (CAP), dinitrophenol (DNP), and impure agar inhibit the growth medium dependent pathway. [From D. A. Youngs, E. Van der Schueren and K. C. Smith, *J. Bacteriol.* **117,** 717–725 (1974).]

DNA.[29] Fibroblasts from patients with the hereditary disease xeroderma pigmentosum were found to exhibit much-reduced levels of repair replication after UV irradiation. It was suggested that this deficiency in excision repair might be related to the induction of the fatal skin cancers that patients with this disease develop upon exposure to sunlight. More recently, it has been reported that skin cells from other patients with xeroderma pigmentosum (so-called "XP variants") have a normal amount of excision repair, but are somewhat deficient in postreplication repair (defined below). (See reference 17, p. 617.)

3. *The damage, while not being directly repaired, is either ignored or bypassed, and the missing genetic information is supplied by redundant information within the cell.*[17,19,25] Several lines of evidence have suggested that the excision mode of repair is not the only mechanism by which cells can repair radiation damage to their DNA in the dark. The first indication was that bacterial cells deficient both in excision repair *(uvr)* and in genetic recombination *(rec)* were much more sensitive to killing by UV radiation than were cells carrying either mutation alone (Fig. 5-12). This suggested that certain steps in genetic recombination might be important in the repair of radiation damage. (In genetic recombination, DNA from a male bacterium is injected into a female bacterium. The donor's DNA is then combined with the recipient's DNA such that subsequent daughter cells contain genetic information from both parents.) Second, the fact that *uvr* cells show a large recovery of viability when plated on minimal

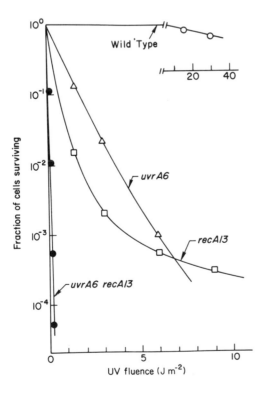

Fig. 5-12. UV radiation survival curves of mutants of *E. coli* K-12. The mutations are defined in Fig. 5-9. [Modified from P. Howard-Flanders and R. P. Boyce, *Radiat. Res. Suppl.* **6,** 156 (1966).]

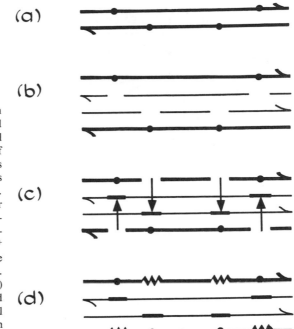

Fig. 5-13. A model for postreplication repair of UV radiation-damaged DNA. (a) Dots indicate photochemical lesions produced in the two strands of DNA. (b) DNA synthesis proceeds past the lesions in the parental strands leaving gaps in the daughter strands. (c) Filling of the gaps in the daughter strands with material from the parental strands by a recombinational process (depends upon functional $recA^+$ genes). (d) Repair of the gaps in the parental strands by repair replication. The reader is cautioned that steps (c) and (d) are highly schematized, and will probably be modified as additional data become available. (Modified from reference 19.)

medium as compared to plating on complex medium (i.e., minimal medium recovery) suggests that excision-deficient cells are still able to repair radiation damage. Cells must be rec^+ to show minimal medium recovery. Third, it has been demonstrated that, contrary to the interpretations of earlier data, photoproducts such as pyrimidine dimers do *not* permanently block DNA synthesis in cells that are deficient in the excision mode of repair. Fourth, the DNA that is synthesized immediately after UV irradiation in cells of *E. coli* K-12 has discontinuities when assayed in alkaline sucrose gradients. In excision-deficient cells the mean length of newly synthesized DNA approximates the distance between pyrimidine dimers in the parental strand. With further incubation of the cells, however, these discontinuities disappear, and the DNA approximates the molecular size of that from unirradiated control cells. A postreplication repair mode is thus indicated that appears to be mediated by some of the enzymes involved in genetic recombination. This process is shown schematically in Fig. 5-13.

Postreplication repair can now be divided into several pathways (Fig. 5-14). The filling of the gaps in newly synthesized daughter strands of DNA does not occur in *recA* mutants. The *recB*, *exrA (lexA)*, and *uvrD* strains are only partially deficient (and additive) in their ability to repair the gaps. Chloramphenicol (an inhibitor of protein synthesis) partially blocks the filling of gaps, but not if the cells are mutant at either *exrA (lexA)*, *uvrD*, or *recB*, suggesting that in addition to their independent functions, there must be another pathway that requires the cooperation of all three gene products. The *recF* mutation has been shown to act

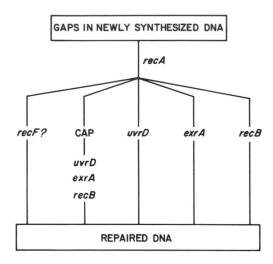

Fig. 5-14. The genetic control of the different pathways of postreplication repair in *Escherichia coli*. While these pathways have been delineated on the basis of the requirements of only a few gene products, it is clear that many more gene products than are shown will be required to complete the repair in each pathway. The mutants are defined in Fig. 5-9. Postirradiation treatment with chloramphenicol (CAP) inhibits a minor pathway of postreplication repair that may produce mutations due to inaccurate repair. [From D. A. Youngs and K. C. Smith, *J. Bacteriol.* **125**, 102–110 (1976)].

independently of *recB* in postreplication repair; whether or not it is independent of *exrA (lexA)* and *uvrD* has not yet been determined. In addition to the products of the genes shown in Fig. 5-14, undoubtedly the action of nucleases, polymerases, ligase, etc. will also be required to complete the repair via each of these pathways. The specifics of these pathways remains to be elucidated.

When plotted as a number of mutants per survivor after a given fluence of UV radiation, the *uvrA,B* strains appear to be more mutable than wild-type cells. This has led to the suggestion that postreplication repair (apparently the only type of dark repair remaining in a *uvrA,B* strain) leads to mutations due to inaccurate repair. To support his concept is the observation that *recA* and *exrA (lexA)* strains, strains deficient in postreplication repair, cannot be further mutated by UV radiation or certain chemical mutagens.

5.4.3. Conclusions

One cannot help but be impressed with the multitude and sophistication of the repair pathways possessed by cells. One system can repair damage in parental strands under nongrowth conditions, while another repairs gaps in daughter strands that can only occur in complete growth medium. A significant percentage of the energy of a cell is spent in synthesizing enzymes to repair and maintain the integrity of the genetic code in its DNA. There is evidence, however, to suggest that these repair systems were not developed just to repair radiation or chemical damage to DNA, but appear to have a necessary function in the everyday life of a cell. This conclusion is based upon the observation that while certain repair-deficient mutants grow normally in the absence of radiation, mutants deficient in two repair systems (e.g., *polA recA*) are not viable. Thus, normal growth processes may produce breaks or other lesions in DNA that must be repaired in order to maintain viability.

The discovery of DNA repair processes has profoundly stimulated research in molecular biology, genetics, cancer biology, and aging. The discovery and elucidation of DNA repair has probably been the most important contribution that UV photobiologists and radiation biologists have made to science.

5.5. MUTAGENESIS[17,30–34]

UV radiation is a very efficient mutagen. For low fluences of UV radiation the mutation frequency can be increased 10^3- to 10^6-fold over the spontaneous mutation level in bacteria, and for high UV fluences, nearly all of the survivors may be mutants. It is not at all surprising that the action spectrum for UV radiation-induced mutations in bacteria mimics the absorption spectrum for nucleic acid (DNA). Thus, to understand the molecular basis of mutagenesis one needs to understand the biochemistry of DNA, both in terms of its replication and its repair.

5.5.1. Alterations in DNA That Lead to Mutations

The first type of alteration is the substitution of one base pair by another. These substitutions can be of two kinds, *transitions* and *transversions*. In transitions a purine base on one of the DNA strands is replaced by another purine (e.g., G→A), or a pyrimidine by another pyrimidine (e.g., C→T). In transversions a purine is replaced by a pyrimidine or vice versa [i.e., (G or A) → (C or T)].

The second type of mutagenic change is the deletion or insertion of a base pair to yield a *frame shift* mutation. The addition or removal of a base pair would alter the continuity of the genetic code for making a protein such that either the polypeptide chain would be terminated by a *nonsense* codon (i.e., a triplet of nucleotides that does not code for an amino acid[33], or incorrect amino acids would be inserted into the enzyme making it inactive *(missense).*

5.5.2. Molecular Mechanisms of Mutagenesis

Two general molecular mechanisms exist for mutation production. A surviving cell may undergo a mutation because an unrepaired DNA lesion has generated an error during the replication of DNA. We have already discussed (Section 5.2.3.) that the photohydrate of cytosine codes for adenine rather than guanine in an *in vitro* enzyme system. Such photochemical lesions leading directly to mutations must be rare events *in vivo,* however, since mutations are not induced in *recA* and *lex (exrA)* strains by UV radiation (see below), although mutations can be induced in these strains by treatment with certain chemicals.

A second mechanism that is currently under active study in several laboratories is based upon the hypothesis that an inducible error-prone repair system changes the base sequence of DNA in the course of restoring a viable DNA structure. This concept is based, in part, upon the observation that certain repair-deficient strains of *E. coli* [i.e., *recA* and *lex (exrA)*] cannot be further mutated by UV irradiation, and that these strains are deficient in other radiation-inducible functions such as filament formation and the induction of λ phage.[17,30]

Error-prone repair leading to mutations is considered to occur via two basic pathways. In one process, mutations appear to arise due to the inaccurate repair, by a minor repair system controlled by the *recA* and *lex (exrA)* genes, of gaps left in the DNA after the excision of pyrimidine dimers and other photoproducts. The other mutation process apparently depends upon the persistence of unexcised photoproducts in the DNA until the time of replication, when mutations result from the error-prone repair of gaps produced in newly synthesized daughter strand DNA opposite damaged regions in the parental strands of DNA [i.e., postreplication repair that is also controlled by the *recA* and *lex (exrA)* genes].

The UV fluence response curve for the production of mutations best fits a fluence squared relationship. This has been interpreted to suggested that two lesions in the cell are required in order to produce a mutation; one of which may be required to induce the error-prone repair system.[17,30]

5.5.3. Procedures Used to Study Mutagenesis

Bacteria have been popular subjects for the study of mutagenesis because they grow rapidly, and can easily be obtained in large numbers. The types of mutations that have been studied have been those that are amenable to an easy assay procedure. Thus, most studies have followed the acquisition by bacteria of resistance to killing by bacteriophage or by antibiotics, and the reversion to prototrophy (i.e., an auxotrophic bacterium may require an amino acid for growth but a prototrophic revertant would not).

The reversion to prototrophy has been the most popular assay because low frequencies of mutations can easily be detected, and the same medium (partially enriched medium containing ~1 μg/ml of the required amino acid) can be used to score both mutations and survival. When plating for survival at high dilution, auxotrophs make small but visible colonies. When plated at high cell densities for mutation scoring, the auxotrophic population forms a limited "lawn" of cells against which prototrophic mutants form large colonies. To score the frequency of existing spontaneous mutations (i.e., in the absence of UV radiation), cells must be plated in the complete absence of the required amino acid.

An auxotroph carrying a new mutation specifying prototrophy will still be phenotypically auxotrophic, and unable to form a colony on minimal medium if it does not first have a chance to synthesize the active enzyme before being plated on minimal medium. This explains the need for the partially enriched plates described above.

This procedure for determining mutations is based upon the assumption that survival is the same on an uncrowded plate (to score survival) and on a crowded plate (to score mutations). For some strains, especially those that form filaments after exposure to the mutagen, this situation does not hold true. Filamenting cells can show a 100% survival on a crowded plate ("neighbor restoration"), and a 5% survival on an uncrowded plate. Such a response would have a profound affect on the "apparent" mutation yield.

There is a classic and unresolved problem in radiation biology as to whether it is better to compare, at equal fluence or at equal survival, some radiation response of two cell lines that have different sensitivities to killing by radiation. This problem also exists when one wants to compare the mutation frequencies of radiation-sensitive and -resistant strains.[35]

5.5.4. Suppressor Mutations

Alterations in DNA within specific genes give rise to primary mutations. When a mutant bacterium reverts to wild-type as a result of another mutation at a different locus or site, the second mutation is called a *suppressor* of the first. Thus, the enzyme tryptophan synthetase is inactive if tyrosine at position 174 is replaced by cysteine. But, via a second mutation, if the glycine at position 210 is replaced by glutamic acid, the activity of the enzyme is restored. This is an example of intragenic suppression. An example of intergenic suppression is the case in which a minor component transfer RNA (i.e., there are major and minor types of transfer RNAs that code for a single amino acid) is mutated so that its anticodon codes for a different amino acid than it carries.[33] Thus, in tryptophan synthetase, the replacement of glycine 210 by arginine yields an inactive enzyme. If, however, a mutant transfer RNA inserts glycine instead of arginine, an active enzyme will result. The efficiency of suppression of this type is low, and in the presence of the suppressor gene both active and inactive forms of the enzyme are made.

5.5.5. Temperature-Sensitive Mutants

Recently, a number of temperature-sensitive mutants have been isolated that have proved very valuable in studies on DNA repair (e.g., *polC,* deficient in DNA polymerase III; *lig*$_{ts7}$, deficient in polynucleotide ligase). These are mutants that behave as wild-type under permissive conditions (i.e., at ~30°C), but are mutant under *restrictive* conditions (i.e., at higher temperatures such as 42°C). If a cell were completely deficient in DNA polymerase III, it could not replicate its DNA, and the mutation would be lethal. Thus, the *polC* mutation is termed a *conditional lethal* mutation, i.e., conditional upon the temperature at which it is grown. The temperature sensitivity resides in the mutant enzyme molecule. Apparently at the higher temperature the configuration of the enzyme is altered (denatured?) to such an extent as to be inactive.

5.5.6. Conclusions

The recent correlation of mutagenesis with DNA repair has rekindled general interest in mutagenesis, and should provide the basis for the rapid advancement of the field.

Another stimulatory factor is the recent correlation between carcinogenesis and mutagenesis. It is now appreciated that many carcinogenic agents must first be metabolized to form the ultimate carcinogen. Almost all ultimate carcinogens have now been shown to be mutagens.[36] Thus, if we can understand the basis of mutagenesis, we should have a better understanding of the molecular basis of carcinogenesis.

5.6. EXPERIMENTS

5.6.1. Photochemistry of Nucleic Acids

Hydrate Formation. Determine the absorption spectrum (200–300 nm) of uridylic acid (30 μg/ml). Irradiate at 254 nm (15-W GE germicidal lamp) in a quartz cuvet and follow the disappearance of absorption with time at 260 nm. When the absorbance at 260 nm is less than about 0.2, determine the full absorption spectrum. Adjust the irradiated solution to ~pH 1 with concentrated HC1 (~0.1 ml), cover, and let stand at room temperature for 24 h, mix, and determine the absorption spectrum [R. L. Sinsheimer, *Radiat. Res.* **1**, 505–513 (1954)].

Thymine Dimer Formation. Determine the absorption spectrum (230–300 nm) of thymine (25 μg/ml). Place 250 μl of a solution of thymine (500 μg/ml) into each of several 10-ml beakers, freeze and irradiate (254 nm) in the freezer (you must start lamp while at room temperature and then put in freezer). Take samples at various times, thaw, and quantitatively transfer to a 5-ml volumetric flask. Determine the absorbance at 260 nm; when it has dropped to about 0.3, determine the absorption spectrum of the sample to confirm the loss of the characteristic absorption spectrum of thymine. Then irradiate this solution at room temperature and follow the return of absorbance at 260 nm with time. When the maximum increase in absorption has been achieved, determine a complete absorption spectrum to confirm the formation of thymine. (While frozen, the thymines are juxtaposed and the photochemical equilibrium between the formation and splitting of dimers favors dimer formation. In solution, however, the dimers are split by the reirradiation and then the monomers diffuse away and cannot be redimerized.)

The thymine dimers (after irradiation while frozen) can be chromatographed and separated from residual thymine [K. C. Smith, *Photochem. Photobiol.* **2**, 503–517 (1963)]. While the use of radioactive thymine would greatly facilitate these studies, the chromatogram can be cut into 1-cm strips (crosswise), and eluted in water, and the absorbance checked at 260 nm. The solutions suspected of containing the thymine dimer can be reirradiated as above to split the dimer, and yield thymine that will then absorb at 260 nm.

5.6.2. UV Survival Curves and Photoreactivation

Survival. Strains AB1157 (wild-type), AB1886 *(uvrA6)*, AB2463 *(recA13)*, and AB2480 *(uvrA6 recA13)* may be obtained from the Coli Genetic Stock Center, Department of Human Genetics, Yale University School of Medicine, New Haven, CT 06510. This experiment requires some expertise in microbiological techniques and a calibrated UV lamp at 254 nm. Approximate calibration can be achieved using 10^{-4} M uridylic acid (UMP) as a chemical actinometer (reference 8, p. 193). The UV fluence rate (J m^{-2} s^{-1}) is equal to

$$\frac{\left[\dfrac{A_I - A_F}{A_I}\right] \left[\dfrac{\text{moles (UMP)}}{\text{ml}}\right] \left[\text{ml (UMP)}\right] \left[\dfrac{11.9 \times 10^8 \text{ JE}^{-1}}{2537 \text{ (Å)}}\right]}{[1.9 \times 10^{-2} \text{ mol E}^{-1} (\Phi)] [(\text{sample surface area}) \text{ m}^2] [(\text{irradiation time}) \text{ s}]}$$

where A_I and A_F are the initial and final absorbances of UMP at 260 nm.

The survival curves in Fig. 5-12 can be approximated as follows: The cells were harvested after overnight growth in yeast extract nutrient broth and diluted 1:10 in buffer to give ~1×10^8 cells/ml. About 10 ml of cells in a 10-cm Petri dish on a shaker platform were UV-irradiated (8-W GE germicidal lamp at ~47 cm above the cells, giving a fluence rate of ~1 J m^{-2} s^{-1}) with various fluences, and appropriately diluted (estimated from Fig. 5-12) to yield about 200 survivors per plate when 0.1 ml is spread on a 10-cm Petri plate. Colonies may be counted after 24–48 h of growth.

Photoreactivation. Spread about 2×10^6 cells of strain AB2480 (*uvrA6 recA13*) per plate. Prepare 6 plates. Immediately UV irradiate 4 of the plates with 0.4 J m^2 (survival is approximately 7×10^{-6}). Place a Pyrex Petri dish lid full of water, as a UV and infrared filter, on top of 3 of the UV-irradiated plates and 1 non-UV-irradiated plate (without their plastic lids). Place two daylight fluorescent bulbs 5 cm above the agar surface of the plates and irradiate the UV-irradiated plates for 1, 5, and 10 min, and the non-UV-irradiated plate for 10 min. Incubate all plates at 37°C for 24–48 h and determine the effect of photoreactivation on survival.

5.7. REFERENCES

1. K. C. Smith and P. C. Hanawalt, *Molecular Photobiology* (Inactivation and Recovery), Academic Press, New York (1969).
2. J. Jagger, *Introduction to Research in Ultraviolet Photobiology*, Prentice–Hall, Englewood Cliffs, N.J. (1967).
3. A. C. Giese, *Living With Our Sun's Ultraviolet Rays*, Plenum Press, New York (1976).
4. S. Y. Wang (ed.), *Photochemistry and Photobiology of Nucleic Acids*, Academic Press, New York (1976).
5. A. J. Varghese, Photochemistry of nucleic acids and their constituents, *Photophysiology* **7**, 207–274 (1972).
6. R. O. Rahn, Ultraviolet irradiation of DNA, in: *Concepts in Radiation Cell Biology* (G. L. Whitson, ed.), pp. 1–56, Academic Press, New York (1972).
7. K. C. Smith (ed.), *Aging, Carcinogenesis and Radiation Biology* (The Role of Nucleic Acid Addition Reactions), Plenum Press, New York (1976).
8. A. D. McLaren and D. Shugar, *Photochemistry of Proteins and Nucleic Acids*, Pergamon Press, Oxford (1964).
9. F. Hutchinson, The lesions produced by ultraviolet light in DNA containing 5-bromouracil, *Q. Rev. Biophys.* **6**, 201–246 (1973).
10. R. O. Rahn, L. C. Landry, and W. L. Carrier, Formation of chain breaks and thymine dimers in DNA upon photosensitization at 313 nm with acetophenone, acetone or benzophenone, *Photochem. Photobiol.* **19**, 75–78 (1974).

11. D. Lenov, J. Salomon, and D. Elad, Ultraviolet- and γ-ray-induced reactions of nucleic acid constituents with alcohols. On the selectivity of the reactions for purines, *Photochem. Photobiol.* **17,** 465–468 (1973).

12. G. R. Banks, D. M. Brown, D. G. Streeter, and L. Grossman, Mutagenic analogues of cytosine: RNA polymerase template and substrate studies, *J. Mol. Biol.* **60,** 425–439 (1971).

13. Yu. A. Vladimirov, D. I. Roschupkin, and E. E. Fesenko, Photochemical reactions of amino acid residues and inactivation of enzymes during UV-irradiation. A review, *Photochem. Photobiol.* **11,** 227–246 (1970).

14. L. I. Grossweiner, Photochemical inactivation of enzymes, *Curr. Top Radiat. Res.* **11,** 141–199 (1976).

15. A. D. McLaren and O. Hidalgo-Salvatierra, Quantum yields for enzyme inactivation and the amino acid composition of proteins, *Photochem. Photobiol.* **3,** 349–352 (1964).

16. L. Augenstein and P. Riley, The inactivation of enzymes by ultraviolet light. IV. The nature and involvement of cystine disruption, *Photochem. Photobiol.* **3,** 353–367 (1964); see also, *ibid.* **6,** 423–436 (1967).

17. P. C. Hanawalt and R. B. Setlow (eds.), *Molecular Mechanisms For Repair Of DNA,* Plenum Press, New York (1975).

18. R. B. Roberts and E. Aldous, Recovery from ultraviolet irradiation in *Escherichia coli, J. Bacteriol.* **57,** 363–375 (1949).

19. K. C. Smith, The roles of genetic recombination and DNA polymerase in the repair of damaged DNA, *Photophysiology* **6,** 209–278 (1971).

20. P. A. Swenson, Physiological responses of *Escherichia coli* to far-ultraviolet radiation, in: *Photochemical and Photobiological Reviews* (K. C. Smith, ed.), Vol. 1, pp. 269–387, Plenum Press, New York (1976).

21. D. E. Lea, *Actions of Radiation on Living Cells,* Cambridge University Press, London (1955) (Reprinted 1962).

22. R. H. Haynes, The influence of repair processes on radiobiological survival curves, in: *Cell Survival After Low Doses of Radiation: Theoretical and Clinical Implications* (T. Alper, ed.), pp. 197–208 Wiley, New York (1975).

23. W. Harm, C. S. Rupert, and H. Harm, The study of photoenzymatic repair of UV lesions in DNA by flash photolysis, *Photophysiology* **6,** 279–324 (1971).

24. E. C. Friedberg, K. H. Cook, J. Duncan, and K. Mortelmans, DNA repair enzymes in mammalian cells, in: *Photochemical and Photobiological Reviews* (K. C. Smith, ed.), Vol. 2, pp. 263–322, Plenum Press, New York (1977).

25. P. Howard-Flanders, DNA repair, *Annu. Rev. Biochem.* **37,** 175–200 (1968).

26. R. B. Setlow and J. K. Setlow, Effects of radiation on polynucleotides, *Annu. Rev. Biophys. Bioeng.* **1,** 293–346 (1972).

27. L. Grossman, Enzymes involved in the repair of DNA, *Adv. Radiat. Biol.* **4,** 77–129 (1974).

28. C. D. Town, K. C. Smith, and H. S. Kaplan, Repair of X-ray damage to bacterial DNA, *Curr. Top. Radiat. Res.* **8,** 351–399 (1973).

29. J. E. Cleaver, Repair processes for photochemical damage in mammalian cells, *Adv. Radiat. Biol.* **4,** 1–75 (1974).

30. E. Witkin, Ultraviolet mutagenesis and inducible DNA repair in *Escherichia coli, Bacteriol. Rev.* **40,** 869–907 (1976).

31. B. A. Bridges, Mechanisms of radiation mutagenesis in cellular and subcellular systems, *Annu. Rev. Nuclear Sci.* **19,** 139–177 (1969).

32. C. O. Doudney, Ultraviolet light effects on the bacterial cell, *Curr. Top. Microbiol. Immunol.* **46,** 116–175 (1968).

33. J. D. Watson, *Molecular Biology of the Gene* (3rd ed.), Benjamin, New York (1976).

34. W. Hayes, *The Genetics of Bacteria and their Viruses* (2nd ed.), Wiley, New York (1968).

35. K. C. Smith, Ultraviolet radiation-induced mutability of *uvrD3* strains of *Escherichia coli* B/r and K-12: A problem in analyzing mutagenesis data, *Photochem. Photobiol.* **24,** 433–437 (1976).

36. J. McCann and B. N. Ames, Detection of carcinogens as mutagens in the *Salmonella*/microsome test: Assay of 300 chemicals: Discussion, *Proc. Natl. Acad. Sci. USA* **73,** 950–954 (1976).

6

Environmental Photobiology

6.1. INTRODUCTION

In many areas of photobiology, interest is usually focused on the mechanism of a photochemical reaction. In the more complex systems such as the photosynthetic membranes and the phytochromes, structural aspects and biochemical regulation must also be considered. The distinguishing feature of *environmental photobiology* is in the attempt to integrate the photobiological mechanisms first, into the complete organism, and second, into the relationships among the organisms making up an ecosystem or a portion of an ecosystem. In the aquatic ecosystem, environmental photobiology might be concerned with the fact that an increase in water turbidity due to increased delivery of sediment to a tributary estuary might eliminate rooted aquatic plants because of lack of sufficient

Howard H. Seliger • McCollum-Pratt Institute and Department of Biology, The Johns Hopkins University, Baltimore, Maryland

sunlight. The photobiological processes of photosynthesis have not changed, but because certain specific thresholds have not been reached—in this case, the light intensity compensation point for photosynthesis, i.e, the underwater light intensity for the aquatic plants at which the uptake of free energy from sunlight is equal to the energy expended by the plant in keeping alive (lumped into the general term respiration)—the plant will die or will not reproduce. This in turn may reduce the number of higher-trophic-level organisms, fish or crabs, that utilize these plants for food or refuge.

As a second example, an increase in nutrients delivered to an aquatic ecosystem (eutrophication) will usually result in a change in the phytoplankton species distribution. Owing to specific size requirements for prey or to other behavioral aspects, the zooplankton or herbivores may also change their species distribution. Eventually this may affect the species distribution of the fish populations that feed on the zooplankton and of the larger fish that eat the smaller fish.

Thus, environmental photobiology is an area very precisely named. It attempts to examine photobiological processes and their relationships with the complete ecosystem.

6.2. THE SUN AS THE SOURCE OF FREE ENERGY FOR LIFE ON EARTH

Since the formation of our solar system, the free energy for all life on earth has been delivered in 2 kg of sunlight falling on the surface of the earth each second. The solar constant is 2.0 cal cm^{-2} min^{-1} incident on the upper atmosphere, equivalent to an annual energy delivery of 5.61×10^{24} J. In equivalent units of mass and energy this is equal to 68,600 tons per year or 2 kg s^{-1}. The spectral intensity distribution of this incident sunlight is shown in Fig. 6-1, curve 1, in units of photons s^{-1} cm^{-1} nm^{-1} as a function of wavelength from 200 nm in the ultraviolet (UV) region to 1800 nm in the infrared region. Atmospheric absorption of UV and visible radiation by ozone and of red and infrared radiation by water vapor modify the spectral distribution of curve 1 to that shown in curve 2, the spectral intensity of sunlight presently incident on the surface of the earth. At ~290 nm, the short wavelength limit for sunlight penetrating to the earth's surface, the spectral intensity has been attenuated by the atmospheric ozone layer and by molecular scattering by a factor of 2×10^6. In curve 3 of Fig. 6-1, the sunlight UV irradiance at the surface of the earth is plotted in logarithmic units. The superimposed curve d shows the relative action spectrum for UV-induced skin erythema in the same wavelength region, to which we shall return later.

6.2.1. Spectral Intensities for Evolution

Curves 4 and 5 of Fig. 6-1 illustrate the absorption effects of two different types of seawater on the spectral intensities of underwater sunlight. Curve 4 is

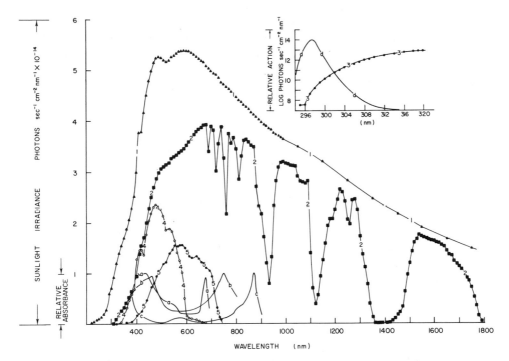

Fig. 6-1. Sunlight spectral irradiance in units of photons per second-square centimeter-nanometer ×
10^{-14}: *Curve 1*. Distribution of solar irradiance outside the earth's atmosphere at a distance of 55 km.
Data recalculated from Table 813 in reference 1. *Curve 2*. Distribution of solar irradiance at the surface
of the earth calculated for a sun zenith angle of 60°, equivalent to two air masses. (Data recalculated
from Table 815 in reference 1.) *Curve 3*. Logarithm of solar irradiance at the surface of the earth in
the wavelength region of the action spectrum for skin erythema. (Data recalculated from Table 815 in
reference 1.) *Curve 4*. Distribution of downwelling irradiance at a depth of 10 m in clear blue tropical
waters near Islas Tres Marias. (From reference 2.) (The data below 360 nm were calculated from
Table 559 and 563 in reference 1 for the transmissions of water and the atmosphere, respectively,
for UV radiation and from Table 813 in reference 1 for the incident spectral irradiance.) *Curve 5*.
Distribution of downwelling irradiance at a depth of 1.0 m in a eutrophic freshwater lake. (From
reference 2.)

    Relative absorbance or action spectra: *Curve a*. Relative absorbance of the brown alga *Gonyaulax
polyedra*. (Data recalculated from reference 3.) *Curve b*. Relative absorbance of the green photosynthetic bacterium *Cholorobium* sp. (Abstracted from reference 4.) *Curve c*. Relative absorbance of
the blue-green photosynthetic bacterium *Rhodospirillum spheroides*. (From reference 5.) *Curve d*.
Action spectrum for UV-induced skin erythema. (Abstracted from references 6 and 7).

the downwelling spectral irradiance over a 2π solid angle at the depth of 10 m in clear "blue" tropical ocean water,[2] while curve 5 is the downwelling spectral irradiance in a eutrophic lake at a depth of only 1 m.

It is generally agreed that the primitive atmosphere was essentially devoid of oxygen and its photodissociation products, including ozone. Life, therefore, is assumed to have originated and evolved on the bottom of shallow pools of depth greater than 10 m, a water thickness sufficient to attenuate the potentially lethal UV radiation with wavelengths below 300 nm. A more complete discussion with references can be found in reference 8. However, it is also quite likely that the primitive "soup" of organic and inorganic compounds synthesized by the UV radiation[9] would afford a protective UV-absorbing layer as the result of a photostationary process, and that eutrophic pools shallower than 10 m were accessible to the evolution of primitive life forms. One of the difficulties in the laboratory duplication of the UV-induced synthesis of primitive biological molecules is the deposition of strongly UV-absorbing films of synthesized molecules on the inside walls of the silica irradiation vessels. This primordial slick,[10] rather than a more homogeneous soup, would be more consistent with the implications of the spectral irradiance distributions shown in curves 4 and 5 of Fig. 6-1. At a depth of 10 m, even in clearest ocean water, incident sunlight intensities are reduced to negligible levels above 620 nm compared with curve 2, which is the surface irradiance. In shallow eutrophic waters, strong absorption of the short wavelengths extends to ~500 nm, suitable only for organisms containing yellow-absorbing (phycoerythrin) and red-absorbing (chlorophyll) pigments.

The relative absorbance spectrum of the dinoflagellate *Gonyaulax polyedra* is shown in curve a of Fig. 6-1. The major absorptions that give rise to the spectral irradiance curve 5 are in the blue region and in the red region below 700 nm. These correspond to the major absorptions of photosynthetic pigments.

The relative absorption spectra of the green sulfur (*Chlorobium*) and blue-green (*Rhodospirillum*) photosynthetic bacteria are shown in curves b and c, respectively, of Fig. 6-1. From inspection of curve c it is apparent that *Rhodospirillum* cannot grow well at any significant depth in water due to the infrared absorption of the latter. Therefore, if photosynthetic bacteria developed early in evolution (before oxygen evolution) they would have been confined to the undersides of the primordial slicks, thick enough to absorb lethal UV radiation, but thin enough to transmit 900-nm light. If they are more recent (subsequent to green plants), they would be confined to nutrient-rich anoxic surface waters that can receive light up to 900 nm.

6.2.2. The Fortuitous Coincidence of the UV Absorptions of Ozone, Nucleic Acids, and Proteins

The UV absorption spectrum due to the effective 3 mm thickness of the atmospheric ozone layer is shown in Fig. 6-2, together with the sunlight irradiance on the surface of the primitive anoxic earth. At 260 and 278 nm, the maxima of the absorption bands of nucleic acids and proteins, respectively, the incident

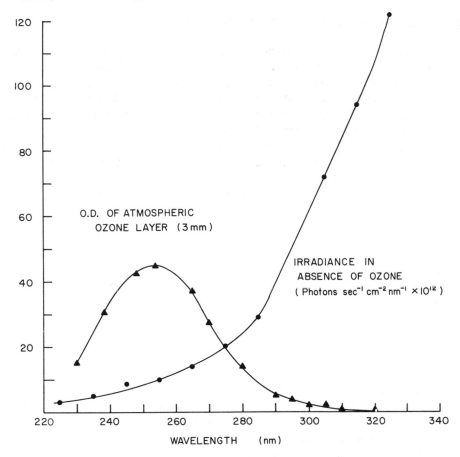

Fig. 6-2. Solar irradiance at the surface of the earth in the absence of ozone absorption of UV radiation (●), and the optical density of the present 0.3 atm-cm layer of ozone (▲). (From reference 8.)

sunlight intensities are reduced by factors of 10^{40} and 10^{17}. A significant necrosis in mammalian cells in culture can be observed subsequent to a fluence of UV radiation (254 nm) from a mercury germicidal lamp as low as 10 J m^{-2}, corresponding to a fluence of 1.3×10^{15} photons cm^{-2}. The action spectrum for this necrosis has a broad peak around 270 nm with a half-width of 50 nm. It is possible to estimate from Fig. 6-2 that in the *absence of ozone* the total irradiance within this action spectrum would be ~1.3×10^{15} photons cm^{-2} s^{-1}, certainly a lethal exposure rate even for a 1-s exposure.

Figure 6-2 shows that the fortuitous location of the ozone absorption band and the 3-mm effective thickness of the atmospheric layer of ozone are sufficient to permit the emergence of simple life forms to surface waters, and thence to the land.

6.2.3. Possible Perturbations to the Ozone Layer

In view of the major importance of the ozone layer to surface organisms, it is of some importance to understand the chemical and photochemical mechanisms giving rise to this very thin protective layer, which in temperate regions averages 0.3 atm-cm. Between the equator and 20°N latitude the thickness is ~0.25 atm-cm, while above 60°N latitude the layer can be as thick as 0.48 atm-cm. A range of 0.197–0.373 atm-cm of ozone has been observed at Davos, Switzerland (elevation 1575 m). These changes give rise to large spectral intensity variations in the UV region, particularly in the range 290–300 nm, the peak of the action spectrum for skin erythema (curve d, Fig. 6-1).

The absorbance of ozone is exponential, the optical density (O.D.) at any wavelength being given by

$$O.D._\lambda = \log \frac{I_{0_\lambda}}{I_\lambda} = \epsilon_\lambda cd \tag{6-1}$$

where ϵ_λ is the decadic extinction coefficient at wavelength λ, in units of reciprocal atm-cm, and cd is the thickness of the ozone layer in atm-cm.

From this it is seen that

$$\frac{dI_\lambda}{I_\lambda} = 2.30 \; O.D._\lambda \frac{d(cd)}{(cd)} \tag{6-2}$$

At 290 nm, $O.D._\lambda = 5$. A 10% decrease in the ozone layer will increase the surface irradiance at 290 nm by as much as 115%; at 300 nm the increase is 46%. From Eq. (6-2) and curve d of Fig. 6-1 it becomes apparent that small relative losses in the atmospheric ozone layer may increase significantly the intensities of UV radiation in the skin erythemal range.

Until recently it was believed that a major effect of air pollution would be increased atmospheric turbidity and, more specifically, increased absorption of UV radiation in the spectral region below 310 nm. However, there are several forms of air pollution that may produce losses in the ozone layer, and thus may *increase* the UV irradiance in the skin erythemal range. A catalytic cycle by which nitric oxide destroys ozone, including the photolytic dissociation of NO_2 and O_3, is shown in Scheme (6-3):

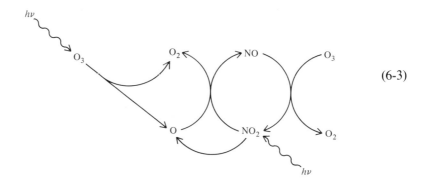

$$(6-3)$$

A second and even more efficient catalytic cycle involves free chlorine atoms.[11] A parallel set of reactions to the NO cycle is shown in Scheme (6-4):

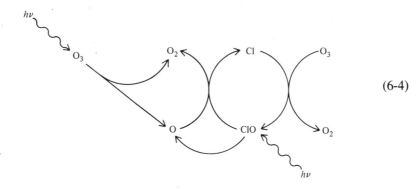

$$(6\text{-}4)$$

In the presence of both NO and ClO we can also have the conversion of ClO to Cl:

$$ClO + NO \rightarrow Cl + NO_2 \qquad (6\text{-}5)$$

In nature: (1) Nitrous oxide (N_2O) produced by soil bacteria, e.g., slowly diffuses into the stratosphere where it reacts with photolytically produced oxygen atoms to form nitric oxide

$$N_2O + O^* \rightarrow 2NO \qquad (6\text{-}6)$$

where O^* is an excited state of the oxygen atom. (2) Ammonia can be oxidized by OH radicals (OH·) or photolyzed by light to produce NH_2 radicals ($NH_2^·$) that can react further to form NO. (3) Ionization by cosmic rays can produce excited state nitrogen atoms N^* that react with molecular oxygen to produce NO + O:

$$N^* + O_2 \rightarrow NO + O \qquad (6\text{-}7)$$

Oxidation of N_2O and NH_3 are the major sources of the natural nitric oxides in the stratosphere controlling the steady-state concentrations of O_3. Cosmic ray ionization produces an average of 10^7 NO molecules cm^{-2} s^{-1} in the stratosphere, while estimates of the total stratospheric input from other natural sources range from 3 to 30 $\times$ 10^7 molucules cm^{-2} s^{-1}.[12] It has been proposed that, in the absence of the nitric oxide resulting from biological activity, the ozone layer would be double its normal thickness. These numbers should be used with care, however; the present ozone layer is the result of the oxygen released by biological activity (photosynthesis) that has overcome the Urey self-regulation mechanism[13] for O_2 concentrations in the primitive atmosphere.

6.2.3.1. Nitric Oxides from Stratospheric Aviation

A stratospheric diffusion model now includes 12 photolysis rate equations and 65 chemical rate equations for the reactions associated with the production of ozone and its subsequent catalytic destruction by NO, which would be injected into the stratosphere by high-flying fleets of supersonic transports (SSTs).[14] The NO source from naturally produced N_2O is approximately 10^8 molecules $cm^{-2} s^{-1}$. There is a feedback stability and temperature coupling whereby a decrease in O_3 reduces the NO yield from N_2O above 28 km. An operational fleet of 300 Concordes (the British–French SST) would inject 0.2×10^8 NO molecules $cm^{-2} s^{-1}$ into the atmosphere. If the present natural NO source has truly been responsible for the reduction by a factor of 2 in the ozone layer, this additional injection could conceivably reduce the ozone layer by 10%, increasing the solar irradiance below 300 nm [see Eq. (6-2)] by as much as 50%.

6.2.3.2. Nitric Oxides from Nuclear Bombs

Thermonuclear explosions produce enormous quantities of NO_x in the lower and middle stratosphere.[15] Various estimates of the reduction of the ozone layer by nuclear explosions prior to the 1962 test ban range from 3 to 6%, with half-time for recovery of 2.5 yr. Thus, the long-range results of even a limited nuclear war will be worldwide, and a major increase in solar irradiance below 300 nm may have disastrous consequences for human and other animal populations and for certain plant species.

6.2.3.3. Nitric Oxides from Cosmic Radiation

Supernova explosions may account for the cosmic rays presently incident on the earth.[12] A supernova explosion at a distance of 50 light years may increase the stratospheric cosmic ray intensity by a factor of 40 for almost a century. Under these conditions the nuclear radiation dose rate would be increased by much the same factor to around 2.5 rem yr^{-1}, increasing the probability for radiation-induced cancer incidence. (For a definition of the rem, the acronym for *r*oentgen *e*quivalent *m*an, see reference 16.) The nitric oxide concentrations, by virtue of Eq. (6-7), may be increased 1–10 times present levels due to all sources, and in time would effect a major reduction in the ozone layer. It is possible, therefore, that nearby supernova explosions may exert a significant biological effect through increased UV irradiance under conditions where nuclear radiation doses are nonlethal.

6.2.3.4. Chlorine from Photodissociation of Chlorofluoromethanes

Freon 12 (CCl_2F_2) and Freon 11 (CCl_3F) are used in large quantities as aerosol propellants and in refrigeration. Because of their chemical inertness they have previously been considered to be harmless. However, that portion of the halomethanes diffusing into the stratosphere can be photolytically dissociated to release free chlorine atoms that are even more efficient at catalyzing the destruction of ozone [see Eq. (6-4)] than nitric oxides.[11] Up to the present time, approximately 5×10^9 kg or 4×10^{10} mol halomethanes have been produced and presumably released into the atmosphere. These halomethanes will not reach maximum concentrations in the stratosphere (where photolysis can occur) until around 1990. An atmospheric model[11] predicts that if all halomethane release to the atmosphere were halted immediately, the rate of destruction of ozone due to the present concentrations of halomethanes would increase and level off in 1990 at values comparable to those from natural sources of NO. If the release of halomethanes continues at the present rate ($\sim 10^9$ kg yr^{-1}) the ozone destruction by chlorine will exceed that due to natural sources of NO by 1986.

Chlorofluoromethanes therefore represent a potentially serious UV radiation threat to biota on the surface of the earth. If indeed the naturally occurring sources of NO have reduced the ozone layer thickness by a factor of 2 as compared with that predicted by the reactions with oxygen,

$$h\nu + O_3 \rightarrow O + O_2 \tag{6-8}$$
$$O + O_3 \rightarrow 2O_2 \tag{6-9}$$

there appears to be no chance of avoiding a further 20–50% decrease in the ozone layer that will continue over the next several decades. There could, therefore, be a *doubling* of the present UV irradiance in the skin erythemal region (curve d, Fig. 6-1).

6.2.4. Possible Biological Consequences of Increased UV Radiation

The major concern over the possible decrease in the ozone layer and the projected increase in UV radiation has been the causal relationship between UV exposure and skin carcinogenesis. The action spectrum for skin carcinogenesis peaks at 297 nm. At this wavelength the optical density of Caucasian stratum corneum (the layer of dead surface skin cells) is ~ 0.5, while for heavily pigmented Negroid stratum corneum the value is ~ 2. On the basis of epidemiological studies of the incidence of squamous cell and basal cell carcinomas, which almost always can be treated effectively, and malignant melanomas, which are often fatal, it has been estimated that a 5% reduction in the ozone layer would produce 8,000 additional cases of skin cancer per year for the white population of the United States, leading to 300 extra deaths.[17] There is some evidence that for the period 1966–1970 compared with the period 1951–1955, there has been a major increase (50–100%) in the death rate due to malignant melanoma. The reasons for this are not yet understood.

There are areas of photobiology other than that of human skin cancer possibly affected by an increase in UV irradiation. Insects exhibit visual sensitivity into the near-UV region.[18] Therefore, all other conditions of spectral illumination remaining constant, an increase in UV intensity between 300 and 400 nm would change the hue of flower colors for bees, as an example. However, this would not appear to present a recognition problem, since over the daily time interval and over the seasonal time interval in which the insect operates, the UV illumination relative to the illumination above 400 nm already undergoes wide variations. In the worst possible case, assuming that the insect's recognition behavior is governed specifically by some absolute UV spectral illumination function, a change in this spectral illumination might change the daily phase of the insect's searching behavior or even possibly its geographical range (since UV intensity is also a function of latitude). We obviously do not know precisely the effect of this spectral illumination change on the totality of the insect behavioral pattern. That this may be only a small effect can be extrapolated from the relatively rapid northward migration from South America of a species of African "killer" bees (indiscriminate stinging behavior) accidentally transplanted to South America only a few years ago. In the case of the African bees, the change with latitude in the relative components of UV to the total illumination does not seem to have inhibited the migration from an equatorial spectral distribution (higher relative UV component) to more northern latitudes.

The sun compass mechanism[19] involves in part the ability to analyze planes of polarization as well as intensities of polarized light. The plane of polarization is determined by the sun as a source, the air as a scatterer, and the eye as an analyzer. Some arthropods make use of the polarization of water surface reflections. In these cases an increase in UV radiation would, in general, enhance the polarized signal received by the organism, since Rayleigh scattering varies as λ^{-4}.

The action spectrum for phototaxis in insects also involves near-UV radiation. In a species of white cabbage butterfly there is a very specific *absorption* of near-UV radiation ($\lambda < 400$ nm) by pigments in the wing surfaces of male butterflies and a *reflection* of these wavelengths by the females of the species.[20] This reflectivity by females in the region 380–400 nm is the means by which flying males recognize a virgin female, the latter resting on a plant leaf with the wings closed. Olfactory stimuli are not important. At these wavelengths the extinction coefficient of ozone is decreased significantly from its maximum at 260 nm, and a small fractional decrease in the ozone layer would produce approximately the same fractional increase in UV radiation. If only near-UV intensity is involved in the male response, there might be a beneficial effect in recognition due to an increase in signal-to-noise ratio. If hue recognition is involved, as for bee vision of flower colors, then the same arguments hold for the cabbage butterfly as for the bee.

In the plant kingdom, where the primary function is to do chemical work through electronically excited states, it might be expected that protection mechanisms to prevent detrimental photochemical reactions have been highly evolved. In many plant species the optimal use of sunlight requires exposure to the "normal" near-UV component of the sun's spectrum. Therefore, adaptation to a wide range of intensities due to seasonal and climatic conditions as well as geographical location would be the norm. A wide range of environmental parameters also contributes to the success, and therefore the geographical distributions of terrestrial plants. These include temperature, water, and soil conditions as well as light. Those plants that might be most affected by a change in the UV component of the sun's radiation would be broad-leafed crops already occupying narrow ecological niches, or those plants that are already pressed by other environmental factors. It can be inferred from these examples that the effect of increased UV radiation may be through a number of complex pathways in the ecosystem, rather than by producing acute photochemical damage.

In eutrophic surface waters the absorption of blue light is extremely high. At 400 nm the optical density per meter can be as high as 3, and rises steeply for shorter wavelengths. The phytoplankton in these eutrophic waters are continually mixed within the water column, and so are carried to the surface and to lower depths during the course of the day. They must, therefore, be able to respond maximally to a wide range of spectral intensity distributions, since the blue and UV intensities have relatively higher contributions at the surface than at deeper layers in the water column. In addition, the algae must respond maximally to the different spectral distributions of sunny days as compared with cloudy days. An increase in UV irradiance due to a decrease in the ozone layer might be "interpreted" by the phytoplankton as an increase in sunny days during which the relative UV contribution is greater than for cloudy days. This explanation should also hold for phytoplankton in clear, low-nutrient (oligotrophic) waters. It is not clear whether an increase in UV irradiance might affect positive phototaxis of some algal species.[21] It is also obvious that there is room for significant experimental studies on the effects of UV radiation on aquatic biota.

The major increases in UV irradiance that might occur as the result of nuclear war or supernovae explosions present an altogether different problem. However, it is evident from all that has been discussed that the human race is the species at risk (both photobiologically as well as sociologically). If this fragile species is destroyed it matters little whether the present distribution of plant and animal species that are directed and cultivated to support man continues. Life will continue and evolution may, by a more fortuitous pathway, result in a more ecologically tolerant species.

The discussion of ozone effects should not be left without some summarizing remarks. Unfortunately a controversy has arisen in the area of *risk versus benefit,* as it has for the questions of nuclear power and environmental chemical carcinogenesis. This involves the projection of necessarily incomplete scientific investigational data into political–economic decision-making. At the present time the "accepted" facts are as follows:

1. Oxygen evolved from the photosynthetic splitting of water has been responsible for raising the atmospheric oxygen content to its present atmospheric level (PAL) of 1.51×10^5 atm-cm from the original concentrations of <0.001 PAL on the primitive earth.

2. The photodissociation of molecular oxygen gives rise to oxygen atoms that add to oxygen molecules to form ozone.

3. The present ozone layer (0.3 atm-cm) is a photostationary state modified by chemically catalyzed dissociation of O_3 as well as chemical sinks for O atoms. Some of the chemical catalysts are themselves the products of photolytic reactions. The photolytic and catalytic reactions are functions of the vertical distributions of spectral light intensities, reactants, catalysts, and temperature. This latter dependence gives rise to daily and seasonal variations in the thickness of the ozone layer.

4. Any climatic change that would modify the level of photosynthesis should be reflected in the oxygen balance and consequently in the O_3 steady-state concentration.

5. There are significant natural concentrations of NO in the atmosphere due to the biological production of N_2O.

$$N_2O \overset{h\nu}{\rightarrow} N_2 + O^*$$
$$\overset{h\nu}{\searrow} \quad NO + N$$

$$O^* + N_2O \rightarrow 2NO$$

(6-10)

Estimates of these production rates range from 2.5 to 15×10^7 cm^{-2} s^{-1}.

6. Any man-made addition to the chemical sinks for ozone should be reflected in a reduction of the O_3 steady-state concentration. However, the feedback mechanisms among the parameters are not known precisely, and therefore the magnitudes of the reductions in O_3 steady-state concentrations are not known.

7. The steady-state production of Cl atoms from photolytic dissociation above 25 km of the 3×10^5 tons of $CFCl_3$ and 5×10^5 tons of CF_2Cl_2 produced annually is 5×10^7 cm^{-2} s^{-1}.

8. In laboratory experiments the Cl—ClO chain is considerably more efficient at catalyzing the destruction of O_3 than the NO—NO_2 chain.

As the logical conclusion from these facts it is acknowledged that the increased use of chlorofluoromethanes and the delivery of nitric oxides to the atmosphere will result in a reduction of the protective ozone layer. The question is, how much?

One of the problems in assessing this possible damaging effect by epidemiological studies of skin cancer incidence is that changing customs of dress and vacation patterns (and thus exposure to UV radiation), the increased use of nitrogen fertilizers (adding to NO levels in the atmosphere), the chlorination of sewage (adding to chlorine compounds in the atmosphere), and unknown and possibly synergistic effects of photosensitizing drugs, cosmetics, and environmental pollutants may mask or contribute to the end result. This entire area must be examined much more carefully. The physical measurements of integrated UV irradiance in different environments must be made more precisely; the contributions of the sources of catalytic destruction of ozone must be determined more accurately; specific effects of increased UV radiation on sensitive species must be evaluated; the feedback mechanisms (both biological and physicochemical) must be understood, in order to estimate the biological hazard of a decrease in the thickness of the ozone layer.

One of the considerations in all environmental studies is the concept of risk versus benefit. This concept introduces a strange set of units for the scientist, for it attempts to make an equation where the dimensions or units or quantities on one side of the equation are different from those on the other. For example, in the area of the use of X-rays, it is known that increased radiation causes general life-shortening as well as cancer. However, we acknowledge that in some cases the small additional risk of carcinogenesis from X-rays is compensated by its diagnostic value, i.e., in setting broken bones, in gastrointestinal studies, or in the therapeutic use of X-rays to destroy malignant tumors. While the application of risk versus benefit is clearest in this medical application, it becomes more diffuse and controversial when applied to risk of nuclear power plant accidents versus our need for additional energy; to risk of lung irritation vs. the economics of removing pollutants from industrial plants and automobiles; to the destruction of some wetland areas vs. the economic and recreational needs for boat marinas, golf courses, and shoreline vacation homes. It is obvious that these balances apply also to smoking, drinking, skiing, fertilizing of agricultural crops, and Freon in spray cans.

The role of environmental photobiology is to understand the interaction of light with components of the ecosystem to such a degree that the possible effects of some perturbation can be stated precisely. A more general presentation of the biological effects of UV radiation can be found in reference 22.

6.3. AQUATIC PHOTOBIOLOGY

There have been many papers devoted to mechanisms of photosynthesis and to studies of terrestrial plant species and biomes. The algae, such as *Chlorella,* have been used mainly for mechanism studies in photosynthesis. Until recently, relatively little research has been directed toward the study of marine phytoplankton ecology. A cogent summary of different types of plankton ecosystems comprises[23]: The "blue" water, low productivity, low nutrient (oligotrophic) regions resulting from weak physical circulation patterns, characterized by (1) low advection of the horizontal and vertical directions; (2) low nutrient input; (3) light penetration approaching that of pure seawater; (4) low concentrations of organisms; (5) a high degree of continuity in space and time; (6) a relatively constant rank order of species abundance; and (7) locally regenerated nutrients serving as the principal sources of phytoplankton nutrition. Species having evolved under these conditions could be expected to show adaptations to conserve scarce nutrient salts and effective mechanisms to compete for such salts. Generation times for such plankton would be relatively long; thus, protection against predation would have to be stretched over a concomitantly long period to assure reproductive success. Perhaps bioluminescence in marine dinoflagellates, which can function as a deterrent to copepod (crustacean) predation,[24] can have great selective value over these long generation times. Measures of nutrient excretion by net zooplankton (large enough to be retained in a towing net of mesh size usually around 120 μm) are possible and may provide estimates of nutrient cycling rates between phytoplankton and zooplankton. In this system it is necessary to assess (1) the role of bacterioplankton and bacteriovores in nutrient cycling; (2) the loss of nutrients contained in the mixed layer of sinking particles; and (3) the upward transport of particulate nutrients by migrating organisms.

The opposite end of the pelagic (oceanic) biome spectrum probably occurs in coastal upwelling regions where strong circulation patterns predominate. Here, readily identified characteristics are (1) high advective activity in the horizontal and vertical directions; (2) high nutrient input; (3) decreased light penetration; (4) high concentrations of organisms; and (5) a high degree of variability in space and time. These conditions impose a completely different set of selective pressures on the resident organisms. High nutrient input leads to high algal growth rates. Animal behavior patterns can also be expected to show adaptations to surface and subsurface currents such that their pelagic eggs and larvae remain within the system. These environmental conditions are common enough, and the selective pressures are sufficiently great to suggest that communities characteristic of upwelling regions have undergone selection in response to them.

If oligotrophic and upwelling systems are analogous to different biomes, a considerable amount of ecological theory, based on the study of terrestrial systems, might be used in understanding the oceans. For example, the aqueous medium itself may be analogous to the soils of the terrestrial biomes in that the chemical nature of seawater may be strongly influenced by organisms, and, thus, may be quite different in the two systems. "Conditioning" of water for phytoplankton growth must occur on a time scale that is short compared to the time scale of advection. Thus, in upwelling areas, this event must take place rapidly, whereas in oligotrophic areas it may take considerably longer. Higher latitudes impose greater variations in light intensities upon a variety of advective regimes; e.g., production in the Arctic Ocean results from a low advective regime and chronically low light. Upwelling regimes in the Bering Sea are strongly affected by seasonal variations in light intensities.

A third identifiable biome is the deep sea. It differs from the previous two in lacking light and therefore primary production, and by its isolation from the other biomes; yet it covers more than four-fifths of the ocean's volume. It is a dark and cold environment with little seasonal change of physiochemical conditions, and is characterized by a uniquely high hydrostatic pressure. The evolutionary adaptations of organisms living in this environment are less well understood than those of accessible pelagic regions, but the surprisingly large number of species found in the deep sea shows that the environment is not inimical to life.

6.3.1. The Estuarine Ecosystem

A subbiome of the coastal upwelling regions is the estuarine system. In an estuary, such as the eutrophic Chesapeake Bay, the time scales of water movements (tides, runoff from tributary rivers into the bay) and of changes in climatic conditions preclude any steady-state description of the plankton ecosystem. There is always an approach to a new steady state. In addition, the shallowness of the area and its convoluted shorelines provide stronger and more complex interactions among marsh, benthic (bottom-dwelling), and planktonic communities.

The ultimate goal of many estuarine studies is to identify relationships among the factors affecting the viability of the aquatic biota, particularly those species of economic, recreational, or aesthetic importance, in order to furnish a basis for the most efficient utilization of the system. Suggestions can be made relative to the management of a portion of an ecosystem, i.e., whether or not to direct a chlorinated waste water discharge directly into a spawning area, or to place a cooling water intake in a nursery area. Very often this general knowledge of life-cycle relationships and physiology can provide the proper advice, and will avoid catastrophic consequences. However, where the cause and effect relationship is not so evident, or when there are complex trophic level interactions, a predictive model does not yet exist. This is due in part to the complexity of the life cycles of the predators of major importance (shellfish, finfish, crabs), in part to the complexity of the trophic interactions among all of the species, and in part to the large experimental variances of the natural systems, daily, seasonal, and annual. Owing to the short generation times (hours to days) of phytoplankton and their low motility, one would expect that their growth and physiological state might be strongly coupled to the physical and chemical environmental parameters characteristic of their particular geographic locations. As the consequence of this strong coupling, it should be possible to describe the kinetics of growth and dissipation of phytoplankton standing crops in terms of a strongly damped, quasi-stable system driven by a forcing function corresponding to a defined natural perturbation of the ambient physical and chemical environment. This perturbation could be a delta function of nutrients brought in by heavy rainfall. It may be temperature changes in early spring or late fall, or it may be a specific delivery of toxic materials. The diffusional loss of excess nutrients and organisms as the result of water exchange, and the loss of phytoplankton as the result of predation by zooplankton are included as damping factors.

Assuming a knowledge of the more important physiological rate constants and their functional dependence, including feedback, it should be possible to define the limits of stability of the system, i.e., to predict the maximum steady-state levels of the physical and chemical environmental parameters (nutrients, temperature, light, etc.) such that the natural plankton system will recover from large natural perturbations. The difference between those maximum levels and the observed levels (at any degree of significance) would then be a measure of the stability of the plankton system. We assume that the quantitative relationships among nutrients and nutrient turnover, salinity, temperature, turbidity, species selection and succession, predation, and water exchange can be determined. From these quantitative relationships it follows that specific parameters will emerge that can serve as diagnostic indicators of the physiological state and of the previous history, and permit the prognosis of the stability of the phytoplankton community. These relationships should permit the prediction of the direction of changes in the community in response to proposed nutrient, sediment, or heat loading.

In the study of any natural system the experimenter may remove samples for study in the laboratory under controlled conditions. However, the natural system, with diverse community population, is, at any time, the integral of all of the aperiodic climatic, biotic, and chemical interactions that have occurred.

Thus, experimental reproducibility in the natural system is very difficult to achieve. We are immediately faced with the problem of how to make statistically significant measurements in this variable system. Consider any given natural system on which measurements are to be made. The total measured variance will be composed of the variance associated with the "treatment" or stress whose effect it is desired to assess, and the large natural variation of the system due to daily, seasonal, and annual fluctuations in wind, tide, sunlight, rainfall, etc. In principle, therefore, a "before" and "after" baseline study of a single system will be subject to both of these sources of uncertainty, and only "treatments" that produce sufficiently large mean differences (i.e., before minus after) can be assessed with any degree of statistical significance. A further complication exists because statistical parameters such as standard deviation, tests such as chi square, Student's t, F variance ratio, Chauvenet's Criterion, and levels of significance have implicit in them the assumption that the data are normally distributed. How then might levels of significance be assigned to differences in time averages of those quantities from one season to another, or from one year to the next? It should be possible to choose a second system that is similar to the first in its response to natural fluctuations, and differs from the first in the absence of the particular "treatment." Under these conditions it should be possible to analyze differences between the two systems, and to separate the large natural variations from the treatment effect. By virtue of the comparability of the systems, the expected value of the mean of the *difference,* properly normalized, should be zero. Nonhomogeneities within the individual systems and their varying responses to localized meteorological changes, in addition to measurement error, should give rise to a normally distributed spread of difference values that is amenable to statistical analysis. The trick is to work with comparable systems. There are fundamental limitations to the study of any ecosystem in which the total number of variables and the functional relationships among the variables are not completely known. A number of reasonably intuitive decisions have to be made relating the parameters to be measured, the sampling frequencies, the sampling areas, the species to be sampled, etc.—all within the physical and financial limitations of the program. There is a threshold of data collection both in time rate and spatial gradient below which the research becomes relegated to a data collection exercise.

In this area of environmental research one must be careful to define the specific questions to be answered by any series of experiments. One of these involves the marked change in the spectrum of underwater sunlight between the ocean and the eutrophic estuary. What determines the absorptivity of estuarine waters? How does the resultant change in underwater spectral irradiance *select* for phytoplankton species with accessory (antenna chlorophyll) pigments (see Section 13.2.1.) in the appropriate spectral regions?

6.3.2. Spectral Distributions of Underwater Sunlight

The photic zone in most eutrophic estuaries is quite limited in depth. The depth of visual disappearance of a Secchi disk, a white diffusing disk, corresponding to approximately two-tenths of the surface sunlight intensity, can be as low as 0.5 m. As a rough approximation, the depth at which sunlight intensity is reduced to 1% of its surface water value is three times the depth at which the outlines of the Secchi disk become indistinguishable. Since the method is a visual one, it measures the relative absorption of a range of wavelengths for human photopic vision. The peak sensitivity of photopic vision, 555 nm, lies between the absorption peaks for chlorophyll and most of the blue-absorbing accessory pigments. One must, therefore, make the tenuous assumption that absorption coefficients do not change with depth or with sediment load. Figure 6-3 shows the relative spectral distribution of sunlight at the water surface and at a depth of 0.8 m, for a subestuary of the Chesapeake Bay[25], corresponding on that day to the depth of disappearance of a Secchi disk. The curves show that the extinction coefficient of these waters cannot be represented by a single value, but is much higher in the blue region of the spectrum and is also affected by absorption of light by the phytoplankton.

The major significance of Fig. 6-3 is to demonstrate the marked absorption of wavelengths below 500 nm in the photic zone in the estuary, as compared with the absorption of wavelenths longer than 600 nm in coastal and oceanic waters[26] (see curve 4 of Fig. 6-1). There is a significant distortion of the original sunlight spectrum. The mean path length for absorption (true absorption and scattering) of blue light changes from 33 m in the clearest ocean waters[27] to 0.5 m in the Rhode River, a subestuary of the Chesapeake Bay. Due to sediment loading and

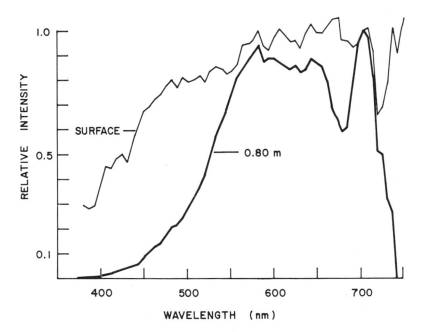

Fig. 6-3. Relative spectral photon intensities of sunlight incident on the surface of the water and at a depth of 0.8 m in the Rhode River on 27 July 1972. The intensity data are normalized at 710 nm. (From reference 25.)

nutrient runoff into the estuary, light may be a limiting factor for phytoplankton growth. Phytoplankton mixed in the water column may spend a significant fraction of their daylight hours at light intensities below their photosynthesis saturation values. A decrease in sediment load may, on occasion, result in an increase in production in the water column. However, the relationships between nutrient delivery, which accompanies sediment delivery, and the direct and indirect effects of sediments (including detritus) in providing nutrients to the phytoplankton and zooplankton have not been completely defined.

6.3.3. Dinoflagellate Accumulation Mechanisms

It might be expected that under conditions where sediment loading is a natural phenomenon, there would be a selection for motile, positively phototactic algae that could to some degree overcome the vertical advective mixing. They would thereby spend a greater fraction of their daylight hours near the surface, and be able to achieve greater yields of photosynthesis than the nonmotile true phytoplankton. This would give rise to a negative gradient or, on occasion, a peak near the surface in vertical profiles of phytoplankton concentrations. These types of vertical profiles are characteristic of dinoflagellate populations in the estuary. Such a peaked distribution in the vertical profile of a dinoflagellate population in the Chesapeake Bay is shown in Fig. 6-4.[28] In this case, the effective phytoplankton concentrations were expressed as concentrations of extractable chlorophyll a, the major photosynthetic pigment of the algae (solid dots in the figure), and as *in vivo* chlorophyll a fluorescence, to be discussed in the next section.

In Fig. 6-4, the peak in chlorophyll a, representative of the phytoplankton

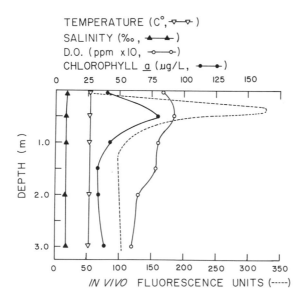

Fig. 6-4. The vertical profiles of temperature, salinity, dissolved 0_2 (D.O.), extractable chlorophyll a, and *in vivo* fluorescence of chlorophyll a in the vicinity of the Bay Bridge of the Chesapeake Bay (38°59′ N. Lat., 76°23′ W. Long.) during the growth of a dinoflagellate population in August 1971. (From reference 28.)

concentration, is very close to the surface, at approximately 0.25–0.50 m. However, on occasion, diatom species (*Cerataulina bergonii* and *Rhizoselenia delicutula*) have been found to be dominant, and to be concentrated at depths between 4 and 8 m in the Chesapeake Bay, where the underwater light intensities were only 1% of surface intensities.[29]

These two cases illustrate the extremes of a generally occurring phenomenon associated with phytoplankton populations in oceanic, coastal, and estuarine waters. The phenomenon is the interaction of the behavioral response to light of a phytoplankter and the physical transport and containment of plankton populations within water masses separated from other water masses by density discontinuities (due to salinity or temperature), and usually moving in different directions. In the case of the phytoplankton population of Fig. 6-4, the peak in chlorophyll concentration at less than 0.5 m is not due to the selection of this depth by an intensity-dependent phototactic search mechanism. (During the morning through bright noon the peak will remain at the same depth.) It is an artifact produced by a surface runoff of lower density river water that does not contain these dinoflagellates. When a vertical profile of water density was measured it was found that the density discontinuity (pycnocline) was also at 0.5 m and that the surface water current vector was different from that of the waters below 0.5 m. This discontinuity occurred within the 0.5-m surface layer, and illustrates the depth resolution that is occasionally required in the shallow estuary compared with coastal and oceanic waters.

A vertical profile of water densities when the diatoms were concentrated between 4 and 8 m showed a pycnocline at 3 m, and a second and even stronger pycnocline at 9 m.[29] Thus, phytoplankton populations may be found within water layers of densities different from upper or lower layers. Unless the water circulation patterns are known, data on phytoplankton distributions can result in incorrect photobiological implications. For example, a plausible though incorrect explanation of the diatom distribution data (in the absence of the water density profile) might be that, having exhausted the nutrients (nitrate and phosphate) in the upper layer, the diatoms sank to a region of higher nutrients. The low light intensity level is the price they had to pay for the higher nutrient concentrations. The lesson to be learned from this example is again that environmental photobiology is involved with the interactions of all parts of the ecosystem.

Under different conditions during calm, sunny days, positive phototaxis results in large accumulations of dinoflagellates into visibly colored patches.[30] This is one of the mechanisms involved in the formation of dinoflagellate blooms or red tides. There are a number of important ramifications of this phenomenon.

1. This relative confinement of populations by water layers will be very selective of those species that can tolerate the ranges of light intensities and nutrient concentrations characteristic of these layers, and can therefore result in the transport of particular phytoplankton species over relatively long distances in subsurface water masses. When and if these water masses are mixed into nutrient-rich and warmer coastal surface waters, it is possible for the transported phytoplankton to bloom and to form dense surface patches, again by virtue of positive phototaxis.

2. There are serious drawbacks to the concept of using the reflectance optical properties of seawater in an effort to monitor phytoplankton populations by airplane or by satellite. Vertical profiles of phytoplankton (chlorophyll) concentrations rarely show the homogeneity in chlorophyll concentration that is assumed in calculations and measurements of the spectral reflectance properties of seawater. The diatom concentrations between 4 and 8 m would not be detectable from spectral reflectance data. The dinoflagellate concentrations of Fig. 6-4 would cause a serious overestimation of the mean chlorophyll concentrations in the surface waters.

3. In the eutrophic estuary, the strong scattering and absorption of blue light might select for phytoplankton whose accessory pigment absorptions fall at longer wavelengths. Therefore, water layer transport mechanisms in estuarine systems should select for phytoplankton that (i) are positively phototactic, (ii) have accessory pigments of photosynthesis that absorb light in the green and yellow regions (peridinin, phycoerythrin), and (iii) can tolerate wide variations in salinity, temperature, and light intensity.

In summary, large, localized accumulations of dinoflagellates can result from the combination of positive phototaxis, the proper accessory pigments, and wide growth tolerances, with (1) confinement within water layers, and lateral movement of those water layers separated by density discontinuities; (2) vertical advection in shallow coastal waters resulting in mixing into surface waters; (3) prevailing on-shore winds producing windrows; (4) reformation of convergences of water masses of different densities; and (5) wind-driven convection cells.

The dinoflagellate *Prorocentrum mariae lebouriae* (*minimum*), which blooms and forms dense visible patches in the northern Chesapeake Bay (39°00′ N. Lat.) during late May and early June, is transported by subsurface waters from the mouth of the bay (37°03′ N. Lat.), a distance of 216 km over a period of approximately 3.5 months. The process is thought to be a combination of processes (1), (2), and (4) above.

In the Chesapeake estuary there is a gradient in nutrient concentrations from the creeks to the rivers to the central bay, and then south to the ocean. Therefore, in a two-layered system where the surface waters are moving generally south and the bottom waters are moving generally north, there would be a major advantage accruing to any organism that could migrate periodically into the upper surface waters, and then down into the bottom waters. In proper phase with the tide, such an organism by a periodic vertical migration could be subjected to a net lateral displacement, north or south or zero. Since phytoplankton are photosynthetic, it would be reasonable to suppose a solar phase photoperiodic rhythm, a positive phototaxis (upward vector) by day, and a negative vector (geo- or chemitaxis) 12 h later. Only organisms able to swim around 1 m h^{-1} (the larger dinoflagellates) in order to take advantage, within reasonable times, of the currents in the upper and lower water layers of the two-layer system, could be successful.

A diurnal migration of the dinoflagellate *Ceratium furca* in surface waters of the Chesapeake Bay can be inferred from Fig. 6-5. The data were taken in the

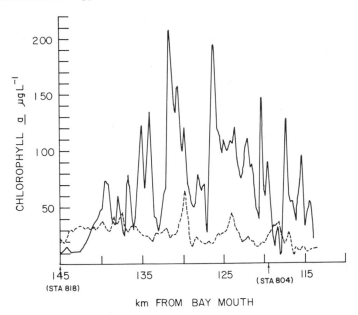

Fig. 6-5. North–south surface water transects of *in vivo* fluorescence of chlorophyll *a* at a depth of 1 m for *C. furca* on August 12, 1974. The dashed curve is the lateral transect at a depth of 1 m during the morning (0850–1050) in the region south of the Patuxent River (Sta. 818). The solid curve is the same lateral transect at a depth of 1 m during the afternoon of the same day (1430–1730). Since the transect extends over the entire accumulation area, the ratio of the integrated transects is a measure of the upward migration of the *C. furca* into the upper meter of surface water. (From reference 31.)

following way: surface waters at a 1 m depth were pumped continuously through a flow cell fluorometer as the research vessel moved at constant speed south along the transect area, a distance of approximately 30 km. In the early morning *C. furca* were uniformly distributed throughout the water column, and therefore the *in vivo* chlorophyll *a* fluorescence was relatively low (dashed line). During the sunny day the organisms migrated to the surface. Therefore, a return trip during the afternoon, again sampling at a 1-m depth, showed much higher concentrations at 1 m (solid line), the result of the positive phototaxis (see, e.g., the peak of chlorophyll *a* concentrations in Fig. 6-4). The oscillations in concentration with distance along the lateral transect are real and show the overall patchiness of the system due to local wind patterns and surface currents. The figure demonstrates the difficulties of making statistically significant measurements of phytoplankton in the estuarine system. Concentrations can vary vertically and in the lateral directions, depending on species, time of day, phase of tide, and climatic conditions. Meaningful measurements in estuarine photobiology can be made, but the approach must be multidisciplinary.

A different dinoflagellate species, *Gymnodinium nelsoni,* in the Rhode River exhibits a downward nighttime migration into interstitial waters, and an upward daytime migration into surface waters. The combination of diurnal upward and downward migration to take advantage at night of higher nutrient concentrations in interstitial waters or in bottom waters and reduce the exchange rate with bay waters of the larger dinoflagellates relative to the uniformly distributed nanno-plankton (less than 10 μm in linear dimension) in the water column can result in much higher *G. nelsoni* populations than would be predicted on the basis of the soluble nutrient concentrations in the surface waters alone.

6.3.4. Phytoplankton Standing Crops

In the tributary estuaries of the Chesapeake Bay and in the central bay proper, between 80 and 90% of the phytoplankton primary productivity and production is carried out by nannoplankton. Stabilities of the water column and turbidities are such that the well-mixed surface waters encompass the entire range of light intensities effective for photosynthesis, i.e., the photic zone. Superimposed on this background, which averages 10–20 μg/liter of extractable chlorophyll *a*, is a seasonal procession of the larger, strongly phototactic dinofla-gellates *Prorocentrum mariae lebouriae* (*minimum*), *Gymnodinium nelsoni*, *G. splendens*, and *Katodinium rotundatum* in the upper bay, and *Ceratium furca* in the lower bay. These dinoflagellates increase in concentration subsequent to an initial cycle of nannoplankton production and herbivore predation on the nanno-plankton. There appears to be a correlation between the conversion of dissolved inorganic nitrogen and phosphorus into organic forms, and the appearance of these species of dinoflagellates.

6.3.4.1. Measurement of Chlorophyll a

One of the more important measurements involves the assessment of the standing crops of phytoplankton. It is obvious that an enumeration of species and number concentrations will have little value of itself because of the large num-bers of species, the large size distributions among the species (2μm blue-green algae, to 6μm chrysophytes, to 70 μm dinoflagellates, to chains of diatoms several hundred micrometers in length), and the occasional large variations among species in photosynthetic efficiency.

Since we are dealing with phytoplankton and their primary photosynthetic production, and since chlorophyll *a* is a primary photosynthetic pigment, a possible measure of phytoplankton standing crops might involve the determina-tion of the volume concentrations of chlorophyll *a*, an assay that is widely used in phytoplankton ecology. The rationale is as follows.

The function of chlorophyll *a* is to sensitize the photoreduction of the pyridine nucleotide NADP to NADPH, and to make ATP. If one assumes a common rate-limiting reaction in the photoreduction system, it follows that at high light intensities (saturation of photosynthesis) the rate of carbon fixation will be directly proportional to the number of primary chlorophyll *a* molecules or reaction centers. Therefore, a measurement of the volume concentration of chlorophyll *a* contained in the phytoplankton[32] should be related to the photo-synthetic potential of the water volume. This maximum rate of carbon fixation can be assumed to be a constant fraction of the total mass of carbon (C), if

$$\frac{1}{C}\frac{dC}{dt} = \text{const} \tag{6-11}$$

At light saturation

$$\left(\frac{dC}{dt}\right)_{max} \sim [\text{chl } a] \tag{6-12}$$

It follows that

$$\frac{[\text{chl } a]}{C} = \text{const} \tag{6-13}$$

Therefore, a measurement of the volume concentration of particulate chlorophyll a should be a reasonable approximation of the phytoplankton carbon biomass. In laboratory cultures of phytoplankton and in natural samples in log phase growth, the mass of chlorophyll a per organism appears to be reasonably constant, and the carbon to chlorophyll a ratio has been measured to be between 50–100 g/g.

A rapid and reproducible method of assaying the chlorophyll concentrations of natural water samples is to pump the water sample through a flow tube in a fluorometer and observe the intensity of the fluorescence of *in vivo* chlorophyll a within the fluorometer active volume.[28] The continuously pumped sampling technique for *in vivo* fluorescence has distinct advantages in time and spatial resolution over the manual extraction technique.

The basic assumption of the *in vivo* chlorophyll a fluorescence technique is that there is a constant ratio between the observed fluorescence intensity of an *in vivo* sample and the extractable chlorophyll a [mg/m^3]:

$$R = \frac{\text{net } in \ vivo \text{ fluorescence intensity [arbitrary units]}}{\text{extractable chl } a \ [\text{mg/m}^3]} \tag{6-14}$$

The efficiency of fluorescence of chlorophyll *in vivo* has long been considered to be an indication of its state in the photosynthetic system. In *Chorella*, inhibition of photosynthesis by cyanide increases the observed fluorescence. An increase in the *in vivo* fluorescence quantum yield has been considered to be a measure of the inefficiency of photosythesis, since fluorescence represents a loss of free energy that might have been channeled into chemical energy of the photosynthetic process.

In spite of its attractiveness as a convenient and rapid assay technique for phytoplankton standing crops, the *in vivo* chlorophyll a fluorescence technique must be approached with care.[29] When the species composition remains relatively constant the determination of the R value [Eq. (6-14)] by extraction of chlorophyll a will be applicable for reasonable time periods and over sufficiently large survey areas to make the method worthwhile.

6.3.4.2. Measurement of Adenosine Triphosphate

Because of the extreme sensitivity of the photon detection technique, and owing to the ubiquitous presence of ATP in all life forms, a method proposed in the NASA program for detection of extraterrestrial life involved the detection of the bioluminescence of a firefly *in vitro* reaction mixture[33] containing excess luciferin, luciferase, and Mg^{2+} ions in a buffered oxygenated solution to which an extraterrestrial grab sample could be added. Light emission would be direct evidence for ATP and presumably of the previous synthesis of ATP by living organisms. With the most sensitive instrumentation, the assay technique can detect as little as 0.01 pg or 2×10^{-5} pmol of ATP. With partially purified luciferase, the presence of "residual light" limits the technique to 1 pg ATP. The kinetics of light emission show an initial rise in light intensity, the peak height of which depends upon the rapidity of injection and subsequent mixing, followed by a complex decay, depending on luciferase purity and concentration, and on ATP concentration. Some investigators prefer to use the flash height, while others prefer to integrate the light intensity over a fixed time.

With present techniques it is possible to detect as few as 10^4 cells of *Escherichia coli*, which contain 10^{-16} g ATP per bacterium.

The ATP assay is now used in marine biology for determining the biomass of heterotrophically growing cells at ocean depths below the euphotic zone, and of phytoplankton in the euphotic zone.[34] In the euphotic zone, depending on whether samples are taken from ocean water, more turbid inshore water, or highly turbid estuarine water, the biomass can vary from 20 to close to 100% of the total organic matter. ATP is present in living cells, diffuses rapidly out of dead cells, and in general is not absorbed appreciably on sediment or detrital particles. Absorption is a problem, however, if large volumes of waters containing high sediment loads are filtered. The ratio of the mass of cellular organic carbon to ATP has been found to be close to 286 g/g for marine and freshwater algae; 250 for bacteria, molds, tumor cells, and a number of zooplanktons. Carbon to chlorophyll *a* ratios of 50–100 have been found for natural phytoplankton samples. The ATP assay furnishes a corroborative check of chlorophyll *a* assays of phytoplankton for the estimation of biomass in surface waters (where phytoplankton are the major source of total organic carbon). It has the further advantage of generating total biomass numbers, and therefore,

$$\text{carbon (from ATP)} - \text{carbon (from chlorophyll } a) = \text{heterotrophic carbon}$$

$$(6\text{-}15)$$

where heterotrophic carbon is the biomass of carbon engaged in nonphotosynthetic, or secondary, production. The mass of phytoplankton carbon divided into primary production (carbon uptake) rates obtained by ^{14}C studies gives the gross generation rate constant for photosynthesis in the water volume. These rate constants, together with carbon:nitrogen:phosphorus ratios in the phytoplankton, and measurements of N and P soluble nutrient concentrations in the sampled waters permit estimates to be made of nutrient turnover rates that are essential in any systematic approach to an ecosystem.

One of the advantages claimed for the C:ATP ratio method is that the carbon derived from the relation (C) [grams] = 250 (ATP) [grams] is representative of living or protoplasmic carbon, as differentiated from detrital carbon (organic particulate carbon from previously living plants or animals). Thus, the ratio of

$$\frac{250(ATP) \, [\text{g/liter}]}{\text{Total particulate organic carbon [g/liter] determined chemically}} \tag{6-16}$$

is the fraction of carbon contained in living organisms in the water sample. This ratio can range from 1 to 10% in oligotrophic or deep water (>1000 m) to close to 100% in eutrophic coastal waters.

The values of 250–300 for C:ATP ratios [g/g] appear to be valid during "normal" exponential growth for a large variety of phytoplankton, zooplankton, and benthic invertebrates in natural waters. However, even for chemostat-grown bacteria, a range of 91–333 has been observed. For phytoplankton cultures subjected to severe nutrient deficiencies (iron, nitrogen, phosphorus) the C:ATP ratio can be as high as 2000. Using the same analogy as in the case of the *in vivo* chlorophyll *a* fluorescence efficiency, the C:ATP ratio and the kinetics of this ratio in light–dark exposures or in test illuminations may be a diagnostic indicator of the physiological state of the organisms.

6.3.5. Light and the Filling of the Photic Zone

The strong correlation between the thickness of the upper mixed layer and the depth of the photic zone has interesting ecological ramifications. The particulate plant pigments appear to account for the major absorption of light in the eutrophic estuary. The presence of sediment scatterers (turbidity) increases the effective geometrical path length for downwelling light. In late August, following a period of minimum rainfall and high insolation, the turbidity in the Chesapeake Bay is at a minimum value. There will be sufficient light intensity throughout the upper mixed layer to support the growth of the phytoplankton. They will continue to increase in concentration until they absorb all the incident light effective for photosynthesis within the upper mixed layer. This is the ideal case. The rate of growth of the phytoplankton less the rate of loss of phytoplankton due to water exchange and predation by zooplankton maintains a steady-state concentration so that all of the available light is used for photosynthesis, and the thickness of the photic zone is equal to the thickness of the upper mixed layer.

When the nutrient concentrations in the upper mixed layer are low (nutrient-limiting case), the growth rate constants for the phytoplankton species will be low, and there will be insufficient phytoplankton concentrations in the upper mixed layer to absorb all of the photosynthetically efficient light. Therefore, the depth at which the light intensity is reduced to 1% of the surface light intensity will be below the upper mixed layer, i.e., the depth of 1% light intensity is greater than the depth of the pycnocline, the thickness of the upper mixed layer.

When the nutrient levels in the upper mixed layer are high due to delivery by

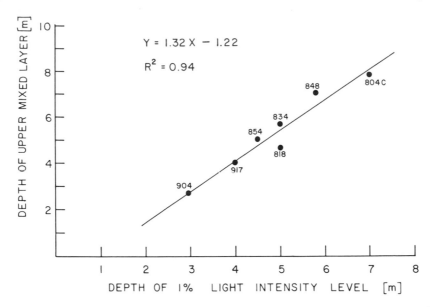

Fig. 6-6. Relationship between the depth of the upper mixed layer in the Chesapeake Bay and the depth of the photic zone (arbitrarily set at 1% of ambient light intensity) during August 1974. The station numbers associated with the data points are shorthand for the latitude, i.e., 848 = 28°48' N. Lat., 808 = 38°08'. C stands for a specific longitude at this latitude. Y and X are the coordinates used in the linear regression, and R^2 is the regression coefficient. (From reference 31.)

runoff, which also carries sediment, the additional scattering by the sediment combined with the added absorption by the increased concentration of photosynthetic pigments compresses the thickness of the photic zone so that it is now less than the depth of the pycnocline. This is the light-limited case.

Both of the above cases are transient phenomena, and both will approach the ideal efficiency (1% light intensity depth equals pycnocline depth) as conditions change seasonally in the estuary. In the light-limited case, as the sediment settles and as predation increases to keep pace with phytoplankton concentrations, the most efficient utilization of light is approached (Fig. 6-6), i.e., the entire mixed upper layer, through which the phytoplankton are transported by the energy delivered by sun and wind, absorbs the total available light energy for photosynthesis.[31] Under these conditions the smaller relatively nonmotile nannoplankton have no need for positive phototaxis.

From this general description it can be seen that simple physical measurements of vertical profiles of light intensities and water densities may be used to monitor extremely complex nutrient–phytoplankton–zooplankton cycles, once the interaction mechanisms are known.

6.3.6. Conclusions

In this complex case of aquatic estuarine photobiology, an attempt has been made to investigate the interaction of light with the primary producers, the phytoplankton. It is necessary to trace the incorporation of inorganic nutrients including carbon through the phytoplankton and to the next higher trophic level, the herbivores.

Even for this small portion of the entire estuarine ecosystem, the number of variables not under the direct control of the experimenter is enormous. The further complication of rapid changes and large seasonal and annual fluctuations in climatic conditions makes the problem a very difficult one. It is relatively easier to study the more placid, less variable ocean systems. However, the estuary is the spawning and nursery area for many ocean species of fish. It is the source of crabs, oysters, mussels, clams, and a wide variety of commercial finfish. Its shorelines are where people live and industry thrives. Its preservation and continued health are thus most important and most threatened.

The examples presented in this chapter touch only briefly upon this complex area of environmental photobiology. Photobiology is already an interdisciplinary field comprising the physics of electronically excited states of molecules, and the chemistry of the reactions of these excited states in complex biological systems, e.g., photosynthesis, vision, and damage and repair of DNA. The expansion into environmental photobiology adds yet another major dimension to be considered; the interaction of the entire photobiological organism with its environment. In the aquatic biology of phytoplankton it is necessary, but not sufficient, to understand the mechanisms of photosynthesis. The student must understand phototaxis, light regulation of metabolic pathways, the accessory pigment distributions, the nutritional physiology and taxonomy of the various classes of phytoplankton, the scattering and absorptive properties of waters in oceanic, coastal, and estuarine systems, the hydrodynamics of water circulation patterns in these systems, the sources of nutrients for the phytoplankton, and the factors influencing predation upon the phytoplankton.

Environmental photobiology is a demanding subject, one that requires specialization in techniques on the one hand, and the capability for broad generalization of concepts on the other. New approaches to systems analysis must be developed. The approach is similar to that taken for human health. The system is assumed to be in some steady state. A reading of blood pressure, a urine or a blood analysis, or a recovery of pulse rate following exercise are diagnostic measurements relating to the overall functioning of the human system. It is not usually necessary to isolate and assay for all of the enzyme cycles, hormones, and antibodies. The physician's diagnostic tests come from an understanding of the body biochemistry and physiology. In an analogous manner in environmental photobiology, it should be possible to understand organism requirements, interactions, biochemistry and physiology, and the external physical and chemical factors due to the environment. From these it should be possible to describe a set of diagnostic measurements for the organism in its environment. For the student who might enjoy making full use of all areas of science: photophysics, photochemistry, biochemical mechanisms, mathematical modeling—as well as working outside of the controlled laboratory with whole organisms in their natural environment—there is the thrill of constant discovery. This is virgin territory, ripe for the multidisciplinary scientist. The problems, in addition to their innate scientific importance, are also very relevant to the proper management of the resources of our increasingly crowded planet.

6.4. EXPERIMENTS

It would be instructive to be able to measure one aspect of the growth rate of algae, the gross rate of uptake of inorganic carbon, by a radioactive tracer technique. The elements of the technique are described in pages 267–274 of reference 32. However, the following suggestions and simplifications should provide the student with a satisfactory grasp of the measurement.

Ten microcuries (μCi) of sodium bicarbonate, ^{14}C-labeled, is available from commercial suppliers under general license and can therefore be ordered for laboratory use without having to set up a specialized radioactivity facility. Most universities already have licenses for radioactive materials for research and the radioactive bicarbonate solution can be ordered through the Radiation Safety Officer. The student should have access to a liquid scintillation counter. However, if this is not available the samples that are collected on the Millipore filters can be placed in liquid scintillation vials containing 3 ml of dioxane and kept indefinitely. The vials can then be taken at any time to a laboratory where liquid scintillation counting facilities are available; any dioxane-based liquid scintillation solution can be added to the vials, and the samples counted. Since the half-life of ^{14}C is 5780 years, the decay of activity before analysis will be negligible.

Any marine or estuarine natural water sample can be used as a source of phytoplankton. A laboratory culture of phytoplankton, if available, will also be perfectly satisfactory. As a rough rule of thumb, take a sample of the phytoplankton solution in a glass test tube into a dark room, wait 15 min to become visually dark-adapted, and irradiate the test tube with a "black light" (UV: 365 nm + 405 nm) lamp. If the red fluorescence of chlorophyll is visible, the sample will contain sufficient algae to give good results in the uptake experiments. Do not use this sample for the experiments as excessive UV radiation can kill some of the algae.

The radioactive bicarbonate should be dissolved in 0.22 or 0.45 μm (pore size) Millipore-filtered natural water or culture medium to a volume such that the activity is approximately 0.5 μCi ml^{-1}. We will be adding exactly 1 ml of this tracer bicarbonate solution to each glass bottle containing 100 ml of the algal sample. In order to keep the irradiation conditions constant, cool white fluorescent lamps should be used for the incubation. All sample bottles including the dark bottles should be filled with 100 ml of the phytoplankton solution, and kept at a fixed distance ($\sim$15 cm) from a bank of two fluorescent lamps for at least 1 h prior to addition of the radioactive tracer bicarbonate solution, in order to acclimate the phytoplankton to the maximum light intensity to which they will be exposed during the experiment. Different light intensities are achieved by wrapping one, two, and three layers of ordinary plastic fly-screening material (as a neutral density filter; Section 1.3.2.) around the "light" bottles. Samples irradiated at 0, 1, 2, and 3 layers of screening and the "dark" bottles should be run in

triplicate. The normalization is obtained as follows: 0.1 ml of the tracer bicarbonate solution is added directly to a liquid scintillation vial to which is also added a
0.22 μm (pore size) Millipore filter through which 100 ml of the cold phytoplankton solution is filtered. This is done so that any "color" added to the liquid
scintillation solution by algal pigments that may "quench" the scintillations
produced by the ^{14}C beta rays can be corrected for. The incubation time should
be between 1 and 2 h (the time of incubation is measured from the time of
addition of the radioactive bicarbonate solution and should be the same for all
samples). The samples should then be placed in dim light and filtered on 0.22-μm
Millipore filters as soon as possible. Those filters containing the phytoplankton
must be washed twice with 50 ml each of 0.22 or 0.45 μm-filtered sample water to
insure that the only activity remaining on the filter is that contained *within* the
algae. When the filters in the scintillation vials are counted in the liquid scintillation counter, the fractional uptake of radioactivity due to photosynthesis in the
light bottle will be

$$f_L = \frac{\text{Net counts/min of incubated sample}}{\text{Net counts/min of normalized sample}} \times 10 \qquad (6\text{-}17)$$

The factor of 10 is due to the addition of 0.1 ml to the normalization scintillation
vial. This value must be corrected by the small amount of "dark" uptake so that
the true fractional uptake is $f_L - f_D$. The true fractional uptake should then be
plotted as a function of the number of absorbing screens, in order to verify that
there is an approach to saturation of photosynthesis with light intensity.

For the more enterprising student the techniques described in reference 32
can be used to measure chlorophyll concentrations and carbonate alkalinity of
the phytoplankton samples. If these are obtained it is possible to specify two
additional parameters. The assimilation number is defined as

$$Z = \frac{\mu\text{g carbon taken up/h}^{-1} \text{ per liter}}{\mu\text{g chlorophyll } a \text{ per liter}} \qquad (6\text{-}18)$$

Using a ratio of 50 (g carbon) /(g chlorophyll a) it is possible to estimate the gross
growth rate constant of the phytoplankton

$$K = \frac{Z}{50} \text{ h}^{-1} \qquad (6\text{-}19)$$

ACKNOWLEDGMENT

This work was supported under U.S. ERDA Contracts E9-76-S-02-3277 and 3278. Contribution
No. 910 of McCollum–Pratt Institute.

6.5. REFERENCES

1. Smithsonian Physical Tables (Prepared by W. E. Forsythe), Smithsonian Institution Press, Washington, D.C. (1959).
2. J. E. Tyler and R. C. Smith, *Measurements of Spectral Irradiance Underwater,* Gordon and Breach, New York (1970).
3. F. T. Haxo, The wavelength dependence of photosynthesis and the role of accessory pigments, in: *Comparative Biochemistry of Photoreactive Systems* (M. B. Allen, ed.), pp. 339–360, Academic Press, New York (1960).
4. A. A. Krasnovsky, The principles of light energy conversion in photosynthesis: Photochemistry of chlorophyll and the state of pigments in organisms, *Prog. Photosyn. Res.* **2,** 709–727 (1969).
5. R. K. Clayton, *Molecular Physics in Photosynthesis,* Blaisdell, New York (1965).
6. H. F. Blum, *Carcinogenesis by Ultraviolet Light,* Princeton University Press, Princeton, N.J. (1959).
7. B. E. Johnson, F. Daniels, Jr., and I. A. Magnus, Response of human skin to ultraviolet light, *Photophysiology* **4,** 139–202 (1968).
8. H. H. Seliger, The origin of bioluminescence, *Photochem. Photobiol.* **21,** 355–361 (1975).
9. R. Buvet and C. Ponnamperuma (eds.), *Chemical Evolution and the Origin of Life,* American Elsevier, New York (1971).
10. A. C. Lasaga, H. D. Holland, and M. J. Dwyer, Primordial oil slick, *Science* **174,** 53–44 (1971).
11. R. J. Cicerone, R. S. Stolarski, and S. Walters, Stratospheric ozone destruction by man-made chlorofluoromethanes, *Science* **185,** 1165–1167 (1974).
12. M. A. Ruderman, Possible consequences of nearby supernova explosions for atmospheric ozone and terrestrial life, *Science* **184,** 1079–1081 (1974).
13. H. C. Urey, Primitive planetary atmospheres and the origin of life, in: *The Origin of Life on Earth* (F. Clark and R. L. M. Synge, eds.), Pergamon, New York pp. 16–22 (1959).
14. M. B. McElroy, S. C. Wolsky, J. E. Penner, and J. C. McConnell, Atmospheric ozone: Possible impact of stratospheric aviation, *J. Atmos. Sci.* **31,** 287–303 (1974).
15. National Academy of Sciences, *Long-term Worldwide Effects of Multiple Nuclear-weapons Detonations,* Washington, D.C. (1975).
16. V. Arena, *Ionizing Radiation and Life,* Mosby, St. Louis, Mo. (1971).
17. National Academy of Sciences, *Biological Impacts of Increased Intensities of Solar Ultraviolet Radiation,* Washington, D.C. (1973).
18. T. H. Goldsmith and G. D. Bernard, The visual system of insects, *Physiol. Insecta,* **2** 165–272 (1974).
19. K. von Frisch, *The Dance Language and Orientation of Bees,* Balknap Press, Cambridge, Mass. (1967).
20. Y. Obara, Studies on the mating behavior of the white cabbage butterfly, *Pieris rapae crucivera Boisduvae. Z. Vergl. Physiol.* **69,** 99–116 (1970).
21. R. B. Forward, Jr., Light and diurnal vertical migration: Photobehavior and photophysiology of plankton, in: *Photochemical and Photobiological Reviews,* Vol. 1 (K. C. Smith, ed.), pp. 157–210, Plenum Press, New York (1976).
22. A. C. Giese, *Living with our Sun's Ultraviolet Rays,* Plenum Press, New York (1976).
23. National Academy of Sciences, *Biological Oceanography: Some Critical Issues, Problems and Recommendations,* Washington, D.C. (1975).
24. H. H. Seliger and W. D. McElroy, *Light, Physical and Biological Action,* Academic Press, New York (1965).
25. H. H. Seliger and M. E. Loftus, Growth and dissipation of phytoplankton in Chesapeake Bay. II. A statistical analysis of phytoplankton standing crops in the Rhode and West Rivers and an adjacent section of the Chesapeake Bay, *Ches. Sci.* **15,** 185–204 (1974).
26. R. C. Smith and J. E. Tyler, Transmission of solar radiation into natural waters, in: *Photochemical and Photobiological Reviews,* Vol. 1 (K. C. Smith, ed.), pp. 117–156, Plenum Press, New York (1976).

27. N. G. Jerlov, *Optical Oceanography,* Elsevier, New York (1968).
28. M. E. Loftus, D. V. SubbaRao, and H. H. Seliger, Growth and dissipation of phytoplankton in Chesapeake Bay. I. Response to a large pulse of rainfall, *Ches. Sci.* **13,** 282–299 (1972).
29. M. E. Loftus and H. H. Seliger, Some limitations of the *in vivo* fluorescence technique, *Ches. Sci.* **16,** 79–92 (1975).
30. H. H. Seliger, H. H. Carpenter, M. E. Loftus, and W. D. McElroy, Mechanisms for the accumulation of high concentrations of dinoflagellates in a bioluminescent bay, *Limnol. Oceanogr.* **15,** 234–245 (1970).
31. H. H. Seliger, M. E. Loftus, and D. V. SubbaRao, Dinoflagellate accumulations in Chesapeake Bay, *in: Proceedings of the First International Conference on Toxic Dinoflagellate Blooms* (V. R. LoCicero, ed.), pp. 181–205, Mass. Sci. and Tech. Found., Wakefield, Mass. (1975).
32. J. D. H. Strickland and T. R. Parsons, *A Practical Handbook of Seawater Analysis,* Fish. Res. Bd. Canada Bull. No. 167 (1968).
33. W. D. McElroy, H. H. Seliger, and E. H. White, Mechanism of bioluminescence, chemiluminescence and enzyme function in the oxidation of firefly luciferin, *Photochem. Photobiol.* **10,** 153–170 (1969).
34. O. Holm-Hanson, Determination of total microbial biomass by measurement of adenosine triphosphate, in: *Estuarine Microbial Ecology* (L. H. Stevenson and R. R. Colwell, eds.), pp. 73–89, University of South Carolina Press, Columbia, S. C. (1973).

7

Photomedicine

7.1. INTRODUCTION

The medical aspects of the science of photobiology are among the most important parts of this discipline, since they are concerned primarily with direct effects

John H. Epstein • Department of Dermatology, University of California, School of Medicine, San Francisco, California

175

of nonionizing radiation on humans. Two organ systems are directly affected by sunlight: the eye and the skin. The eye is discussed in Chapter 10; this chapter will examine primarily cutaneous reactions to the sun.[1-4]

The solar radiation that reaches the earth's surface ranges from around 290 nm (or perhaps 288 nm) in the ultraviolet (UV) region, through the visible and infrared regions, and well beyond. Almost all the photobiological reactions that occur in the skin are induced by radiation between 290 (or 288) and 320 nm (UV-B). These are the wavelengths that inhibit mitosis, DNA, RNA, and protein synthesis; make vitamin D; induce skin cancer; stimulate pigment formation; and produce the erythema response we call "sunburn." These wavelengths comprise what is generally termed the "sunburn spectrum." The longer wavelength UV radiation between 320 and 400 nm (UV-A) causes a few minor photobiological effects, such as immediate pigment darkening (IPD), and the immediate erythema response, but, more critically, UV-A radiation markedly accentuates the acute injury induced by UV-B radiation and enhances its carcinogenic potential. Furthermore, UV-A radiation is responsible for the vast majority of the reactions produced by exogenous photosensitizers. The rest of the sun's spectrum has little specific effect on the skin, although infrared radiation can produce a thermal burn as well as enhance photobiological responses initiated by UV radiation.

7.2. OPTICAL PROPERTIES OF THE SKIN[1,2]

For photobiological reactions to occur, three components are necessary: the biological system, the radiation, and a radiation absorber in the biological system. The biological system with which we are concerned is the skin (Figs. 7-1, 7-2); the radiation emanates from the sun. The radiation-absorbing molecules in skin generally remain to be identified; one important molecule, however, would appear to be DNA.

If a photosensitizing molecule is present in or on the dead cells of the stratum corneum, it will most likely act as a protective filter. If the molecule is in the Malpighian or germinative layers of the epidermis, it may initiate a photoreaction after absorbing radiation of the proper wavelength. It is clear that the radiation must penetrate to the site of the absorbing molecule for the reaction to occur. The penetration of radiation is partly governed by the optical properties of the skin.

Skin reflects a large amount of incident visible and near-infrared radiation. White skin reflects these radiations, as well as the UV-A radiation, much more effectively than does black skin. Very little reflection of UV radiation shorter than 320 nm occurs with either type of pigmentation.

Scattering and absorption play a greater role in the attenuation of sunburn radiation penetration than does reflection. The flattened cells of the stratum corneum, melanosomes, nucleic acids, proteins, lipids, histidine, urocanic acid, peptides, cholesterol, and the like limit penetration of this UV radiation by both scattering and absorption.

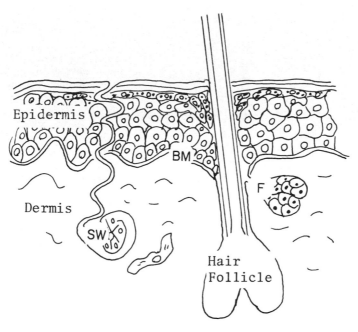

Fig. 7-1. The structure of skin. The skin consists of three main layers. The epidermis is the outermost layer which is separated from the dermis by a basement membrane (BM). The dermis consists primarily of connective tissue, mostly collagen, and some elastic tissue produced by fibrocytes. The dermis also contains blood and lymph vessels, hair follicles, sweat glands (SW), sebaceous glands (F), muscles, and nerves. Below this is a fat layer not shown on the diagram.

Because of the importance of the shorter wavelength UV radiation in the production of photobiological reactions, extensive transmittance studies have been conducted at these wavelengths.[1,2] These investigations have produced conflicting results, but it is generally agreed that a substantial amount of incident radiation in the sunburn or UV-B region (290–320 nm) is absorbed in the stratum corneum. In Caucasian skin at least 20% of this radiation reaches the Malpighian cells, and probably 10% penetrates to the upper dermis. The stratum corneum of black skin absorbs a greater amount of this energy due to the presence of melanin. A large proportion of the longer wavelength UV radiation (UV-A) and of visible light radiation penetrate into, and can be absorbed by, photosensitizers in the dermis.

With the absorption of this energy, an excited state is induced in the chromophore. The excited state may be deactivated by a number of methods, including thermal decay or emission of light energy at a longer wavelength than that absorbed (see Chapter 2). Cutaneous photosensitivity reactions are initiated either by direct alteration in the absorbing molecule or by the transference of this energy to other molecules or cellular components, such as membranes, mitochondria, nucleic acids, and the like.

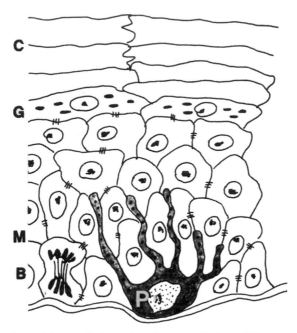

Fig. 7-2. The structure of the epidermis. The epidermis consists of four layers. The basal layer (B) is composed of a single or double layer of germinative cells where cell division occurs for the replacement of the epidermis. The Malpighian layer (M) consists of multiple squamous or prickle cells (keratinocytes) with intercellular bridges. These cells emanate from the basal layer, and eventuate in the outer dead stratum corneum layer. The Malpighian cells become the granular layer (G) where specific proteins are formed, presumably necessary for the formation of keratin. The cells then lose their nuclei, contain large amounts of keratin, and form the protective stratum corneum (C). They are then shed into the environment. The melanocytes (P) are dendritic cells located in the basal layer. Their dendritic (armlike) processes extend up into the Malpighian layer. These cells form the pigment melanin, which they pass into the Malpighian cells through their dendrites. The melanocytes and associated Malpighian cells comprise the epidermal melanin unit.

7.3. ACUTE EFFECTS OF SUNLIGHT ON THE SKIN

7.3.1. Sunburn[1,2]

Sunburn is by far the most common adverse effect produced by sunlight. The cutaneous changes, both morphologic and microscopic, are directly related to the amount of erythemogenic radiation, the degree of melanin pigmentation, and the thickness of the stratum corneum. As may be supposed, even deeply pigmented skin will be "sunburned" if it is exposed to enough radiation of the effective wavelengths.

Erythema, the most visually prominent aspect of the sunburn response, is manifest in a diphasic pattern. There is an immediate faint erythema that occurs during exposure and disappears shortly thereafter. The delayed response appears 2–4 h later, reaches a peak in 14–20 h, and persists for 24–48 h. The final stage is desquamation of the dead epidermal cells.

The action spectrum for sunlight-induced erythema is confined primarily to radiation between 288 (or 290) and 320 nm (Fig. 7-3). The standard erythemal curve formulated by Coblentz and Stair in 1934 showed a maximum efficiency at 297 nm and minimum activity at 280 and 320 nm. Subsequent studies have provided various results with the peak erythemogenic potential ranging between 290 and 294 nm. In addition, Urbach and co-workers, using parameters of

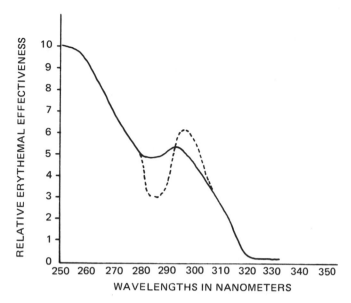

Fig. 7-3. The erythemal action spectrum. The erythemal action spectrum has not been completely agreed upon. This figure represents the different concepts noted in references 1 and 2. It is not clear whether there is a dip (dotted line) or a plateau (solid line) in the effectiveness of wavelengths around 280 nm.

efficiency and available energy, found the most effective wavelength to be about 308 nm (see Section 3.7.). A mild erythema can be produced by radiation between 320 and 400 nm, but this requires 100 times greater exposure. However, UV-A radiation strikingly enhances the erythema induced by UV-B radiation.[4−p.131*,5]

Radiation at wavelengths shorter than the UV-B region, produced by artificial light sources, are also erythemogenic. In truth, UV-C radiation at 254 nm is more efficient in producing erythema than is the UV-B region. Although the various studies have not produced consistent curves, there is agreement that radiation at 254 nm is most efficient with a rapid reduction in erythemal efficacy, especially at 270–280 nm. On this point there is some discrepancy, with some investigators noting a depression and others a plateau at 270–280 nm, relative to 290 nm. The curve continues toward, but not to, zero at 320 nm. The erythema produced by radiation at wavelengths shorter than 290 nm differs from the UV-B response in appearance and latent period, i.e., it is pink rather than deep red, and it reaches a peak at 8 h rather than 14–20 h. Therefore, it may well represent a different phenomenon.

The term "sunburn" stands for a number of complicated cutaneous responses to injury induced primarily by wavelengths between 288 (or 290) and

*Reference and page number therein.

320 nm. Obviously many changes occur simultaneously. In addition, it is difficult to differentiate primary and secondary effects. The nature of the chromophore that absorbs the light energy in order to initiate the primary photochemical responses is not known. Proteins containing aromatic amino acids have been considered likely candidates because of their absorption spectra and the profound effects of UV radiation to these molecules. However, a number of other substances, including nucleoproteins (DNA and RNA), urocanic acid, melanin, and even unsaturated fatty acids of phospholipids may play a role in the initial absorption of light energy. Lipid peroxidation has been demonstrated both *in vitro* and *in vivo*. Since lysosomal membranes are preferentially damaged by lipid peroxides, their formation may play an important part in UV radiation-induced injury. It seems quite likely that multiple chromophores are involved in this complex process.

The pathogenesis of the sunburn erythema perhaps has received the most attention.[1,2,4−p.117,6,7] However, it has not been established conclusively whether this is a primary or secondary phenomenon. Although the latter has been considered more probable, direct UV damage to the upper dermal blood vessels has been demonstrated. Van der Leun's studies suggest that there is a broad action spectrum from 250 nm into the long wavelength UV region that causes a direct effect on blood vessels.[1] In addition, there is a superimposed sharp peak around 300 nm that induces the formation of a diffusible vasodilating substance.

This brings up the question as to the character of the chemical mediator of the sunburn response.[4−p.143] The accumulated evidence indicates that one of the most prevalent vasoactive substances in mammalian tissue, histamine, is not responsible for this process. In addition, although kinins and possibly serotonin are released following exposure to sunburn energy, they are not likely to be related to the characteristic delayed erythema response. More recent evidence suggests that the vasoactive mediator is a prostaglandin.[8] However, further work is needed to establish the validity of this concept, since the information to date is primarily circumstantial.

7.3.2. Microscopic Anatomy

7.3.2.1. Light Microscopy

Histologically, very little is seen in human skin for several hours after exposure to UV radiation. In fair-skinned people, the first recognizable change, noted by 24 h, consists of foci of dyskeratotic cells (sunburn cells) in the epidermal Malpighian layer. Within 48 h damage is observed throughout the epidermis, with abnormal staining (dyskeratotic), vacuolated cells, and shrunken nuclei. The dendrites of the melanocytes also become quite prominent at this time. Regeneration is usually in progress, and the damaged cells form a desquamating (sloughing) layer by 72 h. Hyperplasia (thickening) of the epidermis is quite striking by this time (Fig. 7-4). At 96 h there is a great increase in dihydroxyphenylalanine (Dopa)-positive melanocytes with marked arborization

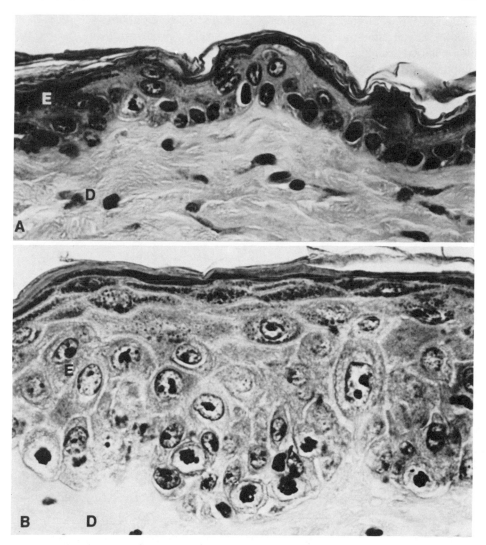

Fig. 7-4. UV-irradiated mouse skin 72 h postirradiation. (A) This shows the normal nonirradiated hairless mouse epidermis that is usually three cell layers thick. There is one mitosis in the basal layer. E, epidermis; D, dermis. (Hematoxylin and eosin; × 600.) (B) Demonstrates epidermal hypertrophy 72 h post-UV irradiation with multiple mitoses in the basal layer. E, epidermis; D, dermis. (Hematoxylin and eosin; × 600.) (From reference 6.)

of their dendrites. These melanocytic changes occur in pigmented skin as well, but the dyskeratotic changes are not nearly as notable, and cellular disorganization is much less prominent.

Histochemical and biochemical techniques have demonstrated a number of other acute epidermal changes following irradiation.[1,2,4–pp.117 and 147] These include lysosomal membrane alterations and rupture, lipid peroxide formation, inhibition of the activity of a number of enzymes, increases in both sulfhydryl and disulfide groups, and the transformation of urocanic acid from a *trans* to a *cis* isomer. This latter change may well play an important protective role in dissipating sunburn energy. Also, glycogen accumulates in the epidermal basal cells 12 h postirradiation, presumably to supply energy for the subsequent burst of mitotic activity.

Dermal changes generally are not prominent. Vasodilatation occurs, and in rodents an inflammatory infiltrate consisting of polymorphonuclear leukocytes can be seen as early as 30 min postirradiation, which becomes maximal between 8 and 24 h.[2] This cellular response is minimal, if present at all, in human skin.

The action spectrum for the microscopic changes is found primarily in the UV-B region. However, as with the erythema response, the long wavelength UV-A radiation markedly accentuates the microscopic damage initiated by the shorter wavelengths (UV-B).[5]

7.3.2.2. Electron Microscopy

Electron-microscopic studies have shown that structural changes occur well before they are demonstrable by light microscopy. The earliest changes noted to date occur in melanocytes associated with the immediate pigment darkening phenomenon, with changes in microtubules, melanosome patterns, and thick filaments.[4−p.195] Within 2–3 h postirradiation, marked changes are also noted in the cytoplasm, nucleus, and nucleolus of the epidermal cells, including vacuole formation, filament aggregation, and apparent loss of lysosomes.[9,10] Thus a number of structural changes that occur shortly after UV radiation exposure seem to result from injury to several sites simultaneously.

7.3.3. Macromolecular Synthesis and Mitosis

The sunburn spectrum has a profound influence on DNA, RNA, and protein synthesis, and mitosis in mammalian skin *in vivo*.[6,11] Autoradiographic, biochemical, and metaphase-arresting techniques show that a moderate erythema dose inhibits the synthesis of these macromolecules and mitosis within the first hour postirradiation and persists for at least 6 h. Functional recovery takes place by 24 h, followed by accelerated activity that reaches a peak at 48–72 h, then gradually subsides. However, increased mitotic activity in mouse skin has been demonstrated up to 40 days after a single exposure.[12] The stage of increased mitotic activity is associated with epidermal hyperplasia, which also gradually subsides over the next 2 months. The mechanism of this hyperplasia and increased mitoses is not established. Recent findings relating to the presence of epidermal mitotic inhibitors (chalone) and the actions of cyclic AMP and cyclic GMP present interesting possibilities.[13,14] In addition, a mitotic-stimulating substance has been identified postirradiation, suggesting that this hyperplasia is due to a combination of the removal of inhibition and the stimulation of growth.

7.3.4. DNA Repair

The inhibition of epidermal cellular DNA synthesis is one of the earliest post-UV irradiation events that occurs in mammalian skin *in vivo,* similar to responses noted in cultured mammalian cells. Among the number of injuries that could occur, pyrimidine dimer formation, primarily between adjacent thymine bases, has received an extraordinary amount of attention. Currently three repair mechanisms for DNA base damage in bacteria have been described: photoreactivation, excision repair, and postreplication repair (see Section 5.4.). In mammalian cells, the excision repair system appears to be the primary mechanism.[15,16] Autoradiographic (Fig. 7-5), density gradient, and chromatographic techniques have been used to demonstrate that thymine dimerization can be induced by sunlight as well as by artificial UV radiation. The DNA can be repaired by excising the dimers and replacing them with normal bases in the proper sequence. This injury and repair occurs in the epidermis *in vivo* and *in vitro,* as well as in cell culture systems. The discovery that the human photosensitivity disease xeroderma pigmentosum is characterized by a defect in this repair

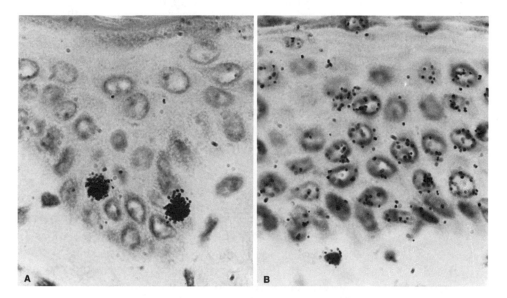

Fig. 7-5. DNA repair 15 min postirradiation. (A) Autoradiographs utilizing tritiated thymidine (^{3}H-TdR) as the radioactive tracer. This figure shows a non-UV-irradiated human epidermis with the labeling (black dots) concentrated in two of the basal cells in a dense labeling pattern, which represents premitotic semiconservative DNA synthesis. (Hematoxylin and eosin; × 720.) (B) Autoradiograph of human skin labeled 15 min post-UV irradiation using ^{3}H-TdR as the radioactive tracer. A sparse labeling pattern is seen throughout the epidermal cell nuclei. This sparse labeling represents unscheduled DNA synthesis or repair replication that is absent or markedly reduced in the skin of patients with xeroderma pigmentosum. In normal human skin postirradiation, the upper dermal fibrocytes and vascular endothelial cells show this unscheduled DNA synthesis, giving visual proof of the penetration of the sunburn rays into the upper dermis. (Hematoxylin and eosin; × 720.)

system has supplied a model for the evaluation of the importance of this repair system in human skin.[4,p.299] It is of interest that bacteria with a defect in excision repair not only are more sensitive to the killing effects of UV radiation, but also form increased numbers of mutations postirradiation. The primary photosensitivity problem in patients with xeroderma pigmentosum is their inordinate sensitivity to the carcinogenic effects of sunlight. Whether this represents cellular mutation or not remains to be determined.

A postreplication repair process in mouse L cells in culture is apparently dependent on *de novo* synthesis and not recombination mechanisms.[4-p.91,17] Recent studies by Lehman indicate that this repair mechanism is defective in cultured cells from four patients with xeroderma pigmentosum (XP variant).[18]

7.3.5. Melanogenesis[1,2,4-pp.165 and 195]

Melanins are black or brownish yellow pigments that appear to be complex polymers of dopa-quinone, 5,6-dihydroxyindole, and 5,6-dihydroxyindole-2-carboxylic acid at various oxidation levels held together by a variety of bond types. This pigment in mammalian skin is produced enzymatically in specialized cells (melanocytes) from tyrosine and Dopa. Light energy can both induce the immediate pigment darkening phenomenon and stimulate new pigment formation.

Immediate pigment darkening is characterized morphologically by a darkening of the skin that occurs almost immediately upon exposure to radiation between 320 and 700 nm. It is presumably caused by the oxidation of reduced or partially reduced melanin already present in epidermal melanocytes (Fig. 7-2). It reaches a maximum at 1–2 h, and decreases between 3 and 24 h postirradiation. Electron-microscopic studies suggest that immediate pigment darkening is related to a distal distribution of the melanosomes toward the tips of the dendrites elicited by the active motor force of microtubules and microfilaments and an apparent transfer to keratinocytes (Malpighian cells) without an obvious effect on the character of the melanosome. There is no significant change in the melanogenic organelles necessary for the synthesis of new melanosomes.

New melanin formation (delayed tanning) is characterized morphologically by a brown pigmentation of the skin starting by 48–72 h postirradiation. It reaches a peak by 13–21 days, and then gradually subsides over the next several months. The action spectrum for this response falls primarily in the sunburn range (290–320 nm). However, long wavelength UV and visible radiation, at least as long as 700 nm, can initiate new melanin formation if enough energy is applied. UV radiation from 250 to 280 nm is also distinctly less effective in inducing new melanin production. Light and electron microscopy correlated with biochemical studies indicate that this new pigment formation occurs in or on cytoplasmic organelles called melanosomes that are formed in the epidermal melanocytes. Subsequently, the pigmented melanosomes are transferred to epidermal keratinocytes, which together with the melanocytes make up the *epidermal melanin unit* (Fig. 7-2). The pigment is then carried to the stratum corneum where it provides the distinctive hyperpigmented appearance. This new pigmentation provides the most significant available natural barrier to sunlight-induced skin damage.

Hyperpigmentation of the skin in "delayed tanning" is associated with (1) proliferation of melanocytes and activation of dormant melanocytes; (2) melanocytic hypertrophy and increased dendritic arborization; (3) increased melanosomal synthesis; (4) increased rate of melanization of melanosomes; (5) increased transfer of melanosomes to keratinocytes that is related to an increase in turnover of keratinocytes; (6) increased size of melanosome complexes (more notable in Caucasoids and Mongoloids); and (7) tyrosinase activation by a direct effect on tyrosinase-inhibiting (sulfhydryl) compounds in the epidermis.[4 – p.165]

7.4. CHRONIC EFFECTS OF SUNLIGHT ON THE SKIN[1,2,17]

7.4.1. Solar (Actinic) Degeneration

Repeated exposures of Caucasoid skin over a number of years result in gross changes, including wrinkling, atrophy, hyper- and hypopigmented spots, dilated blood vessels, yellow raised areas, and actinic keratoses. A distinctive furrowed leathery change may be noted on the back of the neck, especially of very fair-skinned people who have received extensive sun exposure, e.g., sailors and farmers.

Histologically a shortening or flattening of the rete ridges (epidermal downward projections; Fig. 7-1), a possible thinning of the epidermis, and the presence of many abnormal cells in disorderly arrangement are seen in chronically sun-damaged human epidermis. The melanocytes show great variations in size, distribution, and tyrosinase content, although their absolute number generally remains unaltered. In addition, there are greater differences in size and melanin-forming activity among the basal melanocytes, as compared to normal skin. Pigment transfer from melanocytes to keratinocytes also appears to be impaired in areas of chronic solar damage.[19]

More dramatic changes occur after repeated exposure of hairless mouse skin under controlled conditions.[20] These include a striking thickening of all of the layers of the epidermis and the epidermal–dermal basement membrane (Fig. 7-6). In addition, there is an increase in the number of basal cells in DNA synthesis and mitosis, and a decrease in the transit time of cells through the epidermis.

In contrast to the relatively limited demonstrable epidermal changes, there are profound dermal alterations associated with solar degeneration in human skin. Progressive degeneration occurs in the upper portion of the dermis. Specific changes include the development of dilated blood vessels, the accumulation of acid mucopolysaccharides and abnormal-appearing fibrocytes, the loss of mature collagen (but an increase in the soluble component), and a marked increase and degeneration in elastic tissue referred to as *actinic elastosis*. Actinic elastosis, the most prominent and obvious connective tissue alteration due to chronic solar damage, is a dynamic progressive process that has been detected as early as the first decade of life.[4−p.157] The earliest change appears to be a simple increase in numbers of elastic fibers. Subsequent alterations include thickening, curling, and increased branching of the fibers with eventual replacement of the dermis and disorganization of the connective tissue into amorphous masses. Although some questions may still exist as to the origin of the fibers that stain like elastic tissue, biochemical and electron-microscopic studies have confirmed that actinic elastosis is due to the accumulation of elastic tissue.

The action spectrum for the induction of elastosis in experimental animals falls in the sunburn range. It has been postulated that this connective tissue change is the result of photochemically induced alterations in fibroblast function by UV radiation, rather than degradation of connective tissue elements.[1] In support of this hypothesis, the direct injury of human dermal connective tissue cellular DNA *in vivo* has been observed within a few minutes after irradiation with wavelengths shorter than 320 nm.[21] Therefore, the radiation that produces sunburn very likely plays an important role in chronic solar damage to dermal connective tissue. Whether or not the longer wavelengths also have a significant influence must remain a matter of speculation at this time.

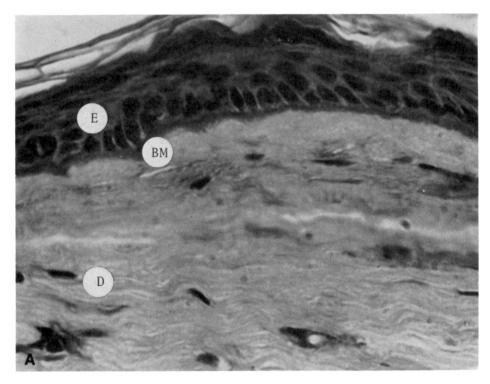

Fig. 7-6. Effects of chronic UV exposures to mouse skin. (A) Hairless mouse skin after repeated UV exposures for one month showing a benign hyperplasia with regular mild epidermal hypertrophy, and moderate but regular thickening of the basement membrane (BM), and thickening of the dermis. E, epidermis; D, dermis. (Hematoxylin and eosin; × 600.) (B) Hairless mouse skin after 4 months of

7.4.2. Carcinogenesis[(4 −p.259,7,20,22,23)]

7.4.2.1. *Human Skin Cancer Formation*

Direct experimentation to verify the sun's role in the production of human skin cancers is not possible. However, extensive empirical observations strongly suggest the etiologic significance of light energy in the induction of these tumors. Skin cancers in Caucasians are generally most prevalent in geographical areas of the greatest insolation (i.e., receiving the greatest amount of solar radiation), and among people who receive the most exposure (i.e., people who work outdoors). They are rare in Negroes and other deeply pigmented individuals, who have the greatest protection against UV-induced injury. Furthermore, the lightest-com-

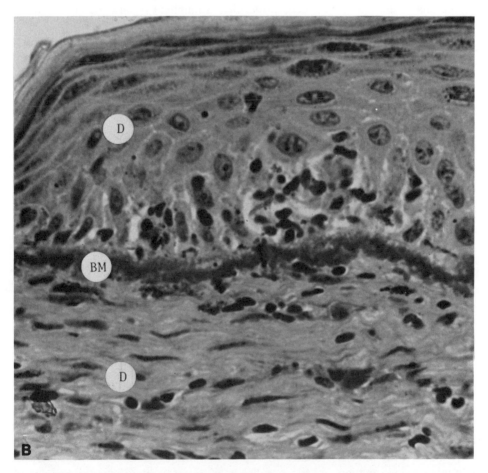

repeated UV irradiation showing a stage of premalignant to malignant change. It is characterized by abnormalities of size, shape, polarity, and chromatin patterns of the basal cells, and a marked thickening and clumping of the basement membrane, which has become discontinuous at several points. (Hematoxylin and eosin; × 600.) (From reference 23.)

plexioned individuals, such as those of Scotch and Irish descent, appear to be most susceptible to skin cancer formation when they live in geographic areas of high UV fluence. When skin cancers do occur in the dark-pigmented races, they are not distributed primarily in the sun-exposed areas as they are in light-skinned people. The tumors in dark-pigmented individuals are most commonly stimulated by other forms of trauma such as chronic leg ulcers, irritation due to the lack of wearing shoes, the use of a kangeri (an earthenware pot that is filled with burning charcoal and strapped to the abdomen for warmth), the wearing of a dhoti (loincloth), and so on. In contrast, the distribution of skin cancer in the Bantu albino, and in patients with xeroderma pigmentosum follows sun-exposure patterns. The arguments supporting the role of sunlight in human skin cancer formation may be summarized as follows: (1) skin cancers do occur predominantly on the sun-exposed parts of the body; (2) skin cancers are more common in regions of the earth that receive the most sunlight; (3) dark-pigmented races are much less susceptible to skin cancer formation than Caucasians.

Although these arguments do not constitute absolute proof, and they are in themselves not wholly established, there is a considerable body of circumstantial evidence supporting the role of sunlight in at least three types of skin cancers: basal cell epitheliomas, squamous cell carcinomas, and melanomas. The formation of the most common skin cancers (and therefore the most common human malignancies), the basal cell epithelioma and squamous cell carcinoma, is influenced by four major factors.[4−p.259]

1. *The total lifetime of sunlight exposure.* Exposure to sunlight for prolonged periods, such as occurs in Texas and Australia, appears to be an important factor in the development of skin cancer. However, a comparison of skin cancer formation in Galway, Ireland, and Philadelphia indicates that hours of exposure do not tell the complete story. Cancer occurred in Philadelphia with much less total exposure.

2. *Intensity and duration of the UV component in sunlight.* Recent studies indicate that skin cancer formation can be correlated with the duration of exposure to the erythemally effective radiation (290–320 nm) more accurately than the duration of total sun exposure. Using this criterion, the discrepancy between cancer formation occurring in rural Galway, with much more total sunlight exposure, and urban Philadelphia becomes easier to understand. The maximum exposure to the sunburn spectrum possible in Galway would be 1.33 months per year, and in Philadelphia it would be 4 months per year (if people were outdoors continuously during daylight hours). On this basis, the "effective" exposure times of the Philadelphia and Galway skin cancer groups become more similar.

3. *Genetic predisposition.* Xeroderma pigmentosum is the most striking example of genetic predisposition to skin cancer formation.[4−p.299] In those individuals who lack the ability to repair UV-damaged DNA all varieties of sunlight-induced malignancies generally develop within the first two decades of life. However, genetic predisposition also appears to influence the formation of basal and squamous cell cancer in the general population. A number of studies have shown a distinct association between these cancers and light eye color, fair complexion, light hair color, poor ability to tan, strong propensity for sunburn,

and a history of repeated sunburn reactions. In addition, individuals of Celtic origin are more prone to this type of carcinogenesis than other light-complexioned individuals.

4. *Factors unrelated to sunlight.* Sunlight exposure to erythemally effective wavelengths appears to be the primary factor in the formation of skin malignancies. Squamous cell carcinomas appear to be directly related to sun exposure. However, the morphologic distribution of the most common cutaneous cancer, basal cell epithelioma, indicates that other factors must also play a role. About one-third of basal cell epitheliomas occur on areas of the skin receiving minimal exposure to sunlight. Thus, although sunlight appears to be a dominant factor in this tumor formation, other, as yet undetermined, influences must participate.

Melanomas are the most dreaded type of skin cancer. Unlike the basal cell and squamous cell cancers, these lesions metastasize readily to other organs in the body, and have a comparatively high mortality rate. Fortunately they are relatively uncommon. Unlike the two common malignancies, the influence of sunlight on melanoma formation is not well established.[23] Melanomas are not found primarily on sun-exposed areas, and the protective effect of melanin pigment is not so obvious as it is in other forms of skin cancer. However, a number of surveys suggest that sunlight does influence the development of at least some of these malignancies. Caucasians with melanomas statistically tend to have light eyes, light hair, fair complexions, and spend more time outdoors when compared to a control group of patients without melanomas. Surveys of the geographic distribution of this cancer, even in genetically similar populations, have demonstrated a much greater melanoma prevalence associated with high insolation as compared to low insolation. The occurrence of solar radiation-induced melanomas in patients with xeroderma pigmentosum and the production of melanomas from benign pigmented lesions by chronic UV exposure in experimental animals further confirms the potential of sunlight to stimulate the production of these tumors under proper circumstances. In addition, the distribution of melanomas developing from circumscribed precancerous melanosis strongly suggests that the sun is at least partly responsible for such lesions. However, the anatomical distribution of melanomas indicates that other factors are more important than sunlight in the etiology of these tumors.

7.4.2.2. Experimental Carcinogenesis

7.4.2.2a. Quantitative Investigations

Traditionally, the skin of the ear of the albino mouse, and, to a lesser extent, the albino rat, was used for the experimental production of tumors by UV radiation, because this area lacks the three main natural inhibiting factors: pigment, a thick stratum corneum, and thick hair growth. The type of tumor produced in this tissue was primarily a sarcoma that was quite adequate for the monumental quantitative studies accomplished by Blum and co-workers.[22] Using the ears of an inbred, albino, haired mouse strain with a UV source primarily emitting radiation shorter than 320 nm, they demonstrated that cancer development follows the law of reciprocity as far as dose–rate dependence is concerned. Increasing the dose or shortening the intervals between exposures accelerated tumor formation, but did not alter the shape of the incidence curve.

A factor in the production of skin cancer by UV radiation that has apparently not been expolored is the influence of protraction of the daily dose on reciprocity. Various experimenters have used exposures varying in length from a few seconds to several hours while exploring such variables as influence of chemicals, action spectrum, etc. There are apparently no published data on whether a radiation source has an equally good chance of producing skin cancer if a given fluence is delivered in a few seconds or is attenuated over several hours.

Under controlled circumstances, skin cancers will develop if enough energy is delivered for a sufficient period of time.[22] However, tumors will not appear no matter how much energy is used if it is not applied long enough. Blum surmised that UV-induced cancer formation is a continuous process that begins with the initial exposure. The appearance of tumors within the lifetime of the animal depends on sufficient acceleration of the growth process. In support of the concept that growth acceleration is an important component of carcinogenesis, the production of squamous cell carcinomas in hairless mouse skin with one UV exposure followed by croton oil promotion has been reported.[23]

The mechanism of tumor growth acceleration remains to be established. The production of mitosis-stimulating substances or perhaps more importantly the removal of mitosis-inhibiting materials, chalones, may well play a role in this process.

7.4.2.2b. Qualitative Investigations[4-p.259,20,23]

The progressive changes occurring in tissue during UV-induced epidermal cancer development were examined in hairless mouse skin; these tumors are identical to primary growths induced by sunlight radiation in human skin. In addition, these animals proved useful in studying the effect of UV radiation on melanoma formation.[23]

A number of investigators have considered dermal influence to be of prime importance in the development of epidermal skin cancers. Traditionally, actinic elastosis was considered to be a formidable agent in this disease process; however, several recent studies have shown that such changes are not essential in humans or experimental animals.[20,23] Many researchers believe that some or all of the following events may be associated with significant dermal influences on the development of epidermal malignancies: dissolution of elastic tissue and collagen; proliferation of young collagen; accumulation of acid mucopolysaccharides, mast cells, and fibroblasts; the formation of new elastic tissue; and alterations in dermal vasculature.

Histochemical and radioactive tracer ([³H]thymidine) techniques used to examine anatomic changes and cellular kinetics[20,23] illuminated the progressive development from benign hyperplasia through stages of actinic keratosis-like lesions to frank invasive malignancy in the hairless mouse system. These studies showed an accumulation of acid mucopolysaccharides, and a loss of insoluble collagen in the upper dermis, and the proliferation of mast cells and fibrocytes. The most striking response occurred in the epidermal–dermal basement membrane (Fig. 7-6), where a progressive thickening of this structure accompanied epidermal hyperplasia. As irregular and abnormal epidermal cell proliferation advanced, the basement membrane became thicker, irregular, clumped, and frayed in appearance, and with frank invasive malignancy, it disappeared altogether. Similar breaks in the basement membrane have been noted with the invasion of human skin cancers and chemically induced experimental tumors. Electron-microscopic studies have suggested that the microprojection of cells through the basal lamina may represent the earliest stage of tumor invasion.

Studies of epidermal cell kinetics showed a progressive increase in the number of germinative basal cells synthesizing DNA and dividing, with a shortening of the DNA synthesis time and G2 period associated with epidermal changes.[20] At the same time, there was a progressive reduction in the cell transit time through the epidermis despite its increasing thickness. With frank malignancy, the germinative cell layer disappeared, and mitoses, many abnormal in appearance, were present throughout the tumor. Thus, the process of carcinogenesis was characterized by the acceleration of cell formation, maturation, and turnover. Furthermore, the epidermal germinative basal cell appeared to be the primary or initial site of abnormal proliferation.

Experimentally, the transformation of benign pigmented growths to malignant melanomas in the skin of hairless mice has been induced with radiation shorter than 320 nm.[23] Repeated exposures over several months resulted in the production of large invasive melanocytic tumors with histologic, autoradiographic, and electron-microscopic characteristics of malignancy. In addition, several of these tumors metastasized to the regional lymph nodes. These results indicate that benign pigmented lesions can be transformed to malignant growths in experimental animals by UV radiation. Whether a similar process occurs in human skin remains to be established.

The enhancing effect of heat on the degree of cutaneous injury and the intensity of erythema response to UV radiation has been well documented.[23] Furthermore, it has been demonstrated that increased temperatures at the time of UV exposure accelerate tumor production. Clinical experience also suggests that heat does, in fact, aggravate UV-induced skin cancer formation in human skin.

The presence of carcinogenic and tumor-promoting chemicals in our environment has made the experimental evaluation of chemical influences on UV-induced carcinogenesis of practical importance.[23] Recent studies have demonstrated that UV radiation and chemical carcinogenic stimuli are additive. Also, repeated applications of the noncarcinogen croton oil results in malignancy following a single UV exposure. These experiments suggest that environmental chemicals may be a significant factor in the development of skin tumors.

Perhaps the most intriguing aspect of UV-induced carcinogenesis is the relationship of the acute responses to the eventual cancer formation.[7,15,22,23] Light-induced carcinogenesis occurs only when acute phototoxic erythema is produced as well. The croton oil experiments also indicate that the process of UV-induced cancer formation begins with the initial exposure. However, the

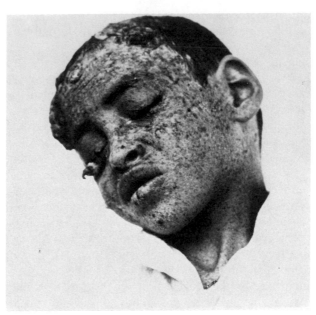

Fig. 7-7. Xeroderma pigmentosum. Clinical picture of a patient with xeroderma pigmentosum, showing actinically damaged skin, multiple actinic keratoses, and tumors. (Courtesy of William Spencer, M.D.)

early changes leading to cancer formation remain undetermined. A number of studies have demonstrated a similar, but not identical, effect of UV radiation and chemical carcinogens on certain vital functions in mammalian epidermis, including DNA and RNA synthesis, cell turnover, and the mitotic rate. The most intriguing findings to date, relating acute photoinjury to cancer formation, have developed from the study of DNA repair systems and the rare genetic disease xeroderma pigmentosum. As noted in Section 7.3.4., in this disease, photosensitivity is expressed as an inordinate susceptibility to the development of sunlight-induced skin cancers[4−p.299,17] (Fig. 7-7). The cells of these patients are defective in their ability to repair UV-damaged DNA. Cultured xeroderma pigmentosum cells do have a higher mutation rate than normal human cells following UV radiation.[24] It has been postulated that this defect may lead to a high mutation rate following sunlight irradiation, and thus result in cancer formation.

7.5. ADVERSE CUTANEOUS REACTIONS TO SUNLIGHT[2,7,25]

Most of the cutaneous effects of sunlight are injurious. The following part of this discussion is designed to define the pathogenesis of some of the adverse effects of this energy. The explanations will of necessity be somewhat crude because of the limitations of our knowledge at this time.

7.5.1. Photosensitivity

Photosensitivity is the broad term used to describe adverse reactions to sunlight or artificial light energy. Two types of photosensitivity reactions may occur, which may be phototoxic or photoallergic in nature.

7.5.1.1. Phototoxicity[25]

Light-induced damage in the skin that is not dependent on an allergic mechanism may be considered to be phototoxic. Theoretically these reactions could occur in everybody if the skin were exposed to enough light energy of the proper wavelengths, and enough molecules that absorb these wavelengths were present. Of course, the radiation must penetrate to the absorbing molecules for the reaction to occur. The sunburn reaction is the classic example of a phototoxic response. Clinically, phototoxic reactions usually are characterized by erythema and at times by edema (swelling) occurring within a few minutes to several hours after exposure, followed by hyperpigmentation and desquamation (peeling) confined to the exposed areas (Fig. 7-8).

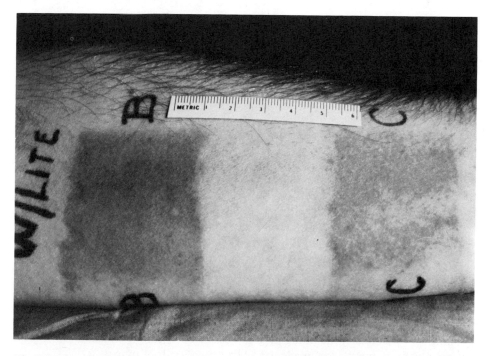

Fig. 7-8. Phototoxic reactions. Demonstrates a phototoxic reaction in two areas on the forearm, B and C, due to the topical application of a psoralen compound followed by exposure to the action spectrum for this chemical (UV radiation between 320 and 400 nm). It is characterized by erythema and edema at this point. The area between B and C received the UV exposure without prior application of the psoralen compound. (Courtesy of H. I. Maibach, M.D. and F. N. Marzulli, M.D.)

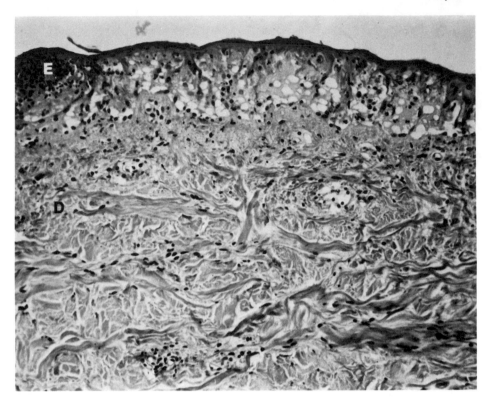

Fig. 7-9. Histology of the phototoxic reaction. Biopsy of the phototoxic reaction shown in Fig. 7-8 demonstrating a marked destruction of the epidermis with intracellular and extracellular edema (swelling), vacuole formation, and cell death. The dermis shows very little cellular infiltrate. E, epidermis; D, dermis. (Hematoxylin and eosin; × 50.) (From reference 7.)

Histologically, epidermal cell degeneration may be prominent when the photosensitizer is in the epidermis. This is most notable after the application of an exogenous photosensitizer to the skin. Edema with a mild-to-moderate inflammatory cell infiltration into the dermis, consisting primarily of polymorphonuclear leukocytes, may be seen (Fig. 7-9).

Despite a remarkable amount of investigation, the mechanisms by which phototoxic responses occur are not well understood. In the case of an exogenous photosensitizer, either the molecule alone or a complex of the chemical and cellular organelles becomes excited by the absorption of light. Triplet states and/or free radicals may thus be formed, and the dissipation of this energy may result in a number of changes, including peroxide formation, cell membrane or lysosomal membrane damage, and nuclear and/or mitochondrial injury.

It is likely that mechanisms vary with the photosensitizer. In support of this idea, there is evidence that the phototoxicity induced by the furocoumarins (see Fig. 4-3) is associated with the formation of adducts with DNA. Chlorpromazine appears to form similar adducts, primarily with RNA. Other chemical photosensitizers, e.g., methylene blue, acriflavine, rose bengal, and porphyrins, require the presence of oxygen. This last type of phototoxicity has been termed "photodynamic action" (see Chapter 4).

7.5.1.2. Photoallergy[25]

Photoallergy can be defined as an acquired, altered capacity of the skin to react to light energy alone or in the presence of a photosensitizer that is presumably dependent on the development of a circulating antibody or a cell-mediated immune response. These reactions are generally uncommon, and the clinical patterns range from an immediate hive response (Fig. 7-10) to delayed itching, scaling, or weeping lesions (Fig. 7-11). The eruption frequently extends beyond the exposed areas, and eruptions in distant, previously involved sites may occur. Under controlled conditions, the general characteristics of allergic responses can be identified, including detection of an incubation period, spontaneous flare, and transfer of the process to normal subjects. Usually, less radiation exposure is required to induce a photoallergic response than a phototoxic response having the same action spectrum.

Photoallergy differs histologically from phototoxicity. The immediate urticarial (hive) lesions show very little microscopically other than some edema and

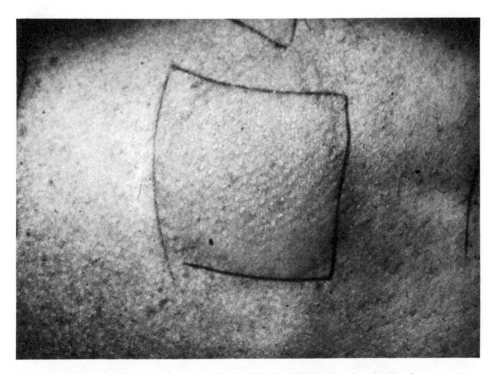

Fig. 7-10. Solar urticaria. Clinical picture of urticarial wheal (hive) occurring shortly after exposure to UV radiation and subsiding within 30 min. (From reference 2.)

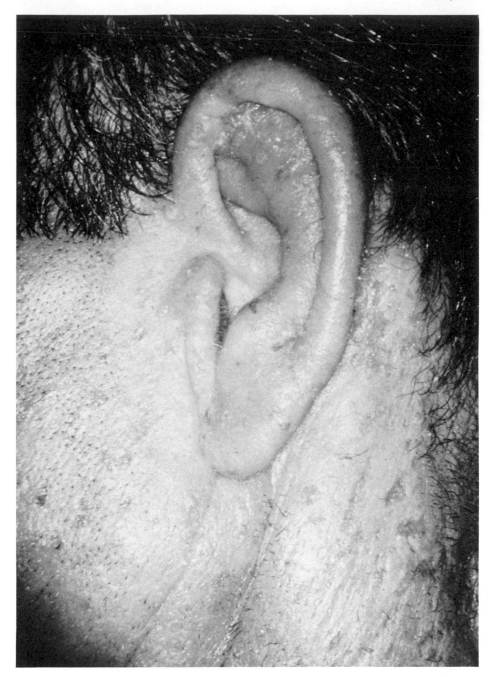

Fig. 7-11. Allergic photocontact dermatitis. Clinical picture of an eczematous photoallergic contact dermatitis induced by utilization of a soap containing tribromosalicylanilide and exposure to the action spectrum (UV radiation between 320 and 400 nm).

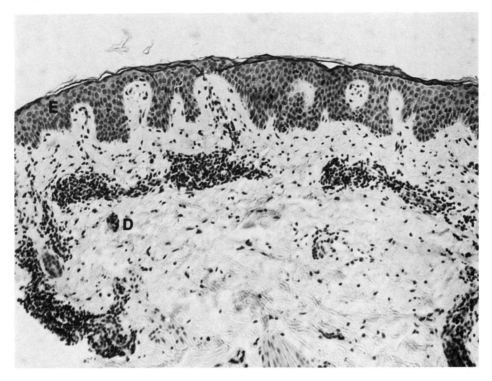

Fig. 7-12. Histology of the allergic photocontact dermatitis. Biopsy of an allergic photocontact reaction showing a dense perivascular round cell (collection of black dots in the dermis) infiltrate in the dermis characteristic of a delayed hypersensitivity response. E, epidermis; D, dermis. (Hematoxylin and eosin; × 80.) (From reference 7.)

blood vessel dilatation. The delayed reactions present a dense round-cell infiltrate around the blood vessels in the dermis that is characteristic, though not diagnostic, of these responses (Fig. 7-12). Similar to phototoxicity, photoallergy can be produced with or without the presence of known exogenous photosensitizers.

7.5.2. Clinical Problems

It is not appropriate for this chapter to dwell on the clinical responses that fall primarily in the field of individuals with training in dermatology. However, an outline of these reactions may serve a useful purpose in putting medical effects of light energy in perspective. The adverse clinical effects may be divided into three large categories: (1) responses due to lack or loss of protection; (2) responses due to the presence of a photosensitizing chemical; and (c) responses not due to a deficiency in protection or to the presence of a known photosensitizer.

7.5.2.1. Lack or Loss of Protection[2,7]

Melanin pigment is the primary natural protective agent in the skin. Therefore, almost all the clinical problems in this category relate to a melanin deficiency. This group includes patients with albinism who have a defect in the enzymatic tyrosinase system and are unable to produce sufficient melanin; vitiligo, in which there is a loss of pigment cells, and thus a loss of melanin; phenylketonuria (PKU), in which pigment dilution occurs due to an inherited error in phenylalanine metabolism; Chidiak-Higashi, a genetic syndrome with pigment dilution, apparently due to a melanosome structural abnormality; and, the most common of all photosensitive conditions, the fair-complexioned individual with light hair and eye color. This category also includes patients with xeroderma pigmentosum, in whom the lack of melanin does not play a role (Fig. 7-7), but rather, relates to an inability to repair UV-damaged DNA.

The clinical reactions in this category consist of acute sunburning, and with chronic repeated injury, skin degeneration, actinic keratoses (Fig. 7-13), and skin cancer formation. The responses are phototoxic in nature, and the action spectrum falls in the sunburn range (290–320 nm).

7.5.2.2. Photosensitized Reactions[2,7]

A photosensitizer in this discussion will be considered to be a substance that, as the result of absorption of the sun's energy, induces an adverse response in the skin. The photosensitizer may be endogenous or exogenous in origin.

The porphyrins are the only well-established photosensitizers made by the human body. These are pyrrole ring structures that are essential to cellular metabolism throughout the body. Among other things, the heme part of hemoglobin is formed by these molecules. However, in the group of diseases named the porphyrias, large amounts of porphyrins are made in the erythropoietic (bone marrow) and/or the hepatic (liver) tissues. In these states, the porphyrins are not associated with iron, and are potent photosensitizers. As such, they induce cutaneous changes ranging from marked photodestruction of the skin and underlying tissues to simple hyperpigmentation, depending on sun exposure and the amount of porphyrin molecules available. The cutaneous changes are due to photodynamic phototoxic responses, and the peak of the action spectrum is around 400 nm.

Exogenous photosensitizers may get to the skin by topical application (contact) or through the bloodstream (systemic).

1. Topical chemicals may cause photosensitivity reactions advertently, as occurs with tars and psoralen compounds (see Fig. 4-3) in the treatment of psoriasis and vitiligo, or inadvertently, as occurs with the psoralen molecules in plants and perfumes, halogenated salicylanilides in deodorant soaps, sulfonamide and phenothiazine medications, and so on.[2,7] The reactions are usually phototoxic, though occasionally they are photoallergic in type. The action spectrum usually falls in the range of 320–400 nm.

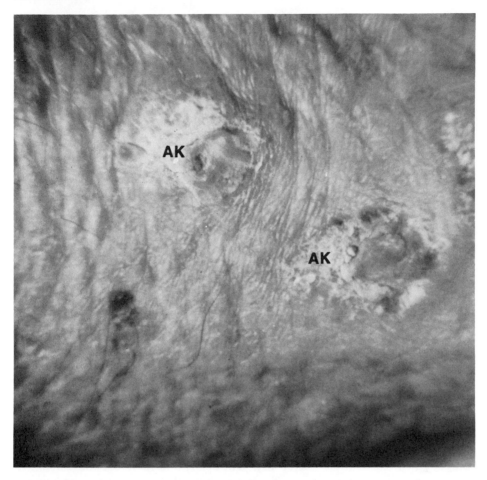

Fig. 7-13. Chronic sun (actinic) damage. Chronic actinic damage of the back of a hand with actinic keratoses (AK).

2. Systemic chemicals also may induce photosensitivity reactions advertently, as occurs with the use of psoralen compounds in the treatment of vitiligo and psoriasis, or inadvertently, which is the usual case. These inadvertent photoreactions are generally induced by commonly used medications, including antibacterial sulfonamides, thiazide diuretics, sulfonylurea antidiabetic drugs, phenothiazines, and the broad-spectrum antibiotic, demethylchlortetracycline.[2,7] The vast majority are phototoxic in nature, although occasionally delayed hypersensitivity responses may occur. The action spectra for both phototoxic and photoallergic reactions usually include long-wavelength UV radiation.

7.5.2.3. Protection Normal and No Known Photosensitizer

Photoreactions not due to deficient protection or the presence of known photosensitizers comprise a catch basket of conditions, including allergic responses such as the immediate antibody-mediated solar urticaria (hive response), and the apparently cell-mediated polymorphous light eruption (PMLE), certain autoimmune diseases, vitamin deficiencies, and a number of genetic problems. In general these are uncommon diseases, except for PMLE. The action spectrum and mechanisms vary with the conditions. Because of the complicated and primarily clinical nature of this category it will not be discussed further.[2,7]

7.6. BENEFICIAL EFFECTS OF SUN AND/OR ARTIFICIAL LIGHT ENERGY

Although adverse effects are much more common than beneficial responses, UV radiation does have therapeutic uses.

7.6.1. Vitamin D Formation[1,2,4−p.247,26]

Perhaps the only completely established beneficial effect of UV radiation on normal human skin is the formation of vitamin D. The D vitamins are a factor in the transport of calcium from the gut and in bone metabolism. These vitamins prevent rickets (softening of bones) in children and osteomalacia (loss of bone substance) in adults by insuring proper bone calcification.

Vitamin D_3 (calciferol) is formed in the skin, most likely in the Malpighian cells (Fig. 7-2), through conversion of 7-dehydrocholesterol by the action of UV radiation. Further hydroxylation to 25-OH calciferol provides a more efficient compound. It has been suggested that skin color, and therefore the amount of protection against UV radiation, has been geographically regulated to prevent naturally occurring vitamin D intoxication.[27] Thus, the light skin of northern Europeans was needed because of the limited amount of natural UV exposure, and the dark skin of peoples who live in the tropics is a necessary protection against producing too much vitamin D. However, as yet no *natural* vitamin D intoxication has been reported in light-complexioned individuals who have migrated southward. Thus, vitamin-D formation by UV radiation remains primarily a beneficial effect.

7.6.2. Treatment of Skin Diseases

Niels Finsen, an outstanding physician and an early pioneer in photobiology, received the Nobel prize in 1903 for his use of UV radiation in the treatment of cutaneous tuberculosis. Though this form of treatment is no longer necessary because of medical advances, the therapeutic value of light energy has been described for a large number of conditions, including acne, eczema, pityriasis rosea, and particularly vitiligo, psoriasis, and herpes simplex.

7.6.2.1. Vitiligo

Vitiligo is a cutaneous disorder characterized by a loss of pigment cells and thus pigment, the etiology of which is speculative at present (Fig. 7-14). Phototherapy represents almost the only available treatment, and success is limited, at best. The procedure consists of the topical or systemic administration of psoralen compounds followed by exposure to the action spectrum of these chemicals, i.e., from 320 to 380 nm.[4–pp.355 and 783] It should be noted that the action spectra for the stimulation of pigmentation and the induction of erythema are quite similar, but they may not be identical. Until recently it was felt that a significant phototoxic erythema effect was necessary for the induction of repigmentation; however, studies now indicate that the pigment response may be separate from the phototoxic injury (as defined by the erythema and visible damage).[28] The basic mechanism for the induction of hyperpigmentation induced by psoralen compounds and long-wavelength UV radiation has not been established. However, the following observations have been noted following application of 4,5′,8-trimethylpsoralen (TMP) and exposure to these wavelengths of radiation: (1) an increased number of functioning epidermal melanocytes occurs due to proliferation and/or activation of inactive cells (mitotic activity has been detected between 48 and 72 h after TMP plus UV-A irradiation); (2) melanocyte hypertrophy and increased extension of dendrites around the keratinocytes; (3) increased number of melanosomes in melanocytes and keratinocytes; (4) increased tyrosinase activity in melanocytes; (5) increased rate of transfer of melanosomes to keratinocytes; and (6) dispersion of melanosomes in keratinocytes in nonaggregated distribution. (This last effect has only been described in Caucasoid skin.) This dispersed distribution of melanosomes may persist for more than 9 months, suggesting that some long-term gene depression may occur, perhaps related to the photoaddition of psoralens to DNA (see Section 4.4.1.).

7.6.2.2. Psoriasis

Psoriasis is a common dermatological disease of unknown etiology that affects 1–3% of the world's population, and between 2 and 8 million people in the United States. The clinical picture varies from localized scaling plaques to generalized exfoliation (Fig. 7-15). The most notable pathological characteristic of this disorder is a marked increase in epidermal cell proliferation, with rapid turnover of the germinative cells.

Treatment of psoriasis has revolved around attempts at the inhibition of this increased cellular proliferation, and phototherapy represents one of the most time-honored, safe, and effective methods in this respect.[4–p.793] Astute clinical observations, and subsequent controlled studies, have established that radiation in the UV-B region (290–320 nm) is therapeutically effective in this disease. In 1925, Goekerman described increased benefits from the topical application of crude coal tar to the sunburn exposure regimen. Since the action spectrum of the coal tar photosensitization is in the UV-A range, the increased effects had to be additive in nature, i.e., the tar and UV-B acted independently. One of the

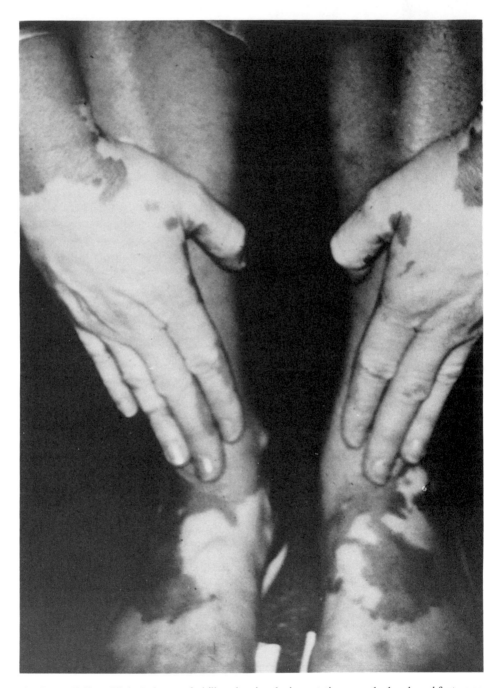

Fig. 7-14. Vitiligo. Clinical picture of vitiligo showing depigmentation over the hands and feet areas caused by loss of pigment cells.

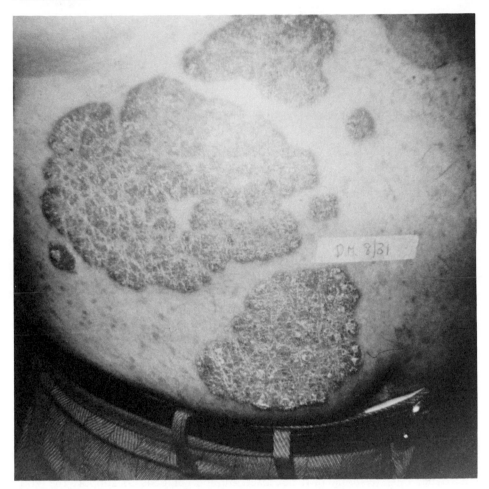

Fig. 7-15. Psoriasis. Clinical picture of psoriasis characterized by large erythematous scaling plaques.

disadvantages of the Goekerman regimen was the messy nature of the tar, which generally had to be used in a hospital setting. Within the past few years, regression of lesions following topical application of psoralen compounds and UV-A has been reported. Most recently, an apparently extremely effective treatment has been described using systemic 8-methoxypsoralen and UV-A emitted from a newly developed high-intensity light system.[29]

Though the mechanism of the effect of phototherapy on psoriasis has not been completely established, a reduction in cellular proliferation most likely plays a part. UV radiation shorter than 320 nm has a profound influence on epidermal cellular DNA synthesis and mitosis, which are inhibited shortly after irradiation, and remain so for several hours.[6] Similarly, psoralens plus long-wavelength UV radiation significantly inhibit these functions, presumably due to photoaddition between the chemical and thymine bases of the DNA. It is also possible that phototherapy influences the dermal vasculature either directly or indirectly, and thus alters the course of the psoriatic lesions. Further study is obviously needed to clarify these concepts.

7.6.2.3. Herpes Simplex Infection[30]

The herpes simplex virus is a common human pathogen that causes primary and recurrent skin and mucosal infections, and occasionally may involve other organs. Recently a possible relationship between this virus and human cancer formation has been described. By far the most common clinical problem is the recurrent herpes simplex infection that is characterized by localized vesicular lesions involving the face, especially the lips, and the genitalia. Two varieties of closely related viruses are responsible for the lesions: those that occur above the waist are usually due to the type 1 virus, and those that appear below the waist are generally due to the type 2 variety. The eruption may occur as frequently as at monthly intervals or less, and thus present a most disturbing and, at times, debilitating problem.

Recently a group of investigators have advocated the use of photoinactivating procedures utilizing photodynamically active dyes, such as proflavine and neutral red, and visible light to inactivate the virus in these lesions. This concept resulted from the fact that photoinactivation of the viruses could be accomplished in tissue culture by this procedure. The following hypothetical mechanism has been considered most reasonable to account for herpes simplex virus inactivation. It is postulated that, during replication, the heterotricyclic dye molecules are inserted between the bases of the viral DNA. The dye DNA complex absorbs the light energy, resulting in an oxidation reaction that disrupts the DNA and leads to a loss of guanine. This leaves holes in the base sequence, and results in breaks in the DNA strands; the viruses are then unable to replicate. An alternative theory suggests that the cells in which the viruses live are destroyed, thus resulting in a deceased number of cells to support the viral particles.

At present the clinical efficacy and safety of this herpes simplex virus inactivation technique has not been established. The clinical results are as yet not convincing owing to the great difficulty in obtaining enough patients to have adequate controls. In addition, certain *in vitro* studies suggest that there may be a risk of inducing cancer by this procedure. Thus, this treatment should be considered with some reservation until further information is available.

7.6.3. Treatment of Systemic Diseases (Neonatal Jaundice)

Historically, phototherapy has been prescribed for a variety of diseases. As noted previously, there is evidence that UV radiation is beneficial for diseases of calcium metabolism, such as rickets and osteomalacia, due to the photochemical formation of vitamin D in the skin. Perhaps the one other disease in which good evidence for benefits is available is neonatal jaundice.[4−p.231,31]

Bilirubin ($C_{33}H_{33}O_6N_4$), the chief pigment in human bile, is formed from heme when hemoglobin from red blood cells is broken down, primarily by the Kupffer cells of the liver. It appears to result from the opening of the porphyrin ring. In the neonatal period, increased amounts of unconjugated bilirubin are present in the circulation. This is due to the shorter survival of red blood cells and to the functional immaturity of the neonatal liver, such that it is limited in its capacity to convert bilirubin from a lipid-soluble to a water-soluble compound (conjugation) that can be excreted in the urine. In addition, increased permeability of the blood–brain barrier makes the newborn infant more susceptible to central nervous system damage by the deposition of bilirubin in brain cells (kernicterus). The danger is most notable in the premature infant. At present, therapy is designed to keep unconjugated bilirubin levels below 10–15 mg%. One of the mechanisms for the reduction in circulating unconjugated bilirubin which has received considerable attention recently is phototherapy.

In vitro studies indicate that visible light will cause the photodegradation of bilirubin. The exposure of experimental animals (Gunn rats) and infants to intense visible light results both in the photochemical alteration of bilirubin, and a striking reduction in the concentration of circulating unconjugated bilirubin. *In vitro* studies have demonstrated that wavelengths in the blue part of the visible spectrum (primarily 450–500 nm) are the most effective in photooxidizing bilirubin. These *in vitro* photooxidation products lack the toxicity of bilirubin. However, although they are similar, the photoderivatives *in vitro* are not identical to those produced *in vivo*. The *in vivo* derivatives are readily excreted and therefore probably will not accumulate in the central nervous system tissue where the damage in kernicterus occurs. Perhaps the most interesting, and as yet unexplained, aspect of this process is that the primary effect of phototherapy may not relate to photodegradation of bilirubin. In animals and infants exposed to blue light, the major portion of the excreted products is unconjugated bilirubin. Thus, phototherapy enhances the excretion of this neurotoxic substance. The mechanism of this increased excretion is unknown at present.

As with all therapies, the potential disadvantages must be considered. Fortunately the photoproducts produced appear to be either nontoxic or so readily excreted that they do not accumulate in the central nervous system. Photodynamic injury to the cutaneous cells must be considered as a possible detrimental effect, though no such response has been observed as yet. In addition, a number of injurious effects on bacteria and purified DNA have been recorded. Whether such findings relate to clinical responses remains to be determined. Perhaps the most potentially disturbing aspect of this phototherapy relates to possible retinal damage that has been induced in newborn piglets exposed to blue fluorescent lights. No such damage has been noted in human infants to date despite extensive exposure.

7.7. INDIRECT EFFECTS OF LIGHT[4−p.231]

A number of miscellaneous indirect effects of light energy on animal functions have been described. These include possible control of biological rhythms

(Chapter 8), gonadal function (Chapter 9), cortisol production, and the like. One of the best characterized indirect effects, other than vision, concerns the inhibition of melatonin formation in the pineal gland by light energy that impinges on the retina. What general influence this response engenders remains to be determined, though recent studies suggest that the pineal gland may play a most important physiologic role in human hormonal and central nervous system function. These indirect effects appear to be mediated through a retinal receptor, as yet unidentified. Other effects include reduction in systolic and diastolic blood pressure, increase in exercise tolerance, reduction in blood cholesterol, and a decrease in serum tyrosine.

7.8. CONCLUSION

Light energy has profound effects on mammalian structure and function. These influences play a key role in the relationship between humans and their environment. The primary effects, both detrimental and beneficial, occur at the site of impact, i.e., the skin. The skin is not only the most common site of photoresponse, but it also is the most available human tissue for examination and investigation. Thus, medical photobiologists have unique opportunities; they can examine cancerous growths as they are being formed, evaluate radiation and chemical interactions as they relate to disease states, study the influence of light energy on immunological processes (see Section 15.5.), and so on. Though the sophistication of methodology for the study of human responses is somewhat limited compared to other disciplines in photobiology, the significance of such studies cannot be overemphasized. The field of medical photobiology is wide open. At present we have just "scratched" the surface of understanding.

ACKNOWLEDGMENTS

This work was supported in part by Grant No. CA 15601 from the National Institutes of Health. Preparation of this chapter was accomplished with extensive aid from the Department of Dermatology Research Unit, University of California Medical Center, San Francisco, California.

7.9. REFERENCES

1. B. E. Johnson, F. Daniels, Jr., and I. A. Magnus, Response of human skin to ultraviolet light, *Photophysiology* **4**, 139–202 (1968).
2. M. A. Pathak and J. H. Epstein, Normal and abnormal reactions of man to light, in: *Dermatology in General Medicine* (T. B. Fitzpatrick, K. A. Arndt, W. H. Clark, A. Z. Eisen, E. J. Van Scott, and J. H. Vaughan, eds.) pp. 977–1036, McGraw–Hill, New York (1971).
3. F. Urbach (ed.), *The Biologic Effects of Ultraviolet Radiation*, Pergamon, New York (1969).
4. T. B. Fitzpatrick, M. A. Pathak, L. C. Harber, M. Seiji, and A. Kukita (eds.), *Sunlight and Man*, University of Tokyo Press, Tokyo (1974).
5. I. Willis, A. Kligman, and J. H. Epstein, Effects of long ultraviolet rays on human skin: Photoprotective or photoaugmentative?, *J. Invest. Dermatol.* **59**, 416–420 (1972).
6. J. H. Epstein, K. Fukuyama, and K. Fye, Effects of ultraviolet radiation on the mitotic cycle and

DNA, RNA and protein synthesis in mammalian epidermis *in vivo, Photochem. Photobiol.* **12**, 57–65 (1970).

7. J. H. Epstein, Adverse cutaneous reactions to the sun, in: *Yearbook of Dermatology* (F. D. Malkinson and R. W. Pearson, eds.), pp. 5–43, Yearbook Medical Publ., Chicago, Ill. (1971).

8. D. S. Snyder and W. H. Eaglstein, Intradermal antiprostaglandin agents and sunburn, *J. Invest. Dermatol.* **62**, 47–50 (1974).

9. G. F. Wilgram, R. L. Kidd, W. S. Krawczyk, and P. L. Cole, Sunburn effect on keratinosomes, *Arch. Dermatol.* **101**, 505–519 (1970).

10. K. Wier, K. Fukuyama, and W. L. Epstein, Nuclear changes during light-induced depression of ribonucleic acid and protein synthesis in human epidermis, *Lab. Invest.* **25**, 451–456 (1971).

11. K. Fukuyama, W. L. Epstein, and J. H. Epstein, The effect of ultraviolet light on RNA and protein synthesis in differentiated epidermal cells, *Nature* **216**, 1031–1032 (1967).

12. G. A. Soffen and H. F. Blum, Quantitative measurements in changes in mouse skin following a single dose of ultraviolet light, *J. Cell Comp. Physiol.* **58**, 81–96 (1961).

13. W. S. Bullough and E. B. Laurence, Tissue homeostasis in adult mammals, in: *Advances in Biology of Skin,* Vol. 7, *Carcinogenesis* (W. Montagna and R. L. Dobson, eds.), pp. 1–36, Pergamon, Oxford (1966).

14. J. J. Voorhees, E. A. Duell, and W. H. Kelsey, Dibutyryl cyclic AMP inhibition of epidermal cell division, *Arch. Dermatol.* **105**, 384–386 (1972).

15. J. E. Cleaver, Repair of damaged DNA in human and other eukaryotic cells, in: *Nucleic Acid-Protein Interactions—Nucleic Acid Synthesis in Viral Infection* (D. W. Ribbons, J. F. Woessner, and J. Schultz, eds.), pp. 87–112, North-Holland, Amsterdam (1971).

16. J. E. Cleaver, Repair processes for photochemical damage in mammalian cells, *Adv. Radiat. Biol.* **4**, 1–75 (1974).

17. J. E. Cleaver and D. Bootsma, Xeroderma pigmentosum: Biochemical and genetic characteristics, *Annu. Rev. Genet.* **9**, 19–38 (1975).

18. A. R. Lehman, S. Kirk-Bell, C. F. Arlett, M. C. Paterson, P. H. M. Lohman, E. A. de Weerd-Kastelein, and D. Bootsma, Xeroderma pigmentosum cells with normal levels of excision repair have a defect in DNA synthesis after UV-irradiation, *Proc. Natl. Acad. Sci. USA* **72**, 219–223 (1975).

19. R. L. Olson, J. Nordquist, and M. A. Everett, The role of epidermal lysosomes in melanin pigmentation, *Br. J. Dermatol.* **83**, 189–199 (1970).

20. J. H. Epstein, K. Fukuyama, and R. L. Dobson, Ultraviolet light carcinogenesis, in: *The Biologic Effects of Ultraviolet Radiation* (F. Urbach, ed.), pp. 551–568, Pergamon, New York (1969).

21. W. L. Epstein, K. Fukuyama, and J. H. Epstein, Early effects of ultraviolet light on DNA synthesis in human skin *in vivo, Arch. Dermatol.* **100**, 84–89 (1969).

22. H. F. Blum, *Carcinogenesis By Ultraviolet Light,* Princeton University Press, Princeton, N.J. (1959).

23. J. H. Epstein, Ultraviolet carcinogenesis, *Photophysiology* **5**, 235–273 (1970).

24. V. M. Maher, L. M. Oullette, R. D. Curren, and J. J. McCormick, Frequency of ultraviolet light-induced mutations is higher in xeroderma pigmentosum variant cells than in normal human cells, *Nature* **261**, 593–595 (1976).

25. J. H. Epstein, Photoallergy: A review, *Arch. Dermatol.* **106**, 741–748 (1972).

26. H. F. DeLuca, Vitamin D: The vitamin and the hormone, *Fed. Proc.* **33**, 2211–2219 (1974).

27. W. F. Loomis, Skin-pigmentation regulation of vitamin D biosynthesis in man, *Science* **157**, 501–506 (1967).

28. K. H. Kaidbey and A. M. Kligman, Photopigmentation with trioxsalan, *Arch. Dermatol.* **109**, 674–677 (1974).

29. J. A. Parrish, T. B. Fitzpatrick, L. Tanenbaum, and M. A. Pathak, Photochemotherapy of psoriasis with oral methoxsalen and long-wave ultraviolet light, *N. Engl. J. Med.* **291**, 1207–1211 (1974).

30. L. E. Bockstahler, C. D. Lytle, and K. B. Hellman, *A Review of Photodynamic Therapy for Herpes Simplex: Benefits and Potential Risks,* DHEW Publication No. (FDA) 75-8013. (Available from the Bureau of Radiological Health, Rockville, MD 20852.) (1974).

31. *Phototherapy in the Newborn,* Final report of the Committee, Division of Medical Sciences, National Research Council (2101 Constitution Ave.) Washington D.C., 20418 (1974).

8

Chronobiology (Circadian Rhythms)

8.1. INTRODUCTION [1-9]

Our earth is spinning, making a complete turn in 24 h. If the axis of this rotation were in the plane of the sun's rays, half the earth would always be in darkness, half in light. Fortunately for us, the earth does not spin in this manner. Instead, its axis of rotation is roughly perpendicular to the sun's rays, so that most of the surface of the earth is alternately illuminated and in shadow. All organisms have evolved in this changing illumination, which we know as day and night. It is not surprising, then, that they have adapted to this situation. For example, some animals are active only during the day, while others are nocturnal. Perhaps only the bacteria disregard the periodic quality of natural illumination.

Animals more complex than bacteria are, with very few exceptions, able to perceive light. It has been tacitly assumed that organisms first sense a change in

Beatrice M. Sweeney • Department of Biological Sciences, University of California, Santa Barbara, California

the environmental light and then respond appropriately. That diurnal and nocturnal behavior may not be so directly controlled by the environment has only recently been appreciated. We know now that the 24-h periodicity observed in complex organisms can be controlled in a more subtle way. Many of these organisms have been observed to continue their cycles when they are not exposed to alternating day and night. This occurs naturally in nocturnal insects and mammals, e.g., the cockroach and the rat, which spend the day in dark burrows.

Chronobiology, the study of how organisms regulate their functions with respect to time of day and other naturally occurring cyclic changes in the environment, began with the studies of De Mairan in 1729. He was an astronomer who became interested in the periodic behavior of plants. It had been known for centuries that some plants, particularly members of the bean family, close their leaves at night and spread them during the day. De Mairan wondered whether or not light was necessary for leaves to open in the morning. He placed plants in darkness for several days, and found that they continued to open and close their leaves. He realized that, since alternating light and darkness were not required, some other timing device must control leaf movement.

In the last fifty years, accurate and flexible instruments have become increasingly available to biologists. Through the use of isotopes and counters, oxygen electrodes, sensitive recorders, photomultipliers, and other devices, biologists have been able to make quantitative measurements of the activities of their experimental subjects. Many have been surprised to find that the rates at which plants or animals carry out their various physiological processes are often not constant with time. Rates were found to fluctuate, sometimes very greatly, as a function of the time of day when the measurements were made. This was particularly true of organisms in their natural environment, or in simulated day and night in the laboratory. The fluctuations were not random, but showed a cyclic pattern that was often correlated with day and night. In many cases, like the movement of bean leaves, these cyclic variations in rate did not require the alternation of day and night. The fluctuating pattern was still observed when the organisms were transferred to constant darkness or continuous light. Maxima in various rate measurements continued to recur roughly every 24 h under constant illumination (or the lack of it) and constant temperature. Since the period of these cycles was about, but not exactly, 24 h in constant conditions, these phenomena came to be called "circadian" rhythms, i.e., rhythms of "about a day." In this chapter, we shall examine the interaction of light with these circadian rhythms.

8.2. EXAMPLES OF CIRCADIAN RHYTHMS[1–9]

The discovery of circadian rhythms, not only in the rather esoteric bean leaf movement, but also in the rates of obviously important physiological processes, raised the intriguing possibility that many, if not most, organisms have at their disposal information regarding time of day, which is innate and independent of

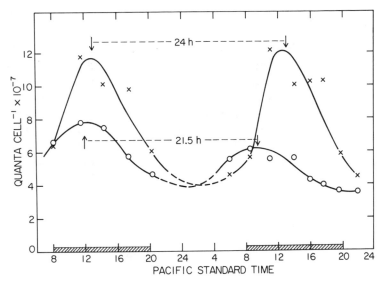

Fig. 8-1. The circadian rhythm of bioluminescence in *Gonyaulax polyedra*. Samples were taken from the same culture, one set left in the light–dark cycle (x), and the other, transferred to continuous cool white fluorescent light (580 lx) 60 h before the beginning of measurements (○). The temperature under both light conditions was 22°C. The environmental dark periods are shown on the abscissa as dark bars. The estimated times of maxima in bioluminescence are indicated by arrows. Note that the period of the culture in a light–dark cycle is 24 h, while that of the culture in continuous light is 21.5 h.

external light and darkness. How widespread is this ability to ''tell time''? How does it work? What is it used for? All these became urgent questions. In this section I shall describe some of these circadian rhythms and how one was discovered.

8.2.1. The *Gonyaulax* Clock[8,9−p.567*]

Gonyaulax polyedra is a marine dinoflagellate, common in the plankton of many coastal waters. It is a photosynthetic unicellular organism, and it also has the interesting property of emitting a flash of blue-green light when given a sharp shake. With the advent of photomultiplier tubes and the accessory circuitry, it became possible to measure the amount of light emitted in this flash. Quite unexpectedly, cells taken from a light–dark cycle gave flashes which varied 100-fold. Bright flashes were observed at night, while during the day cells emitted much less light (Fig. 8-1). If the samples of the *Gonyaulax* culture were now transferred to constant light of an intensity just sufficient to allow photosynthesis to compensate for respiration, the bioluminescent flashes continued to vary in brightness with a cycle of about 24 h (Fig. 8-1). The precise period of the rhythm in bioluminescence was found to depend on the ambient temperature, although

*Reference and page number therein.

the variation in the length of the period with temperature was not great. Interestingly enough, slightly longer periods were found at higher temperatures.

The rhythm in bioluminescence was not the only physiological rhythm observable in *Gonyaulax*. When measurements of the carbon fixation in photosynthesis were made utilizing [^{14}C]bicarbonate, this process also varied with time. The greatest fixation of carbon occurred about midday in a light–dark cycle, and a cycle was still distinct in cells transferred to dim continuous light. This circadian rhythm in photosynthesis could also be measured manometrically by following the capacity of cells to produce oxygen. *Gonyaulax* cells divide only at the end of the night, and this behavior also is not directly dependent on a light–dark cycle. All these rhythms behave as if timed by a single mechanism, one biological clock.

8.2.2. Circadian Rhythms in Other Organisms

So many other organisms have been found to possess circadian rhythms that it is not possible to mention them all. A few examples will suffice to show the diversity of the processes that are cyclic. Adults of the fruit fly *Drosophila* emerge from the pupa, a process called "eclosion," only during the hours near dawn, and not at other times of day. The green flagellate *Euglena* swims into a light beam during the day, but not at night. Some strains of the mold *Neurospora* form spores, called conidia, only for a short time once each day. When this fungus is growing on a solid surface, these dark-colored spores form a visible band daily. Mice are active only at night, as any of you know who have had a pet mouse in a cage with a running wheel that squeaked. All these processes continue for a time to be cyclic in a constant environment, with a period of about 24 h. In general, processes appropriate to a lighted environment are restricted to daytime, while those that are useless or detrimental in light take place only at night.

8.2.3. Circadian Rhythms in Man

Man too is a rhythmic organism. Body temperature, urine excretion, and sleep are all examples of human circadian rhythms. Although humans are not convenient experimental subjects, special rooms providing a constant environment without time cues have been constructed, and human volunteers have lived in these rooms for weeks, even months.[10] The functions of these subjects continued to be cyclic, and the periods were generally considerably longer than 24 h.

Our circadian rhythmicity affects our daily lives more than we sometimes realize. Our body temperature is higher in the afternoon than at other times of the day. Correlated with this is the effectiveness with which we can carry out various tasks, such as solving problems in multiplication and card sorting.[11] We feel pain more intensely, worry more, suffer more from asthma, and are more likely to begin labor or to die in the early hours of the morning. Although few of us rely on the ability to tell time without consulting our watches, many of us have noticed that we often awaken just before the alarm sounds.

Studies with animals have shown that some drugs are much more effective when taken at certain times of day than at other times. By taking advantage of this difference, it may be possible to reduce the dose of many drugs and hence minimize injurious side effects.[11]

As long as travel was relatively slow, we were scarcely aware of crossing time zones. However, jet airplanes now transport us so rapidly that we may experience a complete reversal of day and night in a few hours. On reaching a new time zone, many people feel extreme exhaustion, popularly known as "jet lag," which results from the discrepancy between internal and external time. Not much is known concerning how circadian rhythms in man are brought into synchrony with each other and with time in the environment. Does light reset human rhythms as it does those in other organisms? If we knew more about this, we could plan our trips for the most rapid resynchronization and avoid the inconvenience of jet lag.

8.3. ENTRAINMENT OF CIRCADIAN RHYTHMS TO LIGHT–DARK ENVIRONMENTAL CYCLES[3–5,8]

When circadian rhythms are measured in noncyclic environments, the period is usually found to differ slightly from 24 h. However, in nature or in the laboratory in light–dark cycles 24 h long, the period of circadian rhythms is precisely 24 h. If this were not so, rhythmic processes would quickly become desynchronized with the environment and fail to serve a useful timekeeping function. Thus, the light–dark pattern in the environment must influence the circadian rhythms or "entrain" them, so that they conform to day and night. While circadian rhythms do not require a cyclic environment for their existence, they are clearly not insensitive to such an environment.

Very early in the study of circadian rhythms of various kinds, it became clear that, if an organism was illuminated at night and darkened during the day, its rhythms readjusted to this new light–dark schedule. Processes, which before the transfer had reached their highest values at night, now showed maxima during the new dark period, i.e., during daylight hours in the outside world. Processes normally taking place during the day now occurred during the night, when the organisms were illuminated. The timing of the various rhythmic functions was therefore not unalterably fixed by the earth's rotation, but depended on the light–dark cycles perceived by the organisms.

Experiments in which various rhythmic organisms were exposed to light–dark cycles of lengths shorter or longer than 24 h showed that, under such conditions too, the rhythms readjusted to make their periods exactly match the new environmental cycle, provided that this was not too different from the customary 24 h in nature.

It became apparent that organisms did not learn the new cycles, since, on being returned to continuous conditions, they immediately reverted to their usual approximately 24-h periods, even when the unnatural cycles had been continued for many months. In fact, cycles of light and darkness proved to be unnecessary at any time in the life of an organism for the expression of a circadian rhythm. *Gonyaulax* cultures maintained in bright continuous light for many generations, a condition under which cells are arrhythmic, as we shall see later, become rhythmic as soon as they are transferred to continuous darkness, as if this were the beginning of a normal night. *Drosophila* larvae reared in continuous light also become rhythmic if transferred to continuous darkness. A single short light exposure given to developing eggs of the moth *Pectinophora gossypiella* initiates a circadian rhythm in hatching in the population. We shall return to these points later. Thus, the circadian period is not learned, but is coded in the genetic information of the rhythmic organism. This finding has been confirmed recently; mutations leading to abnormal periods have been induced in the green alga *Chlamydomonas,*[12] in *Drosophila,*[13] and in *Neurospora.*[14]

8.4. SHIFTING THE PHASE OF CIRCADIAN RHYTHMS

8.4.1. Single-Light Exposures

How are circadian rhythms entrained so that they match the light–dark cycles of the environment to which the organism is exposed? Experiments in which organisms received a single light pulse in an otherwise continuously dark environment have provided information regarding this problem. When cultures of *Gonyaulax* are transferred from an environmental light–dark cycle to continuous darkness, the circadian rhythm of bioluminescence continues to be evident for 3–5 days before the organisms die of starvation in the absence of photosynthesis. If a portion of such a darkened culture is exposed to bright light for 3 h while the luminescence is increasing,[9–p.567] subsequent maxima in luminescence will all occur *later* than in cultures unexposed to light (Fig. 8-2). However, if the light pulse is given at the time when luminescence is brightest or when luminescence is decreasing, the subsequent maxima in luminescence will occur *earlier* than in the control in continuous darkness (Fig. 8-2). Light pulses administered to cells during the time when luminescence is low, that part of the cycle normally occurring in the day, will have little or no effect on the time of subsequent maxima in bioluminescence. These findings can be summarized in a graph in which the advance or delay of the rhythm in hours is plotted as a function of the time when the light pulse was given. Such a curve is known as a "phase response curve." Phase response curves for the phase shifts following a single bright light pulse for different organisms are surprisingly similar in general shape (Fig. 8-3). All such curves show little effect of light during the part of the rhythm normally occurring during the day, but show much greater phase shifts at night. The sudden change from delay to advance that occurs in the middle of the part of the

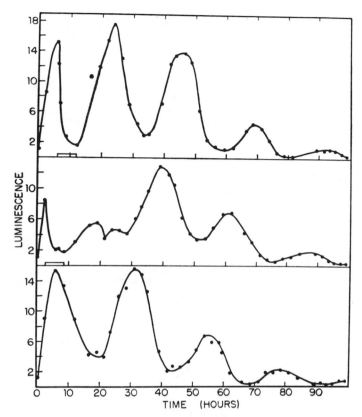

Fig. 8-2. A resetting experiment using the bioluminescence rhythm in *Gonyaulax polyedra*. The rhythm in continuous darkness is shown in the bottom curve. The other curves show the consequence of exposing cells to cool white fluorescent light (14,000 lx) for 6 h when luminescence is increasing (middle curve, and when luminescence is decreasing (top curve). A phase delay is shown in the middle curve, and a phase advance in the upper curve. (From reference 9, p. 567.)

cycle normally associated with night is a common feature of all phase response curves. The similarity between phase response curves in such different creatures as *Drosophila*,[17] the succulent plant *Kalanchoe*,[16] *Gonyaulax*,[8] and *Neurospora*[18] can hardly be a coincidence. It must represent a general property of the timing process.

With regard to phase shifting, different organisms vary widely in their sensitivity to light. An intensity that is dim to *Gonyaulax* may appear very bright to a *Drosophila* larva.

8.4.2. The Spiral-Resetting Surface and the Point of Singularity [20,21]

Phase response curves such as those in Fig. 8-3 each represent the effect of a single exposure to light, usually a relatively high fluence, given at different times. To show the effect of changing the fluence as well as the time of exposure on the

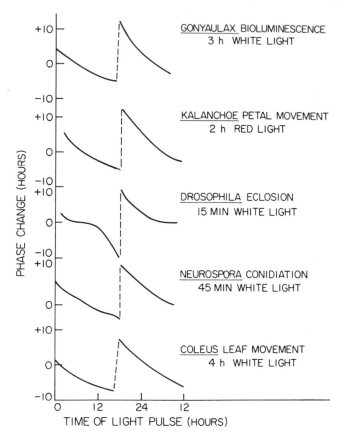

Fig. 8-3. Phase response curves for five circadian rhythms in five different organisms. Note that phase advances are considered as positive and plotted upward, while delays are negative. Dotted lines connect the delaying and advancing portions of the curves. The time at which cells received light (abscissa) is taken as the middle of the light exposure. The data for *Gonyaulax* are replotted from the data of Hastings and Sweeney[8,9–p.567,15], for *Kalanchoe* from Zimmer[16], for *Drosophila* from Pittendrigh[17], for *Neurospora* from Sargent and Briggs[18], and for *Coleus* from Halaban.[19]

subsequent phase of a rhythm, we need a three-dimensional representation. Winfree[20,21] has collected a large amount of information concerning the effects on the *Drosophila* eclosion rhythm of different fluences of light given at different times. This feat is possible because the adult flies emerging from the pupa can be collected every hour automatically in complete darkness. Using his data, Winfree has constructed a three-dimensional model showing the effects on rhythmicity of both light dosage, and the time of the light exposure. On the third axis he plotted the time after the light exposure when the most adult flies emerged. The solid figure that was obtained in this way turned out to be a spiral surface, two side views of which are shown in Fig. 8-4. The axis of this spiral rests on a point representing an S of 50 s and a T of 6.8 h, which Winfree called the "point of singularity." The time, 6.8 h after the beginning of darkness, is the same time

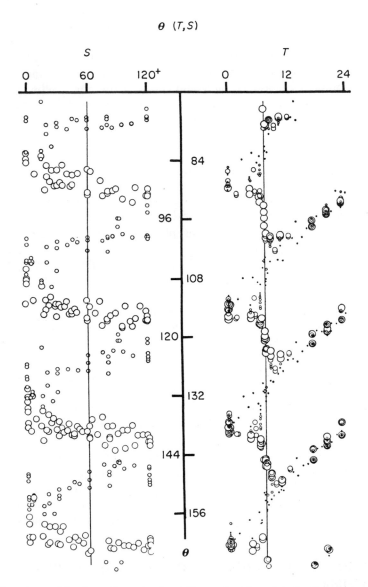

Fig. 8-4. Two views of the three-dimensional resetting surface from data for the eclosion rhythm in populations of *Drosophila pseudoobscura*. The symbols are: S, light dosage; T, time of light exposure; θ, time (h) after light exposure when most adult flies emerged. On the left, the projection shows points $T > 7.3$ h as smaller, giving a three-dimensional effect. T from hour 15 to hour 24 is deleted for clarity. The right-hand projection shows data at $S = 0$–30 s as dots, 31–60 s as tiny circles, 61–90 s as larger circles, and longer values of S as the largest circles. The light used was blue, of about 10 μW cm^{-2} intensity. (From reference 20.)

when the phase shift by a bright light exposure changes from a delay to an advance (see Fig. 8-3). As the spiral-resetting surface winds upward, it does so around this point, making one turn about every 24 h. All the lines drawn along one value for θ converge at this axis at $S = 50$ s. Winfree predicted that if a light exposure of exactly 50 s was given to a population at a time as close as possible to 6.8 h, i.e., the point of singularity, each fly in a population might have any value of θ, and thus the population would appear arrhythmic. When he tried this experiment, he found that rhythmicity was abolished by such a treatment. A point of singularity has since been found in the petal movement rhythm in *Kalanchoe*.[22]

8.4.3. An Explanation of Entrainment

The behavior of rhythms in various environmental light–dark cycles can be understood in terms of the responses to single light exposures. That light falling on a rhythmic organism at night can shift phase by as much as 12 h accounts for the readjustment of rhythmic processes when the day and night are artificially reversed. The different sensitivity to light at different times also explains how circadian rhythms can adjust to different environmental light–dark cycles, including those in which the light and dark portions of the cycle are not of the same length. Unaccustomed light, falling on a rhythmic organism in the last part of the night, causes a phase advance in rhythmicity. Thus, in light–dark schedules with dark periods shorter than 12 h, the phase is brought forward in time in each cycle. This simulates a short period. In environmental cycles with light periods longer than 12 h, if light falls on the organism in the early part of the night phase of the rhythm it delays the next maximum, thus lengthening the apparent period. The combined effect of light early and late in the night phase in a long day–short night schedule restricts the night phase by first delaying, then advancing phase.

When two consecutive light pulses are given within a few hours, the effect of the second pulse shows that the underlying rhythm has been reset rapidly by the first. However, the final phase of a rhythm, particularly in a multicellular organism, may not be obtained at once, but only after several cycles of transients; entrainment to a new light–dark cycle may not occur immediately.

8.5. ACTION SPECTRA FOR SHIFTING THE PHASE OF CIRCADIAN RHYTHMS

It is a general principle of photobiology that, to exert an effect, light must be absorbed. It is sometimes possible, therefore, to identify the photoreceptor molecule by determining an action spectrum for an effect if it is recognizable as the absorption spectrum of a known molecule (see Sections 1.5.4. and 3.7.) It is of great interest to discover which photoreceptors are responsible for shifting the phase of circadian rhythms, particularly because very little is known concerning the biochemistry of rhythmic systems. Do all organisms utilize the same photoreceptor molecule for this purpose? When more than a single process is rhythmic in one organism, do they share the same photoreceptor? Is the same photoreceptor responsible for advancing and delaying phase? Action spectra might provide an answer to these questions. Relatively detailed action spectra have been obtained for only a few rhythms. Phase advances and phase delays of the eclosion rhythm of *Drosophila*[23] are brought about only by blue light of wavelengths shorter than about 500 nm, but the action spectra do not show any distinct peaks between 350 and 500 nm that might suggest the nature of the photoreceptor (Fig. 8-5). The rhythms of egg hatching, eclosion, and egg laying in the cotton moth *Pectino-*

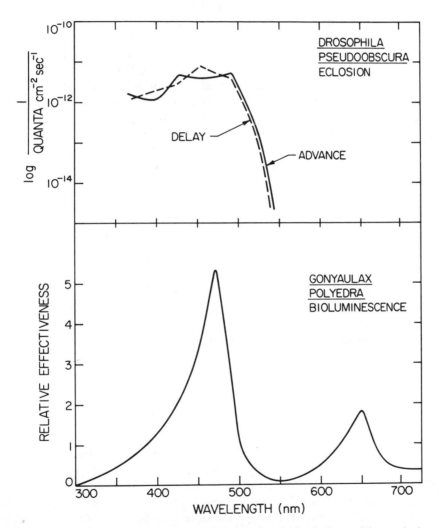

Fig. 8-5. Action spectra for shifting the phase of the eclosion rhythm in *Drosophila pseudoobscura*, and the bioluminescence rhythm in *Gonyaulax polyedra*. While the wavelength specificity and threshold intensity is the same for delaying and advancing phase shifts in *Drosophila*, they differ markedly from the wavelengths active in shifting phase in *Gonyaulax*, indicating a different photoreceptor in these two organisms. The action spectra for *Drosophila* are replotted from the data of Frank and Zimmerman,[23] for *Gonyaulax*, from Hastings and Sweeney.[15]

phora[24] are all very similar. Blue light of wavelengths shorter than 500 nm are effective (Fig. 8-5). In *Pectinophora,* the rhythms of egg hatching, eclosion, and egg laying are all entrained by cylces of blue light (480 nm) and darkness, but not in red light (600 nm). Phase advances of the bioluminescence rhythm in *Gonyaulax*[15] can be brought about by both blue and red light (Fig. 8-5), the action spectrum showing very sharp peaks at 470 and 650 nm. However, shading of the photoreceptor by chlorophyll in both these spectral regions may distort the action spectrum, so that it does not truly represent the absorption spectrum of the photoreceptor. It is clear that chlorophyll is not the photoreceptor. Red, but not blue light advances the phase of the rhythm of CO_2 uptake in leaves of the succulent plant *Bryophyllum.*[25] The leaves of the familiar house plant *Coleus*[19] move up and down rhythmically. Red light is effective during the part of the cycle when light advances phase, while delays are brought about by blue and not red light. There is also a leaf movement rhythm in the garden bean, *Phaseolus*. The phase response curve in this plant can be reproduced by red light irradiation, which causes both delays and advances when the pulvinus at the base of the leaf, where movement takes place, is irradiated.[26] It is clear, then, that the photoreceptors for phase-shifting rhythms are different in different organisms. The effectiveness of blue light in phase shifting all the rhythms in *Pectinophora* suggest that each organism may have only one photoreceptor system for this purpose, but confirmation of this point is required from experiments with other organisms. None of the photoreceptor molecules in rhythms have so far been identified. There is strong evidence, however, that in *Drosophila*[27] the photoreceptor is not a carotenoid, since carotene starvation does not alter light sensitivity.

The rhythms of bioluminescence and cell division in *Gonyaulax* can be phase-shifted by ultraviolet (UV) as well as by visible light.[28] Short exposures (2–4 min) to UV radiation (24 J m^{-2} at 254 nm) are sufficient to reset these rhythms by as much as 9 h if given at night, while during the day part of the circadian cycle only a small phase shift can be induced. All phase shifts after UV irradiation are advances, so that the phase response curve is quite different from that for visible light. The rhythm of sexual fusion in the protozoan *Paramecium*[29] can also be reset by UV radiation.

8.6. LIGHT AND THE INITIATION OF SOME CIRCADIAN RHYTHMS

As mentioned above, some organisms are arrhythmic if they are reared in complete darkness, but rhythmicity begins after a short illumination. This is true of the hatching rhythm in *Pectinophora,*[30] and the conidiation rhythm in *Neurospora crassa* strain "timex."[18] Action spectra for initiation of rhythmicity have been determined in both organisms, and blue light between 400 and 500 nm is effective (Fig. 8-6). These action spectra show some structure, with small peaks at about 420 and 460 nm. In both action spectra, there is activity at wavelengths as short as 350 nm. Action spectra that peak in the blue region are particularly difficult to identify with a specific pigment, since a number of carotenoids and

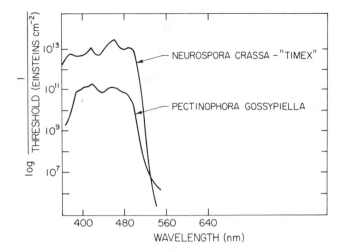

Fig. 8-6. Action spectra for initiating a rhythm of conidiation in the mold *Neurospora crassa* strain "timex," and the rhythm of egg-hatching in the moth *Pectinophora gossypiella*. *The Neurospora* action spectrum is replotted from the data of Sargent and Briggs[18] and that for *Pectinophora* is from the data of Bruce and Minis.[30] The two action spectra are clearly very similar.

flavins absorb in the blue region of the spectrum. Recently, Munoz *et al.*[31] have examined whole cells and extracts of *Neurospora* in the blue region of the spectrum to discover whether or not bleaching of a pigment occurs on illumination. They observed the photoreduction of a *b*-type cytochrome mediated by a flavin photoreceptor, which absorbed at 460, 410, and 360 nm, and could be replaced by flavin mononucleotide or flavin adenine dinucleotide. They therefore suggested that a flavin is the photoreceptor for the responses of rhythms to blue light.

The action spectra for initiation of rhythmicity in *Pectinophora* and *Neurospora* are remarkably similar to that for changing phase in *Drosophila*. However, this may be a coincidence, and may not imply that initiation and rephasing of rhythms have a common mechanism.

8.7. STOPPING CIRCADIAN RHYTHMS WITH BRIGHT LIGHT

Bright light has been observed to stop a number of rhythms, so that this may be a common property of all circadian systems. Not only do rhythms come to a stop in bright continuous light, they do so at a fixed phase point that is associated with the end of a 12-h light period. When cells are removed from this bright illumination, either to darkness or weak light, the rhythm begins at once, as if this time represented the beginning of normal night. The action spectrum for this effect is not known. Thus, we cannot tell whether or not this effect of intense continuous light is an extreme case of the phase delay observed when light pulses are given at the beginning of a subjective night.

8.8. LIGHT EFFECTS ON THE PERIOD OF CIRCADIAN RHYTHMS

8.8.1. Period as a Function of Light Intensity[32-34]

One other effect of light on circadian rhythms remains to be discussed: the intensity dependence of the period in continuous illumination. Since bright light inhibits rhythms altogether, "bright" being a relative term and differing from organism to organism, the range of intensities over which observations of the period in continuous light have been made is short. Despite this limitation, it is known that in many different organisms the period is slightly dependent on light intensity (Fig. 8-7). Rhythms in some organisms have shorter periods in brighter light, while the reverse is true in others.

8.8.2. Autophasing

Brown[35] has advanced the notion that the differential light response, best seen in the phase response curves (Fig. 8-3), is responsible for the observation that the circadian periods in constant environments are usually not exactly 24 h. Brown believes that circadian rhythms are responses to an unknown celestial factor that varies with the earth's rotation with a period of 24 h. Although the environmental light intensity is constant, the sensitivity of the organism to light varies over a cycle, as shown by the phase response curves. The apparent period, then, represents an integration of the phase response curve. If this "autophasing" hypothesis holds, the period at a given temperature in constant light should be shorter than 24 h when the phase response curve contains more advancing than delaying phase shifts, and vice versa. However, in *Gonyaulax*, the period in constant light at 24°C is longer than 24 h, but in the phase response curve at 24°C, and several light intensities, advances dominate delays, i.e., the

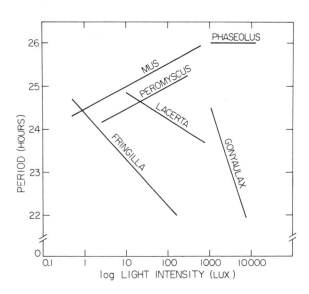

Fig. 8-7. The length of the period of circadian rhythms in a variety of organisms in continuous light of different intensities, illustrating Aschoff's contention that, as light intensity increases, the period of circadian rhythms of diurnal animals, such as the lizard *Lacerta*, shortens, while that of nocturnal animals, such as *Mus* and *Peromyscus*, lengthens. (From Tables 115, 116, and 117 in reference 34.)

integral is positive (R. Christianson, unpublished results in the author's laboratory). If autophasing were taking place, the period should be shorter than 24 h. It therefore seems more likely that periods observed in constant conditions are the natural periods of the biological clocks, and an exactly 24-h period is achieved through phase shifts.

8.9. LIGHT AND THE MEMBRANE MODEL FOR THE CIRCADIAN CLOCK

It would be satisfying if we were able to describe in molecular terms the way in which light stops, starts, and changes the phase of the circadian rhythms. Unfortunately this is not yet possible, since neither the photoreceptor nor the biochemistry of biological clocks is well understood. A recent model, the membrane hypothesis,[36,37] may provide a clue. According to this model, the "clock" consists of a feedback loop with two components: an ion pump incorporated into intracellular membranes, and the transmembrane distribution of the ion being pumped. The pump will function until the difference in the ion level on the two sides of the membrane reaches a critical magnitude. It will then be shut off. Since the membranes allow a certain amount of passive diffusion down a concentration gradient, the ion concentrations on either side of the membrane in question will gradually equalize. Then the pump will be activated again. Illumination of photoreceptors in the plasma membrane has been shown to alter the resting potential of some cells, e.g., *Acetabularia*.[38] This is evidence for a light-induced change in the permeability of the membrane. Any such change would alter the state of the postulated clock feedback loop. A phase advance would be expected when the pump is not operating, while a phase delay would be observed if illumination occurred when the pump was active. Until we know whether pumps sensitive to ion distribution exist, what the ions are, and more about the effect of light in various systems on ion permeability, we will not be able to explain the effects of light on circadian rhythms with any certainty.

8.10. CONCLUSIONS

We have seen that light in the visible region of the spectrum can affect circadian rhythms in at least four distinct ways. A light–dark environmental cycle, either natural or artificially imposed, can entrain these rhythms so that their periods match the environmental periodicities. This is accomplished by a series of single phase shifts in response to light. Single light exposures shift the circadian rhythms differently when the exposures occur at different times. Rhythms in very diverse organisms give the same kinds of phase response curves to light that they perceive as bright, and perhaps also to weak light, although there are insufficient data to be sure of this. Light can start a rhythm in some organisms when they have been reared in darkness, and are arrhythmic. Light can also stop circadian rhythms, and the stopping point is the end of the day. Finally, light can alter the period of circadian rhythms in constant conditions.

Many circadian rhythms respond to blue light, but some, notably in plants, can be phase-shifted by red light. Photoreceptor molecules are apparently not the same in different organisms, but none has been identified. The photoreceptor in *Drosophila* is not a carotene, and in plants it is not a chlorophyll. At least two circadian rhythms can be phase-shifted by UV radiation.

Light interacts powerfully with all circadian systems. Thus, we may be reasonably sure that the photochemical reaction set in motion when photoreceptors are irradiated is closely associated with clock biochemistry. Yet, photoreactions, as probes in studying intracellular circadian oscillators, have hardly been exploited. The identification of the photoreceptor molecules and the reactions initiated by their irradiation may well be the most fruitful approach to understanding this baffling phenomenon.

Understanding circadian rhythms is of theoretical interest, but it also has a practical side. In humans, mental acuity varies with the time of day, just as do body temperature, sleep, wakefulness, and hormone levels.[11] If human rhythms can be phase-shifted efficiently, then pilots might fly during the night without worrying that they might not be at their best to handle an emergency. Workers on night shifts might increase their productivity and contentment. Is it better to change from night shift to day shift often or infrequently? What cues might be employed to hasten readjustment of body rhythms when shifts are reversed? The answers to these and other similar questions may be found from studies of the effects of different light regimes on human circadian rhythms.

Evidence is accumulating that sensitivity to many drugs varies according to a circadian rhythm, much smaller doses being effective at one time of day than at another. This is true of animals, including agricultural pests, and of plants. If sprays are applied at the most effective time, much smaller doses may be used. Hence, environmental pollution by pesticides and herbicides may be reduced. The application of our knowledge of circadian rhythms to problems such as these is just beginning, and may lead to some surprising and unpredicted improvements in the quality of life.

8.11. EXPERIMENTS

Leaf "Sleep Movements" in the String Bean *Phaseolus vulgaris*.

Part I. Use a young (2–3 weeks old) bean plant with the first two leaves well expanded. Since the measurements of a circadian rhythm require more than a few hours' time, take the bean plant home with you, and put it on the window sill or in some propitious place with as much light as possible. Water this plant whenever the pot is dry. Turn it around every day or so in order to prevent too much phototropic bending if the light is unidirectional.

Measure the position of the two primary leaves as often as convenient for 2 days, but at least when you get up, around lunch time, and just before going to bed. Make the measurements as shown below, using a protractor or other device for measuring angles:

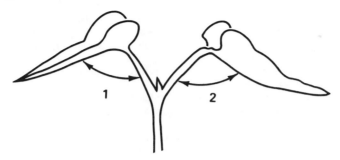

Part II. In the evening, at the end of the 2 days indicated in Part I, place the bean plant in constant darkness or in continuous artificial light for 2 days and continue measurements.

Part III. Try exposing your plant to sound or other environmental changes while in continuous light. Record and plot your measurements as a function of time in hours. Write a short report of your findings (see reference 8, pp. 1–12, 19–21 and reference 9, pp. 507–530).

8.12. REFERENCES

1. J. Aschoff (ed.), *Circadian Clocks,* North-Holland, Amsterdam (1965).
2. F. A. Brown, Jr., J. W. Hastings, and J. D. Palmer, *The Biological Clock: Two Views,* Academic Press, New York (1970).
3. E. Bünning, *The Physiological Clock,* 3rd ed., Springer-Verlag, New York (1973).
4. *Proc. Int. Symp. Circadian Rhythmicity, Wageningen, 1971,* Centre for Agricultural Publication and Documentation, Wageningen, The Netherlands (1972).
5. A. Chovnick (ed.), Biological clocks, *Cold Spring Harbor Symp. Quant. Biol.* **25** (1960).
6. M. Menaker (ed.), *Biochronometry,* National Academy of Science, Washington, D.C. (1971).
7. J. D. Palmer, *Biological Clocks in Marine Organisms,* Wiley-Interscience, New York (1974).
8. B. M. Sweeney, *Rhythmic Phenomena in Plants,* Academic Press, London (1969).
9. R. B. Withrow (ed.), *Photoperiodism and Related Phenomena in Plants and Animals,* Publication No. 55, AAAS, Washington, D.C. (1959).
10. R. Wever, Internal phase angle differences in human circadian rhythms: Reasons for alteration and measurement problems, *Int. J. Chronobiol.* **1,** 371–390 (1973).
11. J. D. Palmer, *An Introduction to Biological Rhythms,* Academic Press New York (1976).
12. V. G. Bruce, Mutants of the biological clock in *Chlamydomonas reinhardi, Genetics* **70,** 537–548 (1972).
13. R. J. Konopka and S. Benzer, Clock mutants of *Drosophila melanogaster, Proc. Natl. Acad. Sci. USA* **68,** 2112–2116 (1971).
14. J. F. Feldman and N. M. Waser, New mutations affecting circadian rhythmicity in *Neurospora,* in: *Biochronometry* (M. Menaker, ed.), pp. 652–656, National Academy of Science, Washington, D.C. (1971).
15. J. W. Hastings and B. M. Sweeney, The action spectrum for shifting the phase of the rhythm of luminescence in *Gonyaulax ployedra, J. Gen. Physiol.* **43,** 697–706 (1960).
16. R. Zimmer, Phasenverschiebung und andere Störlichtwirkungen auf die endogen tagesperiodischen Blütenblattbewegungen von *Kalanchoë blossfeldiana, Planta* **58,** 283–300 (1962).

17. C. S. Pittendrigh, Circadian rhythms and the circadian organization of living systems, *Cold Spring Harbor Symp. Quant. Biol.* **25**, 159–184 (1960).
18. M. L. Sargent and W. R. Briggs, The effects of light on a circadian rhythm of conidiation in *Neurospora, Plant Physiol.* **42**, 1504–1510 (1967).
19. R. Halaban, Effects of light quality on the circadian rhythm on leaf movement of a short-day plant, *Plant Physiol.* **44**, 973–977 (1969).
20. A. T. Winfree, Integrated view of resetting of circadian clock, *J. Theor. Biol.* **28**, 327–374 (1970).
21. A. T. Winfree, On the photosensitivity of the circadian time-sense in *Drosophila pseudoobscura, J. Theor. Biol.* **35**, 159–189 (1972).
22. W. Englemann, H. G. Karlsson, and A. Johnsson, Phase shifts in the *Kalanchoe* petal rhythm, caused by light pulses of different duration, *Int. J. Chronobiol.* **1**, 147–156 (1973).
23. K. D. Frank and W. F. Zimmerman, Action spectra for phase shifts of a circadian rhythm in *Drosophila, Science* **163**, 688–689 (1969).
24. C. S. Pittendrigh, J. H. Eichhorn, D. H. Minis, and V. G. Bruce, Circadian systems. VI. Photoperiodic time measurement in *Pectinophora gossypiella, Proc. Natl. Acad. Sci. USA* **66**, 758–764 (1970).
25. M. B. Wilkins, An endogenous rhythm in the rate of CO_2 output of *Bryophyllum.* II. The effects of light and darkness on the phase and period of the rhythm, *J. Exp. Botany* **11**, 269–288 (1960).
26. E. Bünning and I. Moser, Response-Kurven bei der circadian Rhythmik von *Phaseolus, Planta* **69**, 101–110 (1966).
27. W. F. Zimmerman and T. H. Goldsmith, Photosensitivity of the circadian rhythm and of visual receptors in carotenoid-depleted *Drosophila, Science* **171**, 1167–1169 (1971).
28. B. M. Sweeney, Resetting the biological clock in *Gonyaulax* with ultraviolet light, *Plant Physiol.* **38**, 704–708 (1963).
29. C. F. Ehret, Action spectra and nucleic acid metabolism in circadian rhythms at the cellular level, *Cold Spring Harbor Symp. Quant. Biol.* **25**, 149–158 (1960).
30. V. G. Bruce and D. H. Minis, Circadian clock action spectrum in a photoperiodic moth, *Science* **163**, 583–585 (1969).
31. V. Muñoz and W. L. Butler, Photoreceptor pigment for blue light in *Neurospora crassa, Plant Physiol.* **55**, 421–426 (1975).
32. J. Aschoff, Exogenous and endogenous components in circadian rhythms, *Cold Spring Harbor Symp. Quant. Biol.* **25**, 11–28 (1960).
33. J. Aschoff, Response curves in circadian periodicity, in: *Circadian Clocks* (J. Aschoff, ed.), pp. 95–111, North-Holland, Amsterdam (1965).
34. *Biology Data Book,* 2nd ed., Vol. 2, pp. 1016–1039, Federation of American Societies for Experimental Biology, Bethesda, Md. (1973).
35. F. A. Brown, Jr., The "clocks" timing biological rhythms, *Am. Sci.* **60**, 756–766 (1972).
36. D. Njus, F. M. Sulzman, and J. W. Hastings, Membrane model for the circadian clock, *Nature* **248**, 116–120 (1974).
37. B. M. Sweeney, A physiological model for circadian rhythms derived from the *Acetabularia* rhythm paradoxes, *Int. J. Chronobiol.* **2**, 25–33 (1974).
38. H. D. W. Saddler, The membrane potential of *Acetabularia mediterranea, J. Gen. Physiol.* **55**, 802–821 (1970).

9

Extraretinal Photoreception

9.1. INTRODUCTION

Light plays many roles in the lives of organisms other than that which we normally associate with the visual perception of patterns in the environment. Therefore, it is useful to distinguish between those situations in which (1) light is important to organisms primarily because of its energy content, and (2) light functions as a signal, stimulus, or "trigger" and, from a biological viewpoint, energetic considerations are secondary.

The clearest example of the former is surely photosynthesis (Chapter 13), in which it is specifically the energy of light that the photosynthetic organism is "after"; indeed, the photosynthetic machinery is exquisitely evolved to capture and utilize this energy with maximum efficiency. On *a priori* grounds, one might suppose that the chlorophyll molecules utilized by plants to capture the energy of sunlight are also employed to perceive sunrise and sunset. As we shall see, this assumption, although plausible, is incorrect.

The most familiar signaling function of light is pattern vision. In this activity the amount of light energy involved is of relatively minor importance. Most eyes have not evolved to maximize the amount of light absorbed; in fact, in most

Michael Menaker • Department of Zoology, The University of Texas, Austin, Texas

visual situations, structures in the eye operate to control and often reduce the amount of energy reaching the pigment molecules. Spatial and temporal patterning, small variations in intensity, and differences in wavelength are of much greater importance to the visual process than are large-scale changes in the energy contained in the stimulating light. It is not quite so obvious that these same considerations apply to a variety of other processes in which light plays a signaling, stimulating, or triggering role. These processes include photomorphogenesis (Chapter 11), photomovement (Chapter 12), circadian rhythms (Chapter 8), and reproductive cycles (below), and can reasonably be said to involve photoreception, as distinct from the simple absorption of light quanta by photopigments.

It is characteristic of photoreception that the biological activities initiated by it require the processing of energy in amounts completely out of proportion to the energy in the light stimulus itself. Thus, the number of joules required to escape from the tiger that you perceive to be stalking you is infinitely greater than that involved in conveying his image to your retina and bears no necessary relationship to whether the event is illuminated by the moon or by the noonday sun. It is less evident but equally true that a plant bending toward the light or flowering in response to long days is processing energy in amounts that are unrelated to the energy contained in the triggering light stimuli. On the other hand, the amount of carbon fixed in photosynthesis, the number of cells killed by ultraviolet radiation, and, to a lesser extent, the degree of tanning of human skin, depend much more directly on the number of quanta in the light that produces these effects. All these processes are mediated by photopigments; however, the mechanism of photoreception is mediated by photopigments usually contained within photoreceptors— relatively complex organelles or organs comprising a larger system which functions to amplify the biological response to light.

Having drawn the above distinction, it becomes clear that a single organism might well be expected to possess many photopigments, some of which function within photoreceptors, while others are involved primarily in energy transfer. Hence, there is no reason to assume that green plants must perceive the sunrise with their chlorophyll molecules, and, as described in Section 8.5., they do not. The circadian rhythms of green plants, as of most organisms, are entrained (synchronized) by the natural light cycle, and the photopigment utilized in this response to light, although it has not yet been identified, is clearly not chlorophyll. This is true even for entrainment of the rhythm of the photosynthetic capacity.

The possession of multiple photopigments, and also of multiple photoreceptors that mediate different responses to light, is the rule among organisms. This should not be surprising in view of the above discussion, but our own subjective photoreceptive experience depends so heavily on our eyes that we are usually surprised to learn that most highly evolved organisms with complex image-forming eyes have, in addition, other much less obvious photoreceptors that are used to monitor crucially important aspects of the photic environment. The basic observation that leads to this conclusion is that even though the eyes of an animal have been surgically removed or its optic tracts severed, it still responds in a variety of ways to light in the visible portion of the spectrum. In order to counteract our own pattern vision bias, let us consider the kinds of information about the environment that are available to an organism that has only simple photoreceptors incapable of forming images. The discrimination of day from night enables the entrainment of the circadian system and the consequent adaptive regulation of the timing of many physiological and behavioral activities. The seasons of the year are most unambiguously signaled by the regular annual change in day length, which controls annual reproductive cycles, some aspects of migration, diapause, and other important physiological responses in a wide variety of organisms. (This phenomenon is called photoperiodism.) The intensity of the light to which an organism is exposed will provide cues concerning the cloud cover, whether the organism is in full light or in shadow, the time of day, and even, if moonlight can be perceived, the phase of the lunar cycle. Simple photoreceptors are also capable of detecting the spectral composition of light that impinges on them, and, crudely, the direction from which it comes. These latter capacities improve rapidly with slight increases in complexity of the receptor structure.

Although plants and various simple animals extract many kinds of information from the photic environment, utilizing very simple photoreceptive structures, the recent discovery that vertebrates and complex invertebrates with image-forming eyes also have extraretinal photoreceptors was surprising, even to most photobiologists. After all, granted that it does not require an eye to make such responses, it would seem that once eyes had evolved as a result of selection pressure for the perception of images, they should also serve perfectly well for these other simpler tasks.

9.2. EXTRARETINAL PHOTORECEPTION IN INVERTEBRATES

It has been experimentally demonstrated that many invertebrate species make use of extraretinal photoreception. A few examples will suffice to illustrate the diversity of activities controlled in this way.

Because of their large size and the relative simplicity of performing surgery on them, the giant silk moths have long been a favorite experimental subject of insect endocrinologists. Considerable progress has been made in understanding the hormonal basis of developmental events such as the induction and termination of diapause (a state of greatly reduced metabolic activity that allows the insect to overwinter), which are regulated by environmental light cycles. Several of the hormones involved in the control of diapause and related events are known to be produced and/or stored in the brain and other central nervous system structures. The photoreceptors that mediate these responses are also located in the brain.

In an ingenious experiment, Williams and Adkisson[1] first established that the head end of the diapausing silk moth pupa contained the photoreceptors used by the animal to perceive the long days that trigger the onset of adult development (the termination of diapause). They then showed that simply by transplanting the brain from its normal site in the head to a new location in the tip of the abdomen, photosensitivity was transferred from the anterior to the posterior half of the animal. In a similar experiment, Truman and Riddiford[2,3] demonstrated that the brain of silk moths contains photoreceptors that synchronize the circadian rhythm of eclosion (emergence of the adult moth from the pupal case) with the light–dark cycle, and also the clock that times the event.

There is extensive literature on extraretinal photoreception in the control of insect photoperiodism and circadian rhythms. Indeed there are so many examples of these phenomena in diverse groups of insects that the few well-documented cases in which photoreception involves the exclusive use of the compound eyes and ocelli (e.g., cockroaches and crickets) are of particular interest.[3]

Some time ago, investigators studying the ventral nerve chord of crayfish with electrophysiological techniques observed that changes in the electrical activity of fibers in the chord could be produced by illuminating the 6th abdominal ganglion, but not by illuminating any of the other ganglia. Further work showed that the 6th abdominal ganglion contains a single pair of photoreceptive neurons. The action spectrum for these photoreceptors indicates that the photopigment involved is a rhodopsin very similar to that which mediates photoreception by the crayfish eye.[4] Cave crayfish, which no longer have functional eyes, retain these so-called caudal photoreceptors. Their function remains unknown. Apparently they are not involved in the entrainment of circadian rhythms, although there is reason to believe that they may exert some influence on locomotor behavior. On the other hand, crayfish have other extraretinal photoreceptors, not yet precisely localized (but probably in the supraesophageal ganglion), which are involved in entrainment of the circadian rhythm of locomotion. Page and Larimer[5] have also studied a circadian rhythm in the electroretinogram (ERG) amplitude of the crayfish eye. They have found that even this rhythm, which is, after all, a rhythm of sensitivity to light of the retinal elements themselves, can be synchronized via extraretinal photoreceptors probably located in the brain.

9.3. EXTRARETINAL PHOTORECEPTION IN VERTEBRATES

Extraretinal photoreception by some birds and fish has been documented for over 50 years; however, it is only recently that the extent, complexity, and importance of the phenomenon have become apparent. At least some species in all five vertebrate classes are known to regulate important aspects of their physiology and behavior with reference to extraretinally perceived light. Least is known about this phenomenon in fish (because there has been very little work) and in mammals (because it appears that extraretinal photoreception may be limited to newborn animals[6]). The above is true even if one does not take into account pineal photoreception, about which only a word can be said here.

The pineals and, when present, associated structures such as the parietal eye and frontal organ of all poikilothermic (cold-blooded) vertebrates are probably photoreceptive. Solid electrophysiological evidence that this is so exists for some fish, amphibians, and reptiles. The photoreceptive structures of the pineal complex range from scattered simple outer-segmentlike organelles to highly ordered arrays of photoreceptors that look much like a simple retina and, like the retina of the lateral eye, are served by a lens and are connected with the brain by a substantial nerve tract.[7] We do not yet know whether the avian pineal is photoreceptive, but we are fairly certain that the pineal of most adult mammals is not.

We know more about extraretinal photoreception in the house sparrow *(Passer domesticus)* than in any other vertebrate species. In order to convey some idea of the complexity of this sensory capacity, the remainder of this chapter will be devoted to a detailed description of the phenomenon in that species, with occasional reference to other species for comparative purposes.

9.4. EXTRARETINAL PHOTORECEPTION IN THE HOUSE SPARROW

9.4.1. The Effects of Light on the Circadian Clock: Entrainment

Circadian rhythms of locomotor activity can be most easily assayed in perching birds, such as *Passer,* by continuously recording the switch closures produced when the bird hops on a perch. In this way, one can automatically

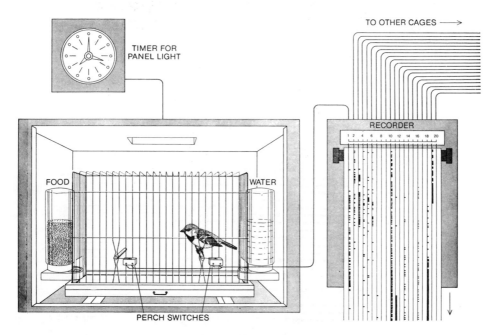

Fig. 9-1. Arrangement for the continuous recording of perching behavior in sparrows. The bird's weight depresses the switch connected to the perch and electrically activates a single pen (of a 20-pen recorder), which writes on a strip of paper moving at 18 inches per day. Each day's record is separated from those of other birds, and each bird's daily records are displayed to yield cumulative data of the kind shown in subsequent figures. While such data are only semiquantitative with respect to amount of activity, they give a very accurate picture of the temporal distribution of activity over long periods of time. (From reference 8; Copyright 1972 by Scientific American, Inc. All rights reserved.)

record the temporal distribution of activity for weeks or months (Fig. 9-1). In the presence of a light cycle, the birds are normally active primarily when the light is on, beginning their activity at very nearly the same time each day (if the light–dark cycle has a period of exactly 24 h). They are said to be entrained (or synchronized) to the light cycle (see Chapter 8). When placed in constant darkness, the birds' locomotor activity remains conspicuously rhythmic, but since it no longer has access to an environmental cue, it "free-runs" with a period close to, but different from, 24 h. In practice, most sparrows have free-running periods longer than 24 h and begin their activity progressively later each day. The difference between entrained and free-running locomotor rhythmicity is dramatically clear in the raw perching data (Fig. 9-2). Since the bird can only entrain to light cycles it can perceive, entrainment provides an unambiguous and easily recorded assay of photoreception.

Sparrows continue to entrain to light cycles even after their eyes and pineal organ have been surgically removed, and many carefully controlled experiments have established that it is visible light rather than some other periodic feature of the regime, such as heat, noise, or electrical fields, to which they are responding.

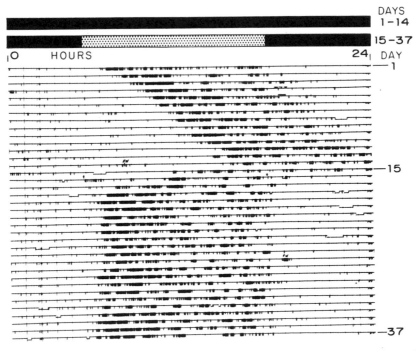

Fig. 9-2. The circadian locomotor rhythm of the house sparrow in its free-running and entrained states. During the first 14 days of the record, the bird was maintained in continuous darkness, and his rhythm free-ran with a period of about 25 h. Beginning on day 15, it was presented with a daily light cycle (diagrammed at the top of the figure—LD 12:12, intensity during the light portion of the cycle = 0.1 lx) to which his rhythm synchronized (entrained). Although this bird was blinded by the removal of both eyes prior to the start of the experiment, his record is indistinguishable from that of a normal sparrow. (From reference 9.)

The threshold intensity to which they will entrain is surprisingly low: about 50% of blinded sparrows will entrain to a cycle consisting of 12 h of darkness alternating with 12 h of light at an intensity of 0.1 lx (approximately the intensity of bright moonlight). This level of illumination can thus be thought of as a crude threshold for extraretinally mediated entrainment.[9] Sparrows with intact eyes will entrain to cycles in which the light portion is considerably less intense, demonstrating that both the eyes and the extraretinal receptors have inputs to the entrainment mechanism, and that these inputs are, to a first approximation, additive (Fig. 9-3). Experiments in which the intensity of light reaching the brain has been either increased by plucking the head feathers, or decreased by injecting opaque substances (e.g., India ink) under the skin of the head, have shown that the extraretinal photoreceptors are located in the brain.[10] More precise localization has been attempted by several workers, but definitive information awaits the development of more refined techniques.

9.4.2. The Effects of Light on the Circadian Clock: Aschoff's Rule

Several different effects of light on circadian rhythms are discussed in Chapter 8. In addition to producing entrainment when it is presented in cycles [light–dark (LD)], light affects the free-running periods of circadian rhythms when it is presented continuously [light–light (LL)]. As the intensity of constant light is increased, the free-running periods of diurnal organisms exposed to it decrease, while those of nocturnal organisms increase. This empirical generalization, and several others related to it, are known collectively as Aschoff's rule (after Dr. Jürgen Aschoff who has been responsible for much of the work in this area). The rule is well supported by data from a variety of very different organisms, although there are exceptions to it.

Sparrows obey Aschoff's rule and continue to do so after blinding. As in extraretinally mediated entrainment, the receptors responsible have been shown to reside in the brain and to act in concert with the image-forming eyes in the intact bird. The free-running period of a bird with both its eyes and its brain photoreceptors intact shows greater changes with varying intensities of LL than does that of a bird in which the eyes have been removed or the brain shielded from light (Fig. 9-4). In sparrows then, extraretinal brain photoreceptors are sufficient for both entrainment and Aschoff's rule, but in nature they probably work together with the eyes (presumably the retinal photoreceptors, although there has been no direct demonstration of this) to mediate the response to light. There is, however, another general response of circadian rhythms to LL for which, at least in sparrows, the brain photoreceptors are not sufficient.

9.4.3. The Effects of Light on the Circadian Clock: Stopping the Clock with Bright Constant Light

In many organisms, the clock is stopped by constant light when its intensity is raised above some threshold, which varies considerably from one species to another. Binkley[12] has shown this to be true for sparrows. One might expect

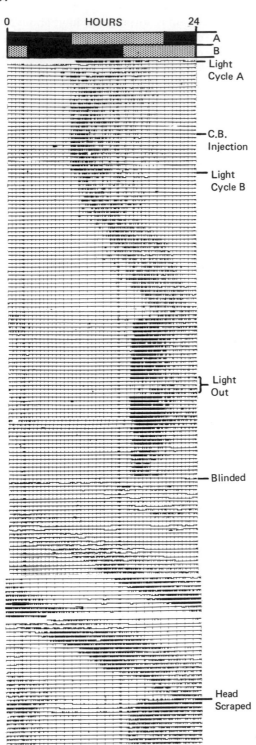

Fig. 9-3. Demonstration of the contribution of both the eyes and brain photoreceptors to entrainment of the circadian locomotor rhythm in the house sparrow. This figure is the activity record of a sparrow exposed to the light regimens (LD 12:12, ~ 0.03 lx) diagrammed at the top. Entrainment to light cycle A continues after carbon black (C.B.) is injected under the skin of the head. Reestablishment of the entrained steady state following the 6-h phase delay (light cycle B) requires 36 days. After blinding, the activity rhythm free-runs through the light cycle twice. When the carbon black is removed, the blind bird reentrains to the light cycle. (From reference 10.)

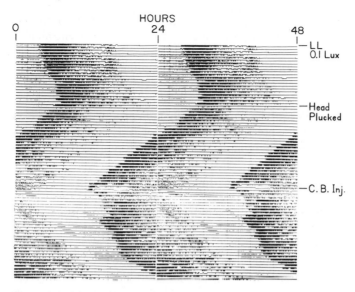

Fig. 9-4. The effects of changing the intensity of light reaching the brain on the free-running period of the locomotor rhythm of a sparrow with intact eyes held in constant light (0.1 lx). The period is initially slightly longer than 24 h. Plucking feathers from the top of the head shortens the period, while subsequent injection of carbon black (C.B.) under the skin of the head reverses this effect. (From reference 11.)

that this effect represents simply an extension of the effects that LL of lower intensities has on the free-running period, but this does not appear to be so in sparrows, at least as far as the photoreceptive mechanism is concerned. Although the intensity of constant light required to produce aperiodic locomotor activity in sparrows (the indication that the clock is stopped) lies between 10 and 100 lx, blind sparrows do not become aperiodic in constant light even when its effective intensity is raised to 20,000 lx. From these experiments, it appears that brain photoreceptors are not sufficient to mediate this effect of light on the sparrow's clock. This conclusion is especially strong in view of the fact that we know from entrainment experiments that the brain photoreceptors are sensitive to light at intensities of about 0.1 lx, so that in exposing them to 20,000 lx we are above threshold by about 2×10^5. Therefore, the eyes are apparently necessary for this particular effect of light on the circadian clock.

To further complicate the issue, it can be shown that, in intact sparrows, the brain photoreceptors are involved in the response to bright constant light, although they are not by themselves sufficient to mediate it. If intact sparrows are brought to aperiodicity by constant light, the intensity of which is just above the required threshold (~10 lx), rhythmicity can be made to recur without changing the light intensity by simply shielding their brain photoreceptors.[13]

The biological meaning of the different patterns of interaction between the eyes and the brain photoreceptors in mediating the effects of light on the circadian system is still obscure, but the experimental results, which have been briefly described above, emphasize the complexity at the photoreceptor level of the response of one system in one species to the photic environment. Although it is beyond the scope of this chapter to explore the comparative aspects of this phenomenon, it is worth mentioning that a similar analysis of the effects of light on the circadian clock has been performed in several species of lizard. While the results of experiments with lizards are generally similar to those with birds, several of the important details, especially concerning the interaction of eyes and brain photoreceptors, are quite different not only when birds and lizards are compared, but even when one lizard species is compared with another.[14]

9.4.4. The Effects of Light on the Reproductive System: Photoperiodism

Many animals and plants and most temperate-zone birds use the information inherent in the cyclic annual change in day length to synchronize their reproductive cycles with the appropriate season (Fig. 9-5). Such responses involve the ability both to perceive and "measure" day length, and are collectively termed photoperiodic. Perhaps the simplest question that can be framed concerning photoperiodic responses is: With what photoreceptors does the organism perceive the photoperiod? When Professor Jacques Benoît posed this question in 1935 with specific reference to the testicular responses of ducks, the answer that he obtained came as a great surprise to the scientific community—so great in fact that 20 or 30 years elapsed before it was fully accepted. By hindsight, and in the context of this chapter, it should not surprise you that this answer was, at least partially, that ducks use extraretinal photoreceptors in their brains to monitor the environmental light cycles that regulate their reproductive physiology.

Since Benoît's pioneering work, photoperiodic photoreception by brain photoreceptors has been directly demonstrated in at least four species of birds not closely related to each other, and probably occurs in most if not all photoperiodic members of the class *Aves*.

Photoperiodic photoreception has been studied extensively in the house sparrow with particular attention to the role, if any, of the eyes. It is perfectly easy to demonstrate that sparrows employ brain photoreceptors to perceive day length. If male birds are brought into the laboratory during the winter when their testes are small and nonfunctional and exposed for several weeks to artificial light cycles, which simulate the long days of spring (e.g., 14 h or more of light per 24 h day), their testes grow dramatically. Not only do the gonads increase in weight by as much as 500-fold, but the testis tubules become packed with spermatids where before there were only spermatogonia. In contrast, the testes of control birds held on short days for the duration of the experiment remain unchanged. If this experiment is repeated using blind birds in both the long- and the short-day groups, the results are identical. In fact, it has not been possible to find any differences between the photoperiodic responses of blind and intact birds on the basis of gonad weight, rates of gonadal growth, or testis histology.[15,16]

These findings led to the speculation that, in this particular aspect of photoreception, the eyes might play no role at all. Some weight was lent to this

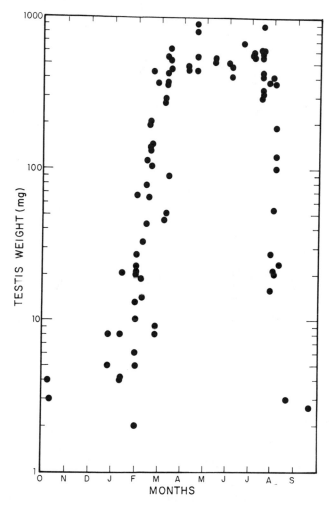

Fig. 9-5. Annual cycle of testis weight in a wild population of *Passer domesticus* in Austin, Texas. Each point represents the combined weight of both testes from a single freshly captured sparrow. Note that testis weight is plotted on a logarithmic scale. (Data obtained by H. Underwood, from reference 8.)

suggestion by the discovery that the threshold intensity required to elicit photoperiodic testis growth in sparrows is of the order of 10 lx, or about 100 times the threshold intensity for extraretinally mediated entrainment, and thousands of times the threshold intensity for vision. A critical experiment was designed in which sparrows with their eyes intact were exposed to long days at a light intensity only slightly above this threshold. Half the birds had feathers plucked from the tops of their heads, thus increasing the amount of light that penetrated to their brains; while the other half were injected with opaque material beneath the skin of their heads, thus decreasing light to their brains. The testes of the

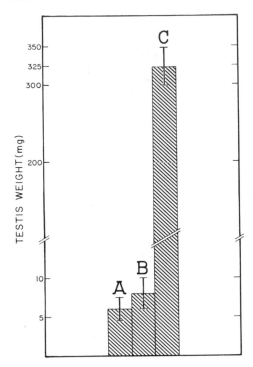

Fig. 9-6. Effects of manipulating the intensity of light reaching the brain on the testis weight of sparrows with intact eyes. A, control group sacrificed at the start of the experiment; B, India ink-injected group; C, group with head feathers plucked. See Section 9.4.4. for additional details. (Figure modified from reference 17.)

plucked birds grew dramatically, while those of the injected birds did not grow at all, in spite of the fact that their eyes were exposed to long days at light intensities above threshold[17,18] (Fig. 9-6). In sparrows at least, photoperiodic photoreception is apparently performed exclusively by brain photoreceptors, perhaps by brain photoreceptors that are different from those that mediate the effects of light on the circadian clock. Note that in the course of investigating four different responses of the house sparrow to the photic environment, three responses of the clock and one of the reproductive system, three different patterns of interaction between the eyes and the brain have been uncovered. Two responses were observed for which brain photoreceptors are sufficient but to which the eyes contribute when present (entrainment and Aschoff's rule); one for which the eyes are necessary (arrhythmicity in bright constant light); and one in which only the brain is involved (photoperiodic reproductive response). Furthermore, there is reason to believe that there are as yet undiscovered responses to light in this as well as other species.

9.4.5. Other Possible Responses to Light

None of the extraretinal responses of sparrows to light that have so far been discussed depends on the presence of the pineal gland; they all proceed unaltered in its absence. Although there is no direct evidence that the sparrow's pineal gland is photoreceptive, indirect evidence suggests that it may well be. One report in the literature describes a behavorial response of young, but not of adult, pigeons that is thought to be mediated by skin photoreceptors.[19] Skin photoreception is fairly common among amphibians and reptiles, and other birds may well have them when juveniles. We should thus tentatively add these two structures (i.e., pineal gland and skin) to our list of photoreceptors in birds that already includes the eyes and an undetermined number of brain photoreceptors.

In looking for photoreceptors that control circadian rhythms and reproduction in birds, it is possible that we have only scratched the surface of the phenomenon. These responses to light are among the easiest to measure. But what of those that are a bit more difficult to assess? Perhaps important aspects of the effects of light on growth and development depend on extraretinal photoreception. Possibly light perceived in this way affects enzymatic activity in the brain and elsewhere in the body as well. The answers to these and a host of other questions await the results of future research.

9.5. CONCLUSIONS

It should be emphasized that as far as extraretinal photoreception is concerned, we do not as yet have any complete cases. Either we have identified and studied the physiology of such receptors but have no clear idea of their function (e.g., the pineals of the poikilothermic vertebrates), or we have identified extraretinally mediated functions, such as entrainment, without any precise localization of or physiological information about the receptors themselves. Not until we know both the function and the identity and physiological characteristics of several such receptor systems will we be able to proceed much further with our analysis.

It is clear from our discussion that many organisms possess and use multiple, discrete photoreceptors to monitor the photic environment. The puzzling question remains: What are the selective advantages of partitioning out photoreceptivity in this way? Although we cannot yet answer this question, we can note that the functional complexity of some of the better-known retinal–extraretinal systems, such as that of the sparrow, argues strongly for some adaptive significance.

In mammals, the processes that have been found to be mediated by extraretinal photoreceptors in the other vertebrate classes seem to be mediated by the eyes. At the present, the only well-documented exception is the effect of light on the level of pineal serotonin in newborn, but not in adult rats.[6]

If extraretinal photosensitivity in mammals is in fact limited to early postnatal life, a whole series of questions is raised concerning the developmental events that lead to its loss. It seems especially important to extend our knowledge in this area in view of the aberrant and occasionally extreme lighting conditions to which newborn human infants are exposed during ordinary hospital routine, and in the course of phototherapy for congenital jaundice (Chapter 7).

9.6. REFERENCES

1. C. M. Williams and P. L. Adkisson, Physiology of insect diapause. XIV. An endocrine mechanism for the photoperiodic control of pupal diapause in the oak silkworm *Antherea pernyi, Biol. Bull.* **127,** 511–525 (1964).
2. J. W. Truman and L. M. Riddiford, Neuroendocrine control of ecdysis in silkmoths, *Science* **167,** 1624–1626 (1970).
3. J. W. Truman, Extraretinal photoreception in insects, *Photochem. Photobiol.* **23,** 215–225 (1976).
4. J. L. Larimer, D. L. Trevino, and E. A. Ashby, A comparison of spectral sensitivities of caudal photoreceptors of epigeal and cavernicolous crayfish, *Comp. Biochem. Physiol.* **19,** 409–415 (1966).
5. T. L. Page and J. L. Larimer, Extraretinal photoreception in entrainment of crustacean circadian rhythms, *Photochem. Photobiol.* **23,** 245–251 (1976).
6. M. Zweig, S. H. Snyder, and J. Axelrod, Evidence for a nonretinal pathway of light to the pineal gland of newborn rats, *Proc. Natl. Acad. Sci. USA* **56,** 515–520 (1966).
7. R. M. Eakin, *The Third Eye,* University of California Press, Berkeley, Calif. (1973).
8. M. Menaker, Nonvisual light reception, *Sci. Am.* **226,** 22–29 (1972).
9. M. Menaker, Extraretinal light perception in the sparrow, I: Entrainment of the biological clock, *Proc. Natl. Acad. Sci. USA* **59,** 414–421 (1968).
10. J. P. McMillan, H. C. Keatts, and M. Menaker, On the role of eyes and brain photoreceptors in the sparrow: Entrainment to light cycles, *J. Comp. Physiol.* **102,** 251–256 (1975).
11. J. P. McMillan, J. A. Elliott, and M. Menaker, On the role of eyes and brain photoreceptors in the sparrow: Aschoff's rule, *J. Comp. Physiol.* **102,** 257–262 (1975).
12. S. Binkley, Constant light: Effects on the circadian locomotion rhythm of the house sparrow, *Physiol. Zool.,* in press (1977).
13. J. P. McMillan, J. A. Elliott, and M. Menaker, On the role of eyes and brain photoreceptors in the sparrow: Arrhythmicity in constant light, *J. Comp. Physiol.* **102,** 263–268 (1975).
14. H. Underwood and M. Menaker, Extraretinal photoreception in lizards, *Photochem. Photobiol.* **23,** 227–243 (1976).
15. M. Menaker and H. C. Keatts, Extraretinal light perception in the sparrow, II: Photoperiodic stimulation of testis growth, *Proc. Natl. Acad. Sci. USA* **60,** 146–151 (1968).
16. H. Underwood and M. Menaker, Photoperiodicially significant photoreception in sparrows: Is the retina involved? *Science* **167,** 298–301 (1970).
17. M. Menaker, R. Roberts, J. A. Elliott, and H. Underwood, Extraretinal light perception in the sparrow, III: The eyes do not participate in photoperiodic photoreception, *Proc. Natl. Acad. Sci. USA* **67,** 320–325 (1970).
18. J. P. McMillan, H. Underwood, J. A. Elliott, M. H. Stetson, and M. Menaker, Extraretinal light perception in the sparrow, IV: Further evidence that the eyes do not participate in photoperiodic photoreception, *J. Comp. Physiol.* **97,** 205–213 (1975).
19. M. B. Heaton and M. S. Harth, Non-visual light responsiveness in the pigeon: Developmental and comparative considerations, *J. Exp. Zool.* **188,** 251–264 (1974).

10

Vision

10.1. INTRODUCTION

10.1.1. The Importance of Vision

A fascination with how vision works is natural. Light provides humans and other animals with very detailed information about their surroundings through the

Edward A. Dratz • Chemistry Board of Studies, Division of Natural Sciences, University of California, Santa Cruz, California

visual sense. Humans are especially visually oriented. There is great practical importance in understanding the eye and its mechanism, so that dysfunctions of this central sensory organ can be avoided or treated more effectively.

How is the eye able to convert sensory information in the form of light into electrical signals to the brain? The major objective of this chapter will be to describe the considerable progress that has been made in understanding this question.

Vertebrate eyes contain two types of photoreceptor cells, the rods and the cones. The rods, responsible for dim-light vision, transmit signals to the brain that we perceive as colorless objects in shades of gray. The cones function in brighter light and enable us to perceive the colors of the spectrum. An eye is said to be "dark-adapted" when it has become well adjusted to the dark. A striking property of the rods can be observed in a dark-adapted eye: If in a cluster of 500 rods as few as 5 to 14 absorb photons, we perceive the light. At this low photon flux, none of the rods absorbs more than one photon. The dark-adapted rod is an efficient single-photon detector! How such a sensitive detection is accomplished is a puzzle that appears to be approaching solution.

The wavelength range of visible light is defined by the wavelengths that can be detected by the human eye. The limits of the visible part of the spectrum are taken to be from about 380 to 700 nm. As will be described in more detail later, the sensitivity of the eye follows the absorption spectra of the visual pigments. The maximum sensitivity of a dark-adapted vertebrate eye (scotopic vision) is near 500 nm, at the maximum absorption of the retinal rod cells. Under bright light conditions (photopic vision) the maximum sensitivity is at about 550 nm, which is near the absorption maximum of the most prevalent cone photoreceptors. The wavelength sensitivity of photopic vision is shown in Fig. 1-2. The sensitivities of the visual systems of all animals fall fairly close to this curve. However, insects can see much deeper into the near-ultraviolet (UV) region with high sensitivity.

The photoreceptors are found in the retina, a thin layer of tissue at the back of the vertebrate eye. The vertebrate retina is of fundamental interest because it is an extension of the brain. Perhaps many general aspects of signal processing in the nervous system can be studied profitably by the use of the relatively accessible and easily stimulated retina.

10.1.2. Phylogenetic Organization of Visual Systems

All vertebrates appear to have extremely similar visual photoreceptor cells. Photoreceptors of animals as diverse as frogs, rats, and cows appear to be comparable, even identical in many aspects. Furthermore, much of the information derived from studies of animal photoreceptors appears to be directly applicable to the understanding of human photoreceptors.

Invertebrates, on the other hand, have evolved quite a variety of morphologically distinct visual systems that are very different from those of the vertebrates.[1,2] In the most primitive photoreceptors, light acts directly on pigmented cells without passing through a focusing apparatus. Rudimentary refractive mechanisms appear in many invertebrates. Cephalopods have a complete cornea–lens system similar to that of the vertebrates (Fig. 10-1). Invertebrate eyes develop from epithelial (skin) tissue, and are arranged in single cells or collections of cells in flat, cup-shaped, and compound or faceted structures. The study of invertebrate vision is an active field. Some invertebrates have extremely large

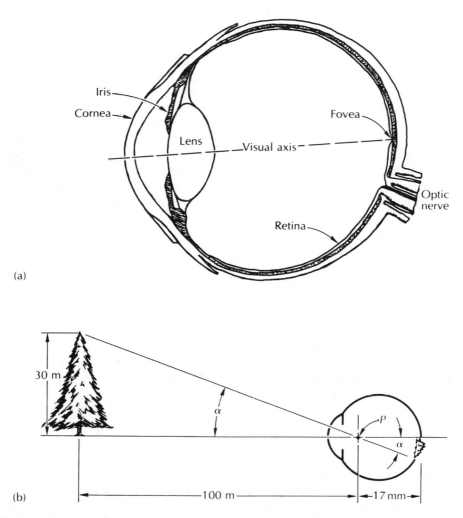

(a)

(b)

Fig. 10-1. (a) Cross section of the human eye. The parts of the eye are discussed in the text. The space inside the eye is filled with a clear jellylike substance called the vitreous humor. The point at which the optic nerve leaves the retina is devoid of photoreceptors, and is responsible for the eye's "blind spot." (b) A simplified optical diagram of the eye looking at a tree. There is a point in the center of any lens system where light rays are not deflected from their incident paths. In the human eye this point, labeled P in the diagram, is about 17 mm in front of the retina. The size of the image on the retina can be determined using simple trigonometry: in this example the image size = 17 mm × 30 m/100 m = 5.1 mm, and the visual angle α = arctan (30/100) $\cong$ 16°. (From reference 62.)

visual cells that are easier to study in detail with microelectrode methods than are the physically smaller visual cells of the vertebrates. Extensive knowledge of the genetics of some invertebrates, such as the fruit fly, provides a potentially powerful research tool. Mutations have been found that alter the development and functional mechanisms of the invertebrate visual system. The availability of such mutants suggests numerous new approaches in research on vision. Invertebrate photoreceptors are not within the scope of this chapter, but leading references to their comparative structural organization and some of the studies of their functional mechanisms are taken into consideration.[1–5]

10.1.3. Overview of the Structural Organization and Functional Mechanisms in the Vertebrate Eye

The optical function of the vertebrate eye can be compared to a modern camera. Figure 10-1 is a diagram of a cross section of the human eye. Light enters through a compound lens that is composed of the cornea and the focusable lens, passes through a variable aperture, the iris, and is focused on the thin light-sensitive layer, the retina. Primates have a tiny specialized portion of the retina, the fovea, that has the most closely spaced photoreceptors, and gives a very high-resolution (fine-grained) image. Figure 10-1 also shows a simplified optical representation of an eye looking at a tree, which is analogous to a camera imaging an object on film.

The eye is, however, functionally much more complex than a camera. It rapidly delivers an image in "real time" to the brain and can control its sensitivity to yield useful signals over a range of light intensities of 10^{10}. The "mechanical" iris aperture of the eye can account for, at most, 10^2 of the intensity range of sensitivity adaptation of the eye. When well dark-adapted, the eye is several orders of magnitude more sensitive than any camera film. Man has not constructed an apparatus to equal the eye in performance, but the analogous device would be an ultrasensitive black and white TV camera combined with a "hi-fi" color TV camera that produces a useful signal over an enormously wide range of light intensities.

The functional mechanisms operating in photoreceptor cells are understood in broad outline, and can be briefly summarized as follows. Light is absorbed by a vitamin A-containing protein, rhodopsin, which is situated in dense stacks of photoreceptor membranes in the retina. Excited rhodopsin appears to cause the release of an unidentified chemical from the photoreceptor membranes inside the receptor cells. The buildup of concentration of the released chemical changes the electrical potential of the receptor cells. This electrical response is in turn processed by a series of interconnected cells in the retina, and is coded into a train of electrical pulses that are passed on to the brain.

Figure 10-2 shows a simplified drawing of the various interconnected cells that receive, process, and transmit signals from the retina. A curious feature of the vertebrate retina is that it is oriented "backward" with respect to the incident light. Light is normally incident from the bottom of the page in this drawing, and passes through several layers of other cellular material before reaching the

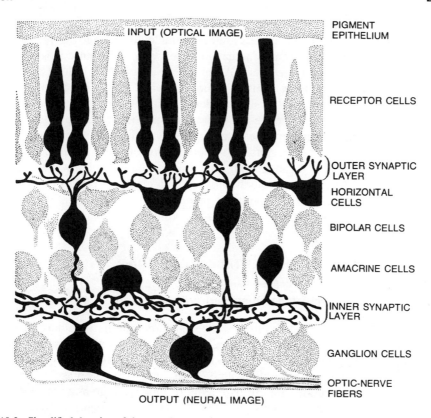

PIGMENT
EPITHELIUM

INPUT (OPTICAL IMAGE)

RECEPTOR CELLS

OUTER SYNAPTIC
LAYER

HORIZONTAL
CELLS

BIPOLAR CELLS

AMACRINE CELLS

INNER SYNAPTIC
LAYER

GANGLION CELLS

OPTIC-NERVE
FIBERS

OUTPUT (NEURAL IMAGE)

Fig. 10-2. Simplified drawing of the vertebrate retina based on a micrograph of a retina prepared by the Golgi method, which selectively stains a few cell bodies and their processes. The various cell types and the major layers of the retina are labeled. Note that light is normally incident from the bottom of the picture, as discussed in the text, and is absorbed by the rod and cone receptor cells. Receptor signals are communicated across the retina by the horizontal cells, while the bipolar cells carry information to the ganglion cell layer. The amacrine cells also communicate across the retina. The ultimate output to the brain is transmitted through the axons of the ganglion cells. (From Werblin, F. S., *Scientific American* **228,** 70–79, 1973.)

receptors. The light receptor cells are at the top of the drawing and are located on the "backside" of the retina. The "backward" orientation of the retina seems to have little consequence for the efficient light detection by the photoreceptor cell. The layers of the retina closest to the front of the eye, and which are traversed first by light, have little or no visible light absorption. Above the photoreceptors are situated the pigment epithelium cells (not shown in this drawing), which supply nutrients to the photoreceptors. The pigment epithelium cells also absorb much of the light that is not captured by the photoreceptor cells, and, therefore, cut down on degradation of the image by light scattering between photoreceptors.

The layer below the receptors contain the bipolar and horizontal cells, each of which is connected to several photoreceptors and to one another. This layer together with the lowest layer comprises the signal-processing "brain" function of the retina. The processed signals are communicated to the lower region of the retina, to the amacrine cells and the ganglion cells that send pulsed signals to the brain through the optic-nerve fibers. After the initial detection of light an extraordinarily varied and complex chain of neurobiological events is activated.

In this chapter we will first consider the structure of the receptor cells; second, the current knowledge of the visual pigment rhodopsin; third, the various measurable light responses; and finally, what is known about the mechanism of action of the system. A useful understanding of visual photoreceptors has come from methods that range from quantum mechanics, photochemistry, and physical chemistry, through biochemistry, and electrophysiology. Much has been learned about photoreceptors in recent years, and this information has revealed the vast territory that yet must be explored.

10.2. STRUCTURE OF THE VERTEBRATE RETINA

10.2.1. Retina Structure as Observed with the Light Microscope

The sizes and shapes of cells and their placement in a tissue may be studied with the light microscope. To prepare cells for microscopic observation, the structural integrity of the tissue is first stabilized by cross-linking it with a variety of chemicals. The tissue is then embedded in a wax or plastic supporting medium, and thin sections are cut with a microtome. A large number of different dyes or stains may be used to enhance the visibility of various parts of the tissues.

Figure 10-3 shows a preparation of monkey retinal tissue. The retina has a layered appearance with different types of cells in each layer. The pigment epithelium, which nourishes the photoreceptors, is the thin layer of cells that extends from the edge of the retina, from 100 to about 105% on the retinal depth scale. The receptors near the top of this micrograph extend from about 45 to 100% of the width of the retina, and the relatively large number of nuclei show them to be the most numerous cell type. There are fewer cells in the layer below the receptors in the region of 20–45% of the width of the retina, so the cells in this layer, the bipolar and horizontal, receive input from several receptors. The amacrine and the ganglion cells are found in the lowest layer from about 0 to 20% of the width of the retina, and they are relatively sparse. The fractional thickness of each of these layers varies from the central to the peripheral retina. The signals detected by the receptors are processed in the lower layers of the retina. The output of the numerous photoreceptors are coded into the firing rate of a smaller number of output channels, the neurons. In experimental animals it has been found that the coding is such that some neurons respond specifically to small moving spots, and light/dark edges of various angular orientations.

The photoreceptor cells have several morphological domains. The so-called outer segments, which are embedded in processes sent out from the pigment epithelium at the top of Fig. 10-3 (back of the eye), are found in the space

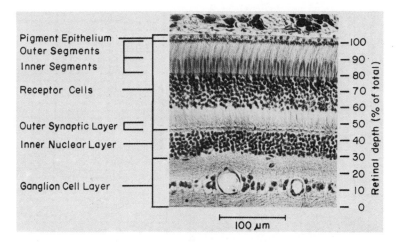

Fig. 10-3. Light micrograph of a thin section of a rhesus monkey peripheral retina stained with hematoxylin and eosin, which makes the cell nuclei appear as dark blobs. The pigment epithelium at the back of the eye is at the top of this picture, and light is normally incident from the bottom as in Fig. 10-2. The identification of the cell layers is discussed in the text. [Adapted from K. T. Brown, K. Watanabe, and M. Murikami, *Cold Spring Harbor Symp. Quant. Biol.* **30**, 457–482, (1965).]

between 90 and 100% of the width of the retina; the so-called inner segments occupy 80–90%; the outer nuclear layer, 60–80%; a fiber layer, 50–60%; and a synaptic portion near 45%. Under the light microscope at least two morphological types of photoreceptors are seen in most portions of the retina in nearly every vertebrate species, although they cannot be distinguished in Fig. 10-3. Cells called rods usually have larger outer segments and thinner inner segments, and cells called cones usually have shorter outer segments and much thicker inner segments.

The distribution of rods and cones, and their morphological appearance, is not constant across the retina. Primate retinas have a central specialized region called the fovea, where the cones are very dense, and no rods are present. Figure 10-4 shows the rod/cone distribution across the human retina. The rods predominate in the other portions of the retina, where only a small number of cones are present. The cones in the fovea are very closely packed and have long, thin outer and inner segments, so that they look similar to the rods in the periphery. The closely packed cells provide high spatial resolution, which produces high visual acuity. Many species show specialized high-resolution regions near the center of the retina, but it is most highly developed in the primates. Every retina has a blind spot, as shown in Fig. 10-4. Here there are no receptors, because this is where the optic nerve leaves the retina. The brain somehow "fills in" the image in the blind spot, so we are normally unaware of it.

There are no blood vessels in the receptor layer. The receptors must get their nutrients from the pigment epithelium above 100% retinal thickness (Fig. 10-3). Most mammals have retinal blood vessels that run through the deep layers of the retina far from the receptors. Two blood vessels can be seen in the section

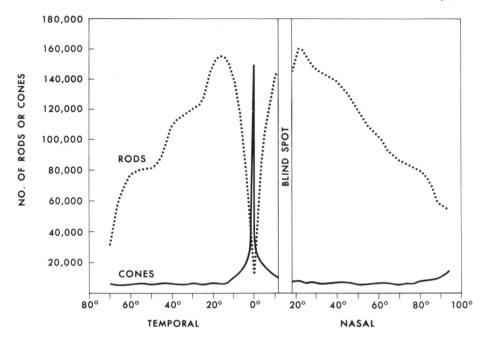

Fig. 10-4. The distribution of rods and cones across the human retina. The central point at 0° is the fovea, which is densely packed with cones. There are few cones in the peripheral retina, which is rich in rods. The blind spot is devoid of photoreceptors, since this is where the optic nerve leaves the retina, and the blood vessels pass through. [After G. Osterberg, *Acta Ophthal. Suppl.* **6**, 1–102 (1935).]

of monkey retina in Fig. 10-3. They appear as large openings in the ganglion cell layer near 10% of the retinal thickness. Many species have no blood vessels in the retina, and obtain all retinal nutrients from the pigment epithelium.

10.2.2. Photoreceptor Ultrastructure as Observed with the Electron Microscope

The electron microscope yields much more highly magnified views of cellular structure than the light microscope. Figure 10-5 shows a thin section of a portion of the photoreceptor layer of the retina. The light-sensitive elements, called the outer segments, are made up of dense stacks of membranes (near the bottom of the figure). The adjoining portion of the photoreceptor cell is called the inner segment and is extremely rich in cellular "power houses," the mitochondria. Numerous mitochondria are seen near the top of Fig. 10-5.

Figure 10-6 shows a higher magnification view of a thin section of a rod outer segment that clearly reveals a stack of membranes surrounded by a plasma membrane. The internal membranes exist in flattened hollow bags called disks. That each flattened disk is hollow can be seen by the presence of open loops at the ends, where the disk membrane turns the corner, and by a thin light space

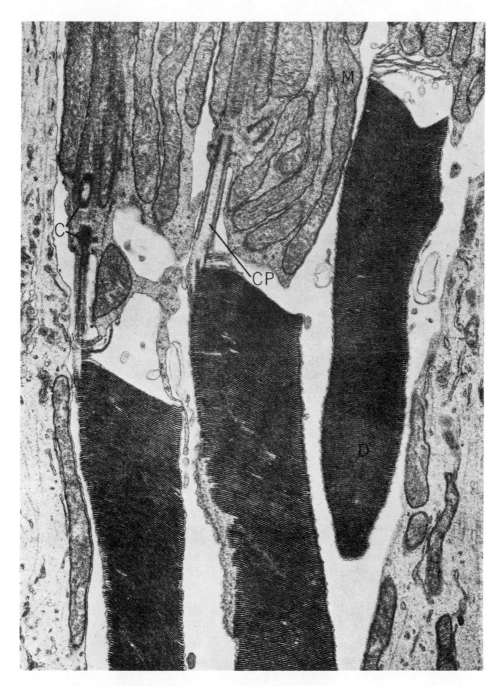

Fig. 10-5. Electron micrograph of an oblique section through a gluteraldehyde-osmium-fixed guinea pig retina showing portions of three outer segments and their attachment to the inner segments. The inner segments contain a pair of centrioles (C), one of which gives rise to a modified cilium composed of a connecting piece (CP), which gives rise to the outer segment during embryological development. The outer segments are made of dense stacks of disk membranes (D), and the inner segments are densely packed with mitochondria (M). [From A. W. Clark, and D. Branton, *Z. Zellforsch. Mikrosk. Anat.* **91**, 586–603 (1968).]

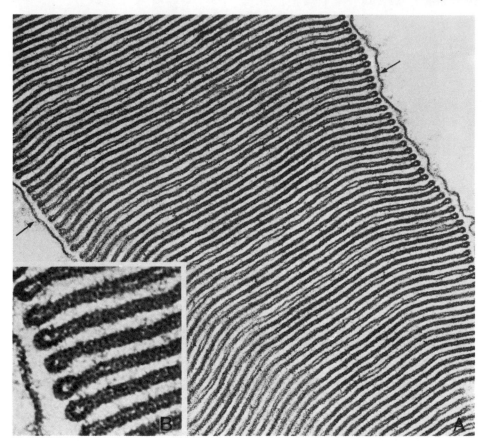

Fig. 10-6. (A) Electron micrograph of a cross section of a portion of a rod outer segment showing the stacked arrangements of the disk membranes. The outer enveloping plasma membrane is indicated by the arrows. (× 72,000.) (B) Magnified view of the edges of the disk membrane and the plasma membrane. Note the loop in the cross section where the disk membranes fold back on themselves to form closed bags. (× 24,000.) [After M. J. Hogan, J. A. Alvarado and J. E. Weddell, *Histology of the Human Eye,* W. B. Saunders and Co., Philadelphia (1971).]

down the center of some of the disks. X-ray diffraction measurements on photoreceptors in living retinas clearly show a space within the disk, but this space has been partially or totally collapsed by dehydration of the specimen during preparation for observation in the electron microscope. The space between the inner surfaces of the disk membrane is about 2.5 nm under isotonic conditions, and has been shown by X-ray diffraction measurements to respond to osmotic variations.[6] Each disk is separated from its neighbor by about 15 nm, and is about 15 nm thick, making the disk–disk repeat distance 30 nm.

The disk membranes are formed by infolding of the plasma membrane. In the rods, most of the disks are pinched off and sealed to form closed bags that are not contiguous with the plasma membrane. Most of the cone disks, however, do not pinch off but remain continuous with the plasma membrane. The mechanisms of assembly and control of the process are not well understood for any membrane system. The work of Young and co-workers[7] using radioactive precursors and autoradiography, revealed the general outlines of the synthesis and turnover of the photoreceptor membranes; this is a very promising system for a more detailed study of membrane assembly. New disks are added to the rods in the region of the junction with the inner segment. Small portions of the tips of the rod outer segments are periodically shed from the cell, and are then phagocytized and digested by the pigment epithelium. The pigment epithelium cells also phagocytize the tips of the cone outer segments.[8]

The rod outer segments are larger than the cone outer segments, and in retinas of most species the rods are much more numerous; consequently, there is much more rod pigment than cone pigment in most retinas. Furthermore, the rod outer segments appear to be more easily isolated and purified than the cones; therefore, nearly all the work on the properties of the photoreceptor outside the intact retina has been done on rod material. Studies of the composition and structure of the color receptors, the cones, remain for the future.

We see that the outer segment portion of the photoreceptor cell is highly membranous, with its membranes arranged in regularly spaced stacks. This organization of regularly stacked membranes is superficially similar to the lamellae of the chloroplasts of green plants, and both types of membranes contain light-absorbing pigment. Each membrane is very thin (~6 nm), and in both the outer segment and the chloroplast lamellae the probability of light absorption has been increased by the evolution of stacks of membranes. The function of the two types of membranes is, of course, very different. The chloroplast converts the energy of absorbed photons to chemical energy, while the outer segment acts as a detector and, as will be seen, uses absorbed photons as signals to control the flow of metabolic energy. Recently a membrane system has been characterized by Stoekenius and co-workers[9] that resembles in some ways both the chloroplast and the outer segment. The membranes are not stacked, but occur as sheets on the surface of the bacterium *Halobacterium halobium*. The light-absorbing molecule, called bacteriorhodopsin, has some similarities to the visual pigment rhodopsin. However, the primary function of bacteriorhodopsin is to carry out photosynthesis, rather than to act as a light sensor.

10.2.3. Molecular Structure as Observed by Chemical Studies and X-Ray Diffraction

The electron microscope falls short of providing structural information at the molecular level. Fortunately, detailed information is obtainable by the use of other tools. The retina may be peeled off the eye cup, and the outer segments broken off the retina. A crude suspension of rod outer segments, which appear to be largely intact and are useful for many experiments, may be obtained by gently shaking the retina in suspension. In this sort of preparation, the outer segments often break off the retina with a portion of the inner segments attached. These preparations contain considerable mitochondrial contamination, since the inner

segments are rich in mitochondria. The purest high-yield preparations of outer
segment material reported to date have been obtained by a more vigorous
homogenization procedure, which breaks the rod outer segments (ROS) into
large fragments. The ROS disk membrane appears to have nearly unique mass
density, so that the fragments can be separated by centrifugation on density
gradients of sucrose or Ficoll (a sucrose polymer with low osmotic pressure).

The ROS are bright pink and "bleach" to a pale yellow if exposed to light
from most of the visible spectrum. The pink color is maintained under infrared or
dim red light. Therefore, most of the preparations and manipulations are carried
out under dim red light. The pigment portion of the ROS cannot be extracted into
aqueous buffers, but remains tightly bound to the membranes. The pigment can,
however, be solubilized in a number of mild detergents without changing its
characteristic absorption spectrum. The color is found to be due to a protein
called rhodopsin. If a very powerful detergent, such as sodium dodecyl sulfate
(SDS), is used to extract the pigment, the visible spectrum of rhodopsin disappears in the dark. In SDS-treated suspensions, the components of the ROS can
be separated by molecular weight using polyacrylamide gel electrophoresis.
Figure 10-7 shows a densitometer scan of such a separation of the proteins of
purified ROS. At least 90% of the protein of the ROS appears in a single dense
band on the gel, and has been shown to be rhodopsin. The identities of the
smaller peaks have not been established. There are several enzyme activities in
the ROS, as will be discussed in Section 10.5.4. Some of the small peaks may
correspond to ROS enzymes. Cattle rhodopsin moves on the gel with an apparent molecular weight of 35,000, a value comparable to that obtained by other
techniques. Numerous minor protein components of larger molecular weight
than rhodopsin are present on the left side of the gel shown in Fig. 10-7. Some of

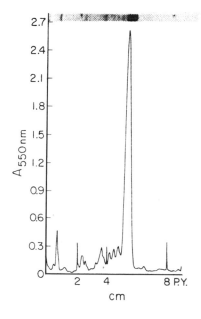

Fig. 10-7. Polyacrylamide gel electrophoresis pattern of
purified bovine rod outer segment membranes dissolved
in sodium dodecyl sulfate (SDS). The gels were stained
with coomassie blue after electrophoresis to visualize
the protein components. A photograph of the stained gel
is shown at the top, and a densitometer scan of the gel
(at the stain absorption maximum) is plotted below. The
major peak is opsin, the protein portion of rhodopsin.
The very sharp peaks at 2, 4, and 8 cm are needle marks
made for length calibration. The tiny peak at 9 cm is the
Pyronine Y tracking dye (P.Y.), which indicates the
distance traveled by this rapidly moving marker. The
other small peaks are protein components, and no single
peak exceeds 1 mole% of the opsin peak. SDS polyacrylamide gels separate by molecular weight, the
smallest components having the highest mobility. Bovine opsin moves with an apparent molecular weight of
35,000, and nearly all of the other minor components
have larger apparent molecular weights. [After D. S.
Papermaster and W. J. Dreyer, *Biochemistry* **13**, 2438–
2444 (1974).]

the minor components may be impurities. The localization and functional identity of the minor components of the ROS have not been established.

The membrane is about 43% protein, 53% lipid, and 4% carbohydrate by dry weight. The lipid portion of the membrane can be extracted with organic solvents and separated by thin-layer chromatography. This treatment also releases the chromophore, a derivative of vitamin A that is discussed in Section 10.3.1. Most of the lipids are phosphatidylcholine, phosphatidylethanolamine, phosphatidylserine, and phosphatidylinositol, which occur in the ratio 40:45:15:2. These lipids each have different, highly polar, phosphodiester "head" groups esterified to a glycerol molecule that is also esterified to two long-chain fatty acids. There is very little cholesterol present, and a trace of sphingomyelin is found. Small amounts of glycolipids may be present, but they have not been characterized.

The fatty acids in the phospholipids of the membrane are strikingly polyunsaturated; in cattle ROS at least 35% of the fatty acids are docosahexenoic acid (22 carbons and 6 double bonds, designated a 22:6 fatty acid). Most of the other fatty acids found are the saturated palmitic (16:0) and stearic (18:0) acids. A typical phospholipid has a saturated fatty acid esterified to one of the hydroxyl groups of glycerol, and a 22:6 fatty acid esterified to the other.[10]

When polyunsaturated fatty acids are removed from a rat's diet, the ROS tenaciously retain their polyunsaturated fatty acids.[11] Most other cells and organs in the animals lose their polyunsaturated fatty acids under these conditions. The exceptionally high level of 22:6 fatty acids and the tenacity with which they are retained by the ROS suggests that these fatty acids may be important for the functioning of the system.[12]

Analyses of X-ray diffraction patterns of the ROS in the intact retina have been carried out by several groups to a resolution of 2.5 nm. These studies indicate that much of the lipid in the disk membrane appears to be arranged in a lipid bilayer. A bilayer consists of two sheets of lipids arranged with their polar "head" groups in contact with water, the hydrocarbons of the fatty acids forming an oily interior. Conflicting conclusions have been reached by X-ray methods as to the localization of the protein in the membrane. A major source of this disagreement appears to have been due to the neglect in the analysis of disorder in the membrane spacing. Recently methods have been developed to account for this disorder and to obtain correct profiles of electron density across disk membranes.[13] Subsequent analysis of the 2.5 nm resolution profile so obtained indicates that a substantial mass of rhodopsin protrudes from the outer surface of the disk membrane, and that the rhodopsin molecule appears to span the disk membrane.

The localization of rhodopsin in the disk membrane has also been studied by several other methods. Membranes have a region of structural weakness where the hydrocarbon chains meet in the center of the two halves of membrane bilayers. Freeze-fracture electron microscopy reveals the textures of the cleavage planes that pass through this region of weakness. Proteins that protrude into the hydrocarbon region are observed as bumps on fracture surfaces. Rhodopsin remains attached to the outer half of the disk membrane when samples are frozen and fractured. These experiments show that rhodopsin is asymmetrically associated with the disk membrane. Chemical labeling with membrane impermeable reagents shows that more than half of the 11 amino groups of rhodopsin are accessible to the aqueous space on the outside of the disk membrane in isolated disks.[14] All of the rhodopsin amino groups, except the one binding the chromophore (see Section 10.3.), are accessible to membrane-permeable analogs of the same reagents. Several laboratories have shown that treatment of the membrane with a number of proteases, which presumably do not penetrate the membrane, cleaves the peptide chain of rhodopsin. Rhodopsin has two short oligosaccharide chains, consisting of mannose and *N*-acetylglucosamine, which protrude into the aqueous region, as shown by binding with oligosaccharide-specific conconavalin A and wheat germ agglutinin.

All of the above information is consistent with rhodopsin being asymmetrically located in the membrane, exposing a portion of its surface and a significant fraction of its mass to the aqueous space outside the disk membrane. Figure 10-8 shows various topological models that have been proposed for the location of rhodopsin in the membrane. Models II and III have now been ruled out. Models IV and V, with rhodopsin spanning the membrane, are supported by chemical labeling, freeze-fracture data, and X-ray diffraction analysis. Labeling with antirhodopsin antibody has also been interpreted to show that rhodopsin spans the membrane. Chemical labeling experiments[14,15] also imply that the lipids are very asymmetrically disposed across the membrane, with most of the phosphatidylethanolamine, and much of the phosphatidylserine distributed toward the outside. The bulk of the phosphatidylcholine is presumably on the inside surface.

The photoreceptor membranes of the rod have a highly specialized function: to signal the absorption of light. Evolution has found a way to do this specialized job with membranes that have a rather simple composition. They contain predominantly one protein—the active light absorber rhodopsin—and have a rather

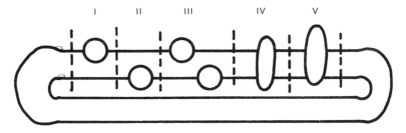

Fig. 10-8. Proposed models for the topographical location of rhodopsin in the rod outer segment disk membrane. Model I shows rhodopsin exposed only on the outer surface of the disk, in contrast to model II where rhodopsin is exposed only to the interior disk space and does not protrude from the exterior surface. Model III shows a symmetrical distribution with rhodopsin molecules protruding from both the interior and exterior surfaces, but single molecules do not span the membrane. In models IV and V, rhodopsin spans the lipid layer, the difference between these being the relative depth of penetration into the interior space. The shape of the rhodopsin is not intended to be taken literally, the schematic drawing only indicates the topological exposure of rhodopsin to the exterior and interior space of the disk. Such models represent a "time-averaged" view, and rhodopsin could be bobbing in the membrane to various depths.

simple lipid composition. As such, these membranes have attracted many workers interested in the more general aspects of membrane structure and function.

If the disk membranes are isolated by shear and density gradient methods at salt concentrations relatively near isotonic, they tend to remain stuck to each other in stacks. The ''sticky'' membranes can be partially separated from each other by treatment with low salt concentrations (of the order of a 100-fold dilution of isotonic).[14] The molecule(s) responsible for the adhesion activity are being characterized and appear to contain protein and polysaccharide.

The disk–disk adhesion molecules may be partially removed by low-salt washes. If the washed membranes are restacked by centrifugation, and the water content reduced to near the native proportion ($\sim$50%), the membranes approach each other much more closely, and are stacked with much less spacing disorder than is present in the intact ROS. Such preparations of washed disk membranes give much higher resolution X-ray diffraction patterns than the intact ROS, and the diffraction patterns have been analyzed, yielding an electron density profile with an 0.8 nm resolution.[16] At this resolution the disk membrane is shown to be more complex than a simple lipid bilayer. The inner surface has a large sharp electron density peak that is probably due to the polar phospholipid head groups, but the outer layer has more structural detail. There appear to be significant changes in the electron density profile when the disk membrane is exposed to visible light. More detailed interpretations of the electron density profiles are in progress, and it is expected that this approach will yield some very detailed information about the disk membrane.

The two-dimensional arrangement of rhodopsin molecules in the plane of the disk membrane appears to be a dynamic one. Spectroscopic measurements on the rhodopsin chromophore have provided information on the rotation and translation of rhodopsin in the membrane (Section 10.3.).

10.3. PIGMENT SPECTRA AND THE CHROMOPHORE

10.3.1. Chromophore Structure and Spectra

Wald showed many years ago that the chromophore in rhodopsin is the 11-*cis* isomer of the polyene vitamin A aldehyde. This molecule is commonly called 11-*cis*-retinal, and its structure (I) together with the structures of some related molecules are shown in Fig. 10-9. When rhodopsin is exposed to light before the chromophore is isolated, retinal is found to be isomerized to the all-*trans* form (structure III). In Wald's laboratory, Bownds found that retinal could be fixed to the ϵ-amino group of a lysine in rhodopsin by sodium borohydride reduction after light exposure, and therefore was probably bound by a Schiff base linkage (structure V, a Schiff base linkage, is equivalent to structure II combined with an R—NH$_2$ group) to the protein before reduction. The reduced Schiff base linkage of all-*trans*-retinal is shown in Fig. 10-9, VII. Borohydride cannot reach the chromophore linkage in the dark, but when rhodopsin is exposed to light, the borohydride reacts with the linkage rapidly.

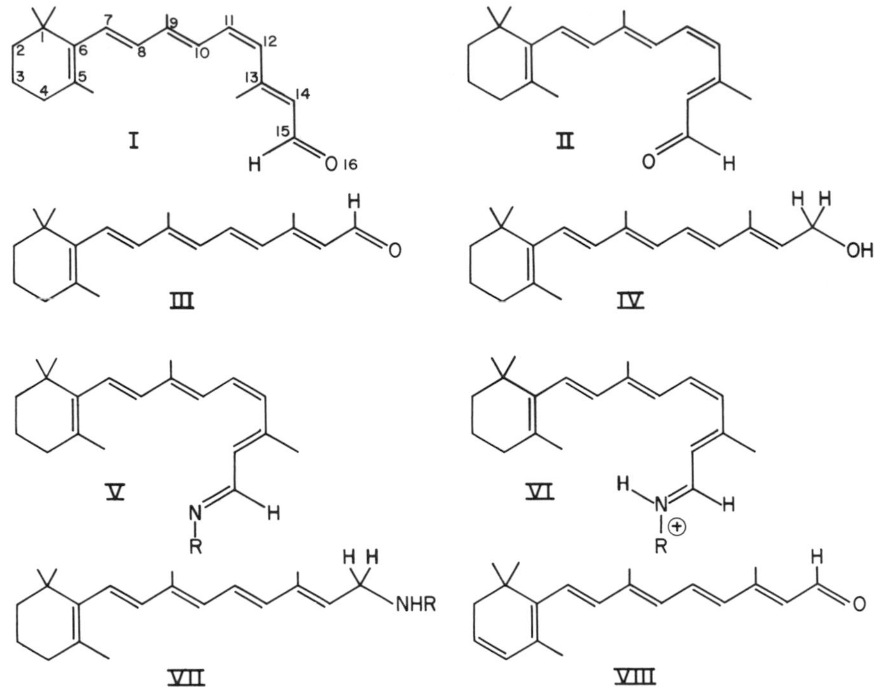

Fig. 10-9. Structures of different vitamin A derivatives important in vision. I, 11-*cis* retinal. II, 11-*cis*-12-s-*cis* retinal. III, all-*trans* retinal. IV, all-*trans*-retinol. V, 11-*cis*-12-s-*cis* retinal Schiff base or 11-*cis*-N-retinylidene-alkyl amine. VI, protonated 11-*cis*-12-s-*cis* retinal Schiff base or protonated 11-*cis*-12-s-*cis* retinylidene-alkyl amine. VII, reduced all-*trans*-retinal Schiff base or all-*trans*-N-retinyl-alkyl amine. VIII, all-*trans* 3-dehydro retinal or vitamin A_2 aldehyde.

Several different geometrical isomers of retinal can be prepared, including 7-*cis*, 9-*cis*, 11-*cis*, 13-*cis*, 9,11-di-*cis*, 9,13-di-*cis*, 11,13-di-*cis*, and 9,11,13-tri-*cis*. The naturally occurring 11-*cis* isomer is energetically the least favorable mono-*cis* isomer, because of considerable steric hindrance between the 13-methyl and the 10-hydrogen (see Fig. 10-9, structure I). If retinal is thermally isomerized to an equilibrium mixture by iodine catalysis, much less 11-*cis* isomer is formed than any other mono-*cis* isomer, and 11-*cis* is exceeded in amount by some of the di-*cis* forms.

Theoretical calculations led to the proposal that the steric strain in 11-*cis* retinal is relieved by twisting about the 12-13 single bond.[17] A twisted nonplanar conformation close to the 12-s-*cis* conformation (Fig. 10-9, structure II) (the s-*cis* refers to the *cis* conformation about a single bond) is predicted to be of the lowest energy, and a nonplanar conformation close to the 12-s-*trans* (Fig. 10-9, structure I) has a slightly higher energy. Both forms were also predicted to be highly twisted about the 6-7 bond and to be closest to a 6-s-*cis* conformation. The crystal structure of 11-*cis* retinal was deduced soon after, and these theoretical predictions were borne out quantitatively in the crystal state. A nonplanar 12-s-cis conformation was found with a 40° twist out of plane, and the 6-7 bond was twisted 135° from the s-*trans* orientation, close to a 6-s-*cis* conformation. Nonplanar conformations about these bonds also persist in solution, as shown by nuclear magnetic resonance (NMR) measurements.

One other closely related polyene chromophore is found in some visual pigments. This chromophore is called 3-dehydroretinal aldehyde or vitamin A_2 aldehyde (Fig. 10-9, structure VIII) and contains an additional double bond in the six-membered ring. A useful shorthand for vitamin A_2 aldehyde is A_2 and, correspondingly, A_1 for vitamin A aldehyde. Visual pigments from hundreds of species have now been analyzed. They all contain either the A_1 or A_2 chromophore. The retinas of some species contain both A_1- and A_2-based visual pigments, and the relative proportion may depend on age, season, or habitat.[18]

Schiff base derivatives of A_1 aldehyde chromophores may be easily prepared with aliphatic amines in organic solvents, and are found to have absorption maxima in the near-UV region. The solid and dashed lines in Fig. 10-10 show spectra for Schiff bases of A_1 in the 11-*cis* and all-*trans* conformation, respectively. Analogous compounds of the A_2 chromophore absorb maximally at somewhat longer wavelengths than those made with the A_1 chromophore. The 11-*cis* chromophores exhibit two peaks on the short wavelength side of the strongest band; these are called *cis* peaks and are found in every *cis* polyene. The all-*trans* chromophore shows a much stronger main band, and weaker short wavelength bands than the *cis* forms. If retinal Schiff bases are protonated by the addition of acid, the spectral maxima are shifted to longer wavelengths. The dotted line in Fig. 10-10 shows a spectrum of the protonated Schiff base of 11-*cis* A_1; its structure is shown in Fig. 10-9, (VI). The chromophore in most visual pigments absorbs even farther to the red than do protonated Schiff bases in solution; the additional shift brought by the protein environment will be discussed in Section 10.3.3.

10.3.2. Visual Pigment Spectra

The absorption spectra of visual pigments (VP) may be measured in three ways. First, VP may be extracted from isolated retinas with a mild detergent

Fig. 10-10. Absorption spectra of retinal Schiff bases made with 1-amino-2-propanol. Dashed curve, all-*trans*-N-retinylidene-1-amino-2-propanol. Solid curve, 11-*cis*-N-retinylidene-1-amino-2-propanol. Dotted curve, protonated Schiff base made with equimolar $HClO_4$ added to 11-*cis*-N-retinylidene-1-amino-2-propanol. All spectra are in 1,2 dichloroethane. [Redrawn from data of J. O. Erickson and P. E. Blatz, *Vision Res.* **8**, 1367–1375, (1968).]

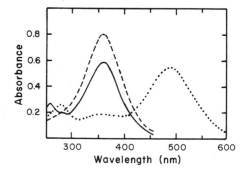

solution. Second, they may be measured *in situ* in single photoreceptor outer segments by direct spectrophotometry using microbeams. Third, they can be estimated in intact animals (including man) with a technique developed by Rushton and co-workers[19]; the amount of light that is reflected from the back of the eye is measured before and after an exposure to bright light, which bleaches a large fraction of the pigment. All VPs show similarly shaped absorption spectra with large extinction coefficients close to 40,000 liters mol^{-1} cm^{-1}. All of the native VPs appear to be bleached with high quantum efficiency (near 0.67).

A striking feature of VP spectra is the great range of the wavelengths of maximum absorption that are observed in photoreceptors of different types in different species. The maxima range from about 435 nm in the blue to about 625 nm in the red. The mechanisms that accomplish this impressive wavelength regulation have not been established, but must involve interactions of the protein and the chromophore. This question of wavelength regulation will be discussed in the next section.

Mixtures of different VPs are present in most intact eyes or in detergent extracts of their retinas. This can be deduced and the absorption maxima estimated by a technique called "partial bleaching," if the maxima of the different pigments are not too close to one another. This method measures the difference in absorbance between visual pigment samples that have been exposed to different sets of light conditions (difference spectra). "Bleaching" is a term that has been widely used to describe the change of color of visual pigments that occurs after exposure to light. In the typical approach, the VPs in the mixture that absorb farthest to the red are first bleached by successive exposures to bright far-red light. Difference spectra are recorded after each bleach until no further change is observed. Then the bleaching light is decreased in wavelength, and the process is repeated. The peaks in the difference spectra are close to the VP absorption maxima if tests are performed in which the wavelengths of the bleaching lights are varied to provide reasonable assurance that predominantly one form of VP is being bleached at each stage.

The spectra of purified bovine rod VP and of the isolated 11-*cis*-retinal are shown in Fig. 10-11. The large peak at 498 nm is called the α peak, the small peak near 340 nm is called the β peak, and the strong peak at 278 nm is called the δ peak. The δ peak is largely due to the aromatic amino acids, tyrosine, and tryptophan in the protein, although there is probably some contribution by the chromophore at these wavelengths. Rhodopsin is bleached with a substantial quantum efficiency by light absorbed in the δ band. We do not see UV light, although some insects do, because it is strongly cut off through absorption by the cornea, lens, and vitreous humor (see Fig. 10-1).

In recent years microspectrophotometry on single isolated photoreceptors has been particularly useful for VP spectral studies. Single cones have distinct spectra; cones with red, green, and blue absorption maxima are present in many species.[20] These observations support the trichromatic theory of color vision; three types of cones exist, each with a VP of a different color. Some species appear to have only one or two types of cones. Some of these species (birds and amphibians) have highly colored oil droplets through which the light must pass to

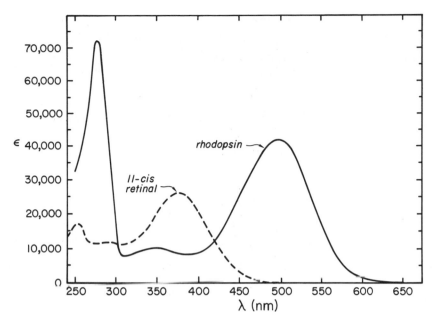

Fig. 10-11. The absorption spectrum of purified bovine rhodopsin solubilized in the detergent Ammonyx LO (solid line) and 11-*cis*-retinal (dashed line). (From reference 53.)

reach the cone outer segments. Different cones in the same species may have different colored oil droplets that apparently act as color filters. Therefore, two types of spectral sensitivity controls are found in cones: wavelength regulation of the VP absorption spectra and, in some cases, an oil drop color filter.

In primates, the central region of the retina is called the macula. This region includes the cone-rich fovea and a surrounding rod-rich area. The macula contains a yellow pigment called lutein. The tissue localization of the macular pigment has not been established. The macular pigment has a series of relatively sharp peaks in the blue and green, and the spectral sensitivity of the human eye shows dips that are due to the screening of the photoreceptor by absorption by this pigment. The function of the macular pigment is not yet understood.

Microspectrophotometry, using polarized light, has shown that the rhodopsin chromophore is highly oriented in the disk membrane.[20] Figure 10-12 shows the geometry of viewing the rod either "end on," as light is incident on the intact eye, or "side on," as can be done experimentally in isolated photoreceptors. If the photoreceptor is viewed "side on" with polarized light, it is found that rhodopsin absorbs light much more strongly if the electric vector is polarized in the plane of the membrane, than if it is polarized perpendicular to the membrane. Viewed visually "side on" in the light microscope, the rods appear nearly colorless when the light is polarized parallel to the long axis of the rod. However, the rods are a distinct red when the light is polarized parallel to the plane of the disk membrane, as diagrammed in Fig. 10-12. The red color results because the

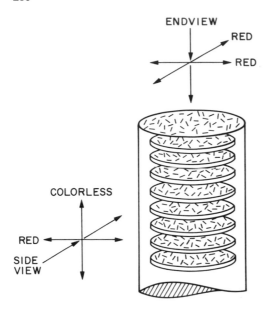

Fig. 10-12. Schematic diagram illustrating the dichroism of outer segments, when viewed from the side with polarized light; and the lack of dichroism, when viewed from the end (see text for more details). The rhodopsin chromophores are highly restricted and must lie in the plane of the disk membrane, but are randomly oriented within that plane.

visual pigment preferentially absorbs the green and blue, and allows the red to pass through. The ratio of the absorbances for the parallel and perpendicular orientation of the polarized light (the dichroic ratio) is 4–6. This large dichroism requires that the chromophore be strongly oriented in the plane of the membrane.

Light incident on an intact eye always hits the photoreceptor "end on," and travels parallel to the long axis of the photoreceptor. Rhodopsin is optimally oriented to absorb this light, for the electric vector of the "end on" incident light lies in the plane of the disk membrane, and has no component perpendicular to it. The rod appears uniformly red, independent of the polarization (i.e., there is no dichroism), when viewed "end on," as shown in Fig. 10-12. The rhodopsin chromophore is not preferentially oriented when viewed "end on." Consequently, vertebrate eyes are not sensitive to the plane of polarization of the light, as are some insect eyes.

If a brief, bright, polarized flash is incident on the rod "end on," a large dichroism is induced that quickly dies away.[21] The initial dichroism is due to photoselection of rhodopsins that happen to be oriented optimally for absorption of the polarized flash and are efficiently bleached. However, the rhodopsin molecules are rapidly rotating in the plane of the membrane, and therefore the dichroism quickly disappears. The half-time for the rotational disorientation is about 20 μs at 5°C. An increase in temperature of 10°C reduces the half-time for disorientation by about a factor of 2.5 (the Q_{10}). The lateral motion of rhodopsin in the plane of the membrane has also been measured spectrophotometrically.[22,23] The lateral diffusion constant is 4×10^{-9} cm²/s at 25°C with a Q_{10} of 2. It is clear that the lipid matrix of the disk membrane presents a relatively fluid medium for rhodopsin movement. The diffusion constants of rhodopsin have been used to estimate an effective viscosity of about 2 P in the plane of the membrane, which is 200 times that of water and is of the order of the bulk viscosity of olive oil.[21-23] Gluteraldehyde fixation stops the rotational and the translational motion of rhodopsin. The receptor potential of the retina is abolished by gluteraldehyde fixation, although rhodopsin can still be bleached, and appears to undergo its normal conformational changes (see the discussion of the early receptor potential in Section 10.4.5.) after fixation. It is tempting to hypothesize that movement of rhodopsin molecules in the plane of the membrane is required for photoexcitation, although other explanations for the above consequences of fixation are possible.

10.3.3. The Question of Wavelength Regulation of Visual Pigment Spectra

The wide range of absorption maxima found in VPs corresponds to a difference of nearly 0.75 eV in the excitation energy of the chromophore. Part of this range is spanned by the use of two slightly different chromophores, A_1 and A_2. However, the A_1 pigments themselves show a range of nearly 0.5 eV (430–575 nm), which is a very large range for a single chromophore. Different protein environments around the chromophore binding site in different VPs somehow cause a change of the absorption maximum of the chromophore from blue to red. This influence of the protein on the chromophore is called wavelength regulation. The mechanism has not been established, but several hypotheses have been proposed to explain this phenomenon.

When a VP protein is coupled with the A_2 chromophore, the VP always absorbs to longer wavelengths than if the same protein is coupled to the A_1 chromophore. If these A_2–A_1 shifts are compared for a number of VPs, it is found that the difference is largest for the pigments that absorb furthest to the red, and gets smaller for pigments that absorb to the blue. In fact, a plot of the A_2–A_1 shift vs. the absorption maxima of either the A_1 or the A_2 VP is smooth and nearly linear, and the A_2–A_1 difference converges to near zero at the short wavelength end. This phenomenon is called "convergence."

In a large number of fish species, the absorption maxima appear to "cluster" near discrete wavelengths.[24] The clustering is shown in the histogram in Fig. 10-13. This clustering suggests discrete changes in the environment of the chromophore in the VPs. The most frequently occurring wavelength maxima is near 500 nm. The maximum of the solar spectrum is at about 520 nm at sea level, decreases to 500 nm at about 20 m water depth, and drops to 480 nm at about a 50 m depth.[25] An absorbance maxima near 500 nm is common in rod pigments of many vertebrates (e.g., 498 nm in cattle, 520 nm in the frog).

A large red shift relative to the isolated chromophore can be obtained by protonation of the Schiff base linkage to rhodopsin (see Fig. 10-9, structure VI, and Fig. 10-10). Resonance Raman spectroscopy experiments have shown that the Schiff base linkage in rhodopsin is protonated. The nature of the counterion to the protonated Schiff base can have a large effect on the spectral maximum.

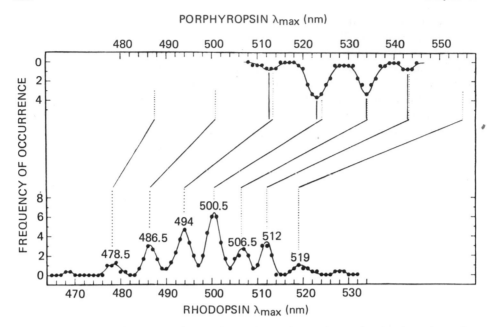

Fig. 10-13. Clustering of the absorption maxima of rhodopsin (A_1 chromophore) and porphyropsins (A_2 chromophore) from teleost fish retina. The smooth, solid lines have been drawn through the data points. The lines, connecting the peaks of the plot of the rhodopsin data with the peaks of the porphyropsin data, connect pairs expected from an empirical relationship between A_1 and A_2 pigment absorption maxima, derived from pigment pairs in a variety of species. (From reference 61.)

Carboxyl groups are probably the only available anions in the protein environment, and variations in the properties of the anion are probably not sufficient to explain the range of absorption maxima.

The absorption maximum of the protonated Schiff base of A_1 is red-shifted to about 500 nm in some solvents, but never as far as the redmost A_1 VP (about 575 nm). It has been shown that the protein could produce a large red shift if it places a second anion (or dipole) near the ring end of retinal (Fig. 10-9).[26] VPs with different colors would then have different distances between the ring of retinal and the second charge.

The VP chromophores have a large circular dichroism (CD) induced in them by the protein in both the α and β bands. This shows that the chromophore is in an asymmetric environment, and that the environment in the protein has a "handedness." The origins of the CD are not understood at present. A combined study of CD and the control of the chromophore absorption maxima (wavelength regulation) may provide a better opportunity for understanding the spectra than does a study of either independently. The α and β band CD are very similar in bovine rhodopsin, and in two of the products that are produced by light absorption, lumirhodopsin and metarhodopsin II[27] (see Section 10.4.1.). The search for explanations for these puzzles is fertile ground for future work. One possible approach is contained in the next section.

10.3.4. Chromophore Analogs

Comparison of the properties of the VPs formed from the naturally occurring A_1 and A_2, provide some testable criteria for theories of VP spectra. The 9-*cis* and 9,13-di-*cis* isomers of these chromophores form pigments with the protein, in addition to the native 11-*cis*. For some time chemists have been making synthetic vitamin A aldehyde analogs that differ in chain length, steric properties, and conformational restrictions. It has been found that the rhodopsin binding site is quite restrictive as to what sorts of alterations of the chromophore it will accept. Longer or shorter polyene molecules do not form VPs, and the six-membered ring must be present. Anywhere between 0 and 3 double bonds in the ring are accepted. None of the methyl groups is required for the formation of VPs; however, some of the analogs missing methyl groups are not bound as tightly as are the native chromophores. The pigment formed from the 9-desmethyl (lacking the 9-methyl in Fig. 10-9, structure I) has a significantly different absorption maximum and CD intensity than the native chromophore. Most of the structural variations of the chromophores produce less stable pigments with opsin and also show a reduced photosensitivity compared with the native pigment. It is beyond the scope of this chapter to discuss these results in greater detail, but much of the work has been reviewed.[28,29] I wish to emphasize that retinal analogs may be very useful in research on the interactions between the protein and the chromophore.

10.3.5. Photochemistry

The quantum efficiency of stimulation of rhodopsin is very high, near 0.67. It was pointed out in Section 10.3.1. that the bleaching of rhodopsin causes the isomerization of the chromophore from the 11-*cis* isomer (Fig. 10-9, structure II) to the all-*trans* conformation (Fig. 10-9, structure III). Conformational changes initiated by photon absorption trigger a series of changes that are discussed in the next section. After a photon is absorbed, all the other reactions in vertebrate photoreceptors occur in the dark. The photochemistry in the system then centers around the isomerization mechanism, and whether the isomerization occurs by a singlet or a triplet pathway. The first detectable spectral change in rhodopsin appears in less than 6 ps (6×10^{-12} s). This observation does not explain the mechanism, but it excludes the involvement of large simultaneous changes in the protein binding site, and suggests that studies of isolated chromophores, as model systems for the early events in the protein, might be useful. Retinals appear to isomerize at least partly by a triplet mechanism. While 11-*cis*-retinal isomerizes predominantly by a triplet mechanism in solution, the protonated Schiff base isomerizes much more rapidly through a singlet mechanism.[30] These observations indicate that retinal may be a poor model system of rhodopsin photochemistry, and that the protonated Schiff base has important, different properties.

10.4. MEASURABLE RESPONSES TO LIGHT STIMULI

10.4.1. Spectra and Properties of "Bleaching" Intermediates

When rhodopsin is exposed to light, it undergoes a series of changes; the color changes from its original pink to yellow in a series of steps. As has been mentioned, this entire process is called "bleaching." Figure 10-14 shows the series of intermediate steps in the bleaching process that have been identified, their absorption maxima, and the approximate half-times of interconversion at 20 or 37°C in the retina or in isolated outer segments. The first change from rhodopsin to prelumirhodopsin is light-induced, all other changes being dark

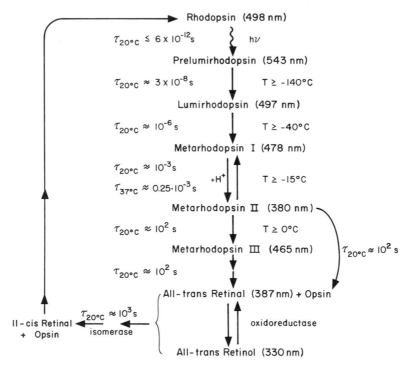

Fig. 10-14. Stages in the bleaching and regeneration of rhodopsin. The initial photoevent is shown as a wavy line, and the subsequent thermal, dark reactions are shown as solid lines. The absorption maximum of each species is shown in parentheses after the name of the species, and is the value observed for bovine rhodopsin. The approximate half-times for interconversion of each species at 20 and 37°C are denoted by $\tau_{20°C}$ and $\tau_{37°C}$, respectively. The approximate temperatures above which the interconversions occur are indicated by a capital T. More than one step may be involved in the interconversions of metarhodopsin I to metarhodopsin II, metarhodopsin III to all-*trans*-retinal + opsin, and the interconversion of all *trans*-retinal to 11-*cis*-retinal.

reactions. All the intermediates may photobackreact to reform rhodopsin; however, it has been claimed that a form of metarhodopsin I (called meta I', and
formed after meta I) does not photobackreact.[31] A single bright flash—no matter
how bright the flash—does not bleach all the rhodopsin because of the photobackreactions. The amount of bleaching in a saturating flash depends upon the
relative absorbance of rhodopsin and the intermediates integrated over the
spectral distribution of the flash, the photobackreaction quantum efficiency, and
the lifetimes of the intermediates relative to the flash length. The lifetimes of the
intermediates, and presumably the photobackreaction quantum efficiency,
depend on the environment of rhodopsin. For example, in many detergents the
meta I → meta II rate is accelerated about 1000-fold. In these detergents nearly
complete bleaching can be obtained in a single 1-ms flash. In contrast, in the lipid
environment of the intact membrane it is difficult to obtain more than a 50%
bleach in a single bright flash.

The kinetics of the spectral changes are rather strongly temperature-dependent. At 37°C meta II appears with a half-time of 0.25 ms in the rat retina,
whereas below 0°C meta II appears very slowly, and meta I may be thermally
trapped below −15°C. Bovine rhodopsin, which has been studied in more detail,
exhibits the same general pattern of temperature-dependent kinetics, with some
quantitative differences. For example, lumirhodopsin can be trapped below
about −40°C, and prelumirhodopsin below −140°C. Prelumirhodopsin can also
be formed at liquid nitrogen temperature (−196°C), and can be photobackreacted
to rhodopsin at this temperature; however, some 9-*cis* pigment (called isorho-
dopsin, λ_{max} = 486 nm in cattle rhodopsin) is also formed in the backreaction
under these conditions. The prelumirhodopsin formed at −196°C completes the
bleaching sequence when it is warmed to room temperature in the dark. The
chicken cone pigment, called iodopsin, behaves differently, and prelumirhodopsin formed at −196°C returns to iodopsin upon warming.

A proton is taken up by rhodopsin during the meta I → meta II transition.
This is surprising because the absorption spectrum of meta II rhodopsin is similar
to that of an unprotonated Schiff base, and that of meta I rhodopsin is similar to
that of a protonated Schiff base. Therefore, the chromophore seems to lose a
proton upon meta II formation, and the protein must take up this proton, as well
as one additional proton during the meta I → meta II transition.

A dependence of the ratio of meta I/meta II on pH can be observed at 5°C,
where bovine meta II has a lifetime of several hours. Meta II predominates at
acid pH, and the ratio shows a simple Henderson–Hasselbalch equilibrium
relationship with a pK near 6.5. At higher temperatures meta III and later
products form. The color of the bleaching product, formed after meta III, is also
pH-dependent in a manner that appears to reflect variable protonation of a retinal
Schiff base linkage.

The meta I/meta II ratio is pressure-dependent at 5°C, when rhodopsin is
located in the lipid environment of the disk membrane.[32] Application of pressure
increases the proportion of meta I at the expense of meta II. Complete reversal of
the meta I ⇆ meta II equilibrium to meta I is obtained at about 2.8×10^6 kg/m².
These observations show that the formation of meta II results in a volume

expansion of the membrane. This volume expansion depends on the lipid–protein interaction in the membrane, because the pressure effect is absent if rhodopsin is extracted into detergent micelles. In detergent extracts, rhodopsin also exposes some additional sulfhydryl groups to reaction with sulfhydryl reagents upon formation of meta II. In contrast in the disk membrane, rhodopsin does not appear to expose additional sulfhydryl groups upon formation of meta II.[33] These observations are some of several indicating that detergent extracts are not adequate models for studying the behavior of rhodopsin upon formation of meta II, and that the membrane environment is very important. The steps up to and including the meta I to meta II interconversion are particularly significant, because spectral changes subsequent to meta II formation are much too slow to account for the brief time delay of visual excitation after light absorption.

10.4.2. The Visual Cycle

The bleaching of rhodopsin leads to the slow liberation of all-*trans*-retinal. In the eye, bleached rhodopsin is regenerated by the enzymatic conversion of all-*trans*-retinal into 11-*cis*-retinal. The 11-*cis*-retinal spontaneously combines with bleached rhodopsin (opsin) to regenerate rhodopsin. There are many intermediate steps in the formation of 11-*cis*-retinal that are diagrammed in Fig. 10-15. The entire process from the bleaching of rhodopsin to its regeneration is called the visual cycle. In isolated retinas, purified ROS, or in detergent extracts, there is little or no spontaneous regeneration. If the bleached rhodopsin is in a membrane environment or in one of a few detergents (e.g., digitonin, Tween-80, or alkylglucoside), regeneration may be accomplished by the addition of 11-*cis*-retinal. Most detergents do not support regeneration from 11-*cis*-retinal. Retinal and other aldehydes make membranes leaky to ions if the membranes contain the lipid phosphatidylethanolamine, but the corresponding alcohols do not.[34] The ROS membranes are rich in phosphatidylethanolamine. Retinal does not build up in the retina upon bleaching, because an oxidoreductase activity (indicated in Fig. 10-14) reduces it rapidly to retinol (Fig. 10-9, structure IV) with the consumption of reduced nicotinamide adenine nucleotide phosphate (NADPH). Some of the retinal-reducing activity survives in isolated ROS, but the NADPH is largely depleted.

The events in the visual cycle, after retinal formation, have not been fully established in detail. If a large amount of pigment is bleached, the retinal is reduced to retinol; the retinol is esterified to a fatty acid and transported to the pigment epithelium. The visual cycle is completed by isomerization to 11-*cis*, return to the ROS, ester hydrolysis, and oxidation to retinal; but the localization and sequences of the events are not entirely clear. There is some evidence for a "short visual cycle" where retinal is isomerized without leaving the ROS, if bleaching of the visual pigments occurs at fluences closer to physiological light levels.[35] Retinas peeled away from the pigment epithelium, and furnished with glucose and simple salts, will not regenerate bleached rhodopsin. However, isolated rat retinas will regenerate slowly if maintained in a very tiny chamber.[36]

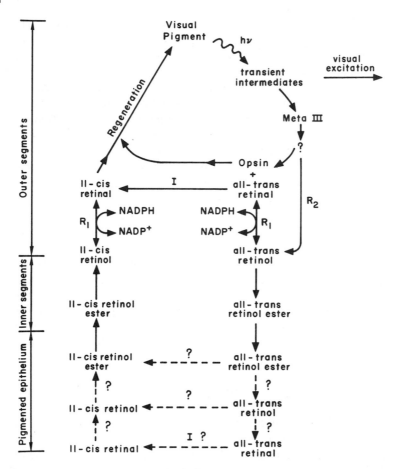

Fig. 10-15. The visual cycle pathway for visual pigment in the living eye. The reactions shown by the solid lines appear to be well established, and the reactions shown by dotted lines are still uncertain under physiological light stimuli. The tissue localization of the reactions is shown on the left, I denotes isomerase enzymes, and R_1 and R_2 denote retinal oxidoreductase enzymes.

Some factor is apparently lost or diluted upon isolation of the retina that appears to be required for regeneration in the visual cycle. Further work is required to deduce the factors involved in the visual cycle, and the role of the pigment epithelium.

10.4.3. Psychophysics and Action Spectra

One powerful way to study the sensory system is to ask an intact animal to report in some way what it detects, or if it detects a carefully controlled stimulus. This method is called psychophysics. Another powerful method, widely used in photobiology, is to measure the quantum efficiency of a response as a function of wavelength of excitation. The resulting curve is called the action spectrum of the response (see Sections 1.5.4. and 3.7.).

Using psychophysical methods, it has been shown in numerous experiments that the action spectrum of vision matches precisely the absorption spectrum of the visual pigments present. These measurements are straightforward in pure rod or pure cone retinas, but are more complex in mixed retinas that have both rods and cones. The sensitivity of a mixed retina matches the absorption of rod VP under dim light, and matches the absorption spectra of cone VP under brighter light stimuli. After stimulation with bright light, the photoreceptors show a marked decrease in sensitivity; this phenomenon is called light adaptation. It is a familiar experience to be momentarily blinded by going from a dim to a brightly lighted environment, then rapidly adapt to seeing well in the bright environment. Similarly, in going from bright to dim light the photoreceptors regain sensitivity. This phenomenon is called dark adaptation, and is also familiar from everyday experience.

Dark adaptation shows complex kinetics. The cones recover sensitivity rapidly; however, while the cones are active, they seem to repress the recovery of the rods.[19] The rods recover sensitivity much more slowly than the cones, with a half-time of about 5–8 min. Complete dark adaptation of the rods takes about 30 min. Understanding the mechanisms of light and dark adaptation will require more knowledge of the internal mechanisms of the photoreceptors (see Section 10.5.)

10.4.4. The Electroretinogram (ERG)

The electrical responses of the photoreceptor cells to light may be measured in a number of ways. A simple and surprisingly informative method is to measure the potential generated across the whole eye in intact animals, excised eyes, or across isolated retinas, in response to light stimuli. These potentials are called the electroretinogram (ERG) and may be measured with gross electrodes (e.g., cotton wicks soaked in saline), or with extracellular microelectrodes. These responses have been separated into components that come from the photoreceptor cells (the a wave or P III), from deeper layers of the retina (b wave), and from the pigment epithelium (c wave).

The a wave can be isolated in retina preparations by the addition of the amino acid sodium aspartate, which serves to inactivate the horizontal cells below the receptors.[37] The a wave is highly sensitive to anoxia, and to the removal of glucose, and therefore requires oxidative energy sources. Furthermore, the a wave is suppressed by the removal of Na^+ or by the addition of Ca^{2+}. The appearance of the a wave shows a several milliseconds delay after a stimulus flash.

10.4.5. Early Receptor Potential (ERP)

Very bright stimulus flashes produce a potential that appears with no detectable delay after the stimulus.[38] This response has been called the early receptor potential (ERP). In contrast with the a wave of the ERG, the ERP is highly resistant to anoxia, and is not dependent on the identity of the salts present. The ERP depends linearly on the amount of rhodopsin bleached, whereas the other potentials generated by the retina are highly nonlinear in the amount of rhodopsin bleached.[39] Utilizing the photobackreactions of the bleaching intermediates, it has been shown that the ERP appears to monitor the interconversion of rhodopsin intermediates.[40]

Hagins and Rüppel[41] have shown that the ERP (which Hagins calls the fast photovoltage, or FPV) appears to correspond to charge displacements across the plasma membranes of the rod. Presumably rhodopsin changes its conformation during interconversion between bleaching intermediates, and these conformational changes lead to a change in the distribution of charges across the membrane. The rhodopsin in the pinched-off disk membranes cannot contribute any net voltage to the ERP, so the ERP is due only to changes of the rhodopsin in the enveloping plasma membrane (including, of course, the disks that have not yet pinched off at the base of the outer segment). The spectral changes after bleaching result primarily from the rhodopsin that is in the disk membrane, since 97–99% of the ROS membrane area is due to the disk. The close similarity of the kinetics of the spectral changes and the ERP changes, in both the forward and backward directions, indicate that rhodopsin in the disk and plasma membranes behave very similarly.

The ERP tells us little about the mechanism of visual excitation, but does indicate that rhodopsin undergoes a conformational change after bleaching, and that a conformational change is particularly evident upon formation of meta II rhodopsin. Conformational changes upon meta II formation have been detected in pressure experiments, as well as in the chemical reactivity experiments described in Section 10.4.1. Meta II rhodopsin is the last spectral intermediate formed that is fast enough kinetically to be involved in visual excitation. All the intermediates up to, but not including, meta II form in a dried system prepared from digitonin micelles in gelatin film. If the light-exposed films are hydrated in the dark, meta II is formed. Meta II is therefore the first intermediate that requires an aqueous environment. All these considerations suggest that the formation of meta II produces a substantial conformational change, and the formation of meta II is a leading candidate for the step that triggers visual excitation.

10.4.6. Intracellular Recording of Photoreceptor Responses

Intracellular recording involves piercing a cell with a glass micropipet electrode of very small diameter, and directly measuring the electrical potential across the cell membrane. Electrical contact is provided by filling the micropipet with a concentrated salt solution. Intracellular recording has been very useful in understanding the electrical properties and ion flows in a variety of cells, the most prominent examples being nerve axons and muscle cells. The relatively small size of photoreceptor cells has made intracellular recording difficult. However, in the last few years, improvements in technique have made it possible to obtain intracellular recordings from several different species of amphibian retinas, which happen to have particularly large photoreceptors. Numerous studies have also been made of the intracellular responses of other cells deeper in the retina that are stimulated by the photoreceptors.[42]

The photoreceptors do not give spike responses like axons, but give smoothly graded responses to graded light stimuli.[5,37,43] This behavior is not unusual for short neurons. The photoreceptors have a negative potential inside, as do all cells; however, they are unusual in that they hyperpolarize (become more negative inside) rather than depolarize in response to stimuli. The graded hyperpolarizing property upon stimulation is also shown by the horizontal and bipolar cells deeper in the retina. The ganglion cells behave like ''standard'' neurons; they depolarize in response to stimuli, and have an output that consists of a train of spikes.

Intracellular recordings have been made from both cones and rods, as shown by spectral sensitivities, characteristic saturation behavior, and intracellular dye marking after the recordings. The rods show much slower electrical kinetics than do cones, particularly in the recovery phase after a brief stimulus. The rods have been shown to receive additive signals from nearby rods. This additive behavior contributes to the high sensitivity of rod vision, although it must cause reduced acuity (spatial resolution) in rod vision. With broad field stimulation of neighboring rods, individual rods give detectable signals at much less than 1 quantum absorbed per rod. Rods give responses that increase with intensity up to about 300 quanta absorbed per rod. Above this intensity, the rod response saturates in amplitude and shows increasingly slow recovery after brighter supersaturating stimuli. The voltage response amplitude of photoreceptors follows the relation $\Delta V / \Delta V_{max} = I^n/(I^n + K^n)$, where ΔV is the voltage response, I is the stimulus intensity, n is a constant that ranges from 0.8 to 2.0 depending on the photoreceptor; K is the stimulus intensity that produces half-saturation, and has a value of about 30 quanta for rods.

Dark-adapted cones give detectable responses from about 100 photons absorbed per cone and increase with intensity up to about 10^4 to 10^5 photons per cone.[43] The half-saturating intensity (K) is about 1200–1300 photons per cone. These numbers are only approximate, but it is clear that the cones are much less sensitive than the rods. Upon exposure to conditioning background lights, cones are desensitized and give detectable responses to flashes above the background, up to about 10^8 quanta absorbed per cone. The two types of receptors, rods and cones, together cover a useful dynamic range of intensities of about 10^{10}.

10.4.7. Extracellular Microelectrode Recording

Hagins and co-workers[44] have developed a method that allows measurement of the spatial localization of currents in slices of retina. This method measures extracellular voltage gradients between two relatively low-resistance microelectrodes. The extracellular medium has a small but significant resistance. Currents between two regions of the tissue produce a small voltage between two electrodes placed in these regions. The spatial distribution of voltage gradients along the photoreceptors has been used to locate sources and sinks of current, and to measure the magnitude of these currents.

In the invertebrate retina there is an increase in Na^+ current into the photoreceptor outer segment in the light,[45] but the opposite result is found in vertebrate photoreceptors.[5,44] For example, in the rat retina there is a large Na^+ current ($\sim 10^7$ Na^+ ions rod^{-1} s^{-1}) from the inner segment to the outer segment in the dark, and this current is decreased by light. One photon absorbed per rod decreases the current by about 1%. The decrease is linear up to about half the maximum response, at approximately 30 photons absorbed per rod. The full response curve to flash illumination follows the relation $V/V_{max} = I/(I + 30)$, which is similar to that found by intracellular recording on other rods.

Yoshikami and Hagins[46] found that a high concentration of Ca^{2+} (about 10 mM) reversibly mimicked the effect of light in repressing the photoreceptor Na^+ current. They proposed that Ca^{2+} was the intracellular signal that controls Na^+ permeability, but that extracellularly applied Ca^{2+} did not get into the cell very effectively. The Na^+ dark current runs down in the presence of ouabain (a Na^+-K^+ ATPase inhibitor) and the rate of rundown is slowed by light exposure or by extracellular Ca^{2+}. That Ca^{2+} may be the intracellular signal for visual excitation is called the "calcium-coupling" hypothesis. More will be said about this in Section 10.5.3.

10.4.8. Rod Outer Segment Ion Currents Measured by Osmotic Responses

Fresh isolated frog or rat ROS shrink rapidly after hyperosmotic shock.[47] With most salt solutions the rods remain shrunken. However, in NaCl solution in the dark, the rods rapidly recover to their native length. If shrunken rods in NaCl are illuminated or exposed to high Ca^{2+} concentrations (10 mM), they do not recover their length. These and other experiments show that light and extracellular Ca^{2+} can control an Na^+ flux into the outer segments. The recovery after a hyperosmotic NaCl shock is easily lost if the preparation is not kept highly concentrated in ROS, or if the ROS age more than about 15 min after the retina is dissected.

The rods act as if they have a "spring" in the 15-nm space between the disks; this space collapses in high salt, but returns to 15 nm in NaCl solutions in the dark, as shown by freeze-fracture measurements of the disk membrane spacing. The ROS plasma membrane appears to be permeable to NaCl in the dark, which allows recovery from the osmotically induced shrinkage. From the rate of recovery, it has been calculated that 10^9 Na^+ ions/s pass through the plasma membrane of each frog rod in the dark, and that the absorption of 1 quanta/rod decreases the Na^+ permeability by 1–3%.[47]

10.4.9. Rhodopsin in Artificial Lipid Membranes

Black lipid membranes (BLM) are macroscopic lipid bilayers formed across a hole in a sheet of Teflon that divides two aqueous chambers. They are so named because the bilayer appears black when viewed with reflected white light. Rhodopsin can be incorporated into BLMs.[48] The membranes are not stable if rhodopsin is placed on both sides of the bilayer; however, substantial stability may be obtained if rhodopsin is placed only on one side of the bilayer. This is an interesting property when compared with the native disk membrane, because, as discussed in Section 10.2.3., freeze-fracture and chemical labeling experiments show that rhodopsin is not symmetrically disposed in the disk membrane, but has an asymmetry toward the outer surface.

The BLM system is particularly useful since the conductivity can easily be measured. Rhodopsin has no large effect on the conductivity of the membrane in the dark. However, after a light stimulus the conductivity of a rhodopsin-containing membrane increases markedly. The ion selectivity of the light-induced conductivity or the single channel conductivity has not yet been reported.

Purified rhodopsin has been reconstituted into lipid vesicles with a variety of phospholipids.[49-51] In these vesicles the native asymmetry of the disk membrane is lost, and rhodopsin is symmetrically disposed on the inside and the outside of the bilayer.[50] If lipids are used that undergo a ''solid'' to ''fluid'' phase transition at an experimentally useful temperature, rhodopsin is squeezed out of the pure ''solid'' lipid phase below the phase transition temperature. However, bleached rhodopsin remains dispersed under some circumstances.[50] Spin-label measurements imply that rhodopsin decreases the lipid hydrocarbon chain mobility.[49] The meta I $\rightarrow$ meta II kinetics are greatly accelerated by detergent, but can be restored to near the native rate in these reconstituted systems.[51] Reconstituted systems provide experimental control over the lipid environment of rhodopsin, which may be useful in many types of experiments.

10.5. TOWARD A MECHANISM OF VISUAL EXCITATION AND ADAPTATION

10.5.1. The Internal Transmitter Hypothesis

The topological organization of the rod cell requires that an intracellular ''transmitter'' must carry a signal from the site of light absorption in the disk membrane to the site of Na^+ conductance change on the outer plasma membrane. Only a few newly synthesized disks at the base of the rod are connected with the plasma membrane, and most of the disks have pinched off and are not connected. The plasma membrane accounts for only about 1–3% of the total outer segment membrane area. The remaining 97–99% of the membrane area is made up of the disks. Therefore, the disks must contain most of the rhodopsin, since rhodopsin could not be much more closely packed in the plasma membrane than it is in the disk. Since the rod is so sensitive, the disks not connected to the plasma membrane must be used to cause the cell to respond. This argument leads to the requirement for a signal or ''transmitter'' between the disks and the plasma membrane. It is highly unlikely that this signal is a brute force electrical coupling, which would require the disks to take up or release sufficient amounts of an ion to make a large contribution to the membrane potential. A more detailed discussion of the implausibility of electrical coupling can be consulted.[5] The high sensitivity of the rod is consistent with the intracellular transmitter being a chemical with specific receptors on the plasma membrane. A specific transmitter could effect the plasma membrane Na^+ permeability by binding to specific sites, even if it were released or taken up in very small amounts.

10.5.2. Requirements for a Transmitter

A transmitter would have to be released or taken up by the disks sufficiently rapidly in the light to change the concentration at the plasma membrane. The time involved is about the time for a peak response to a dim flash (100–200 ms).

After a single photon is absorbed, a transmitter would have to be released or taken up in sufficient quantity to detect the change in concentration above the background leakage of the rod, and above the thermal noise in the conductivity of the plasma membrane. Several lines of theoretical analysis indicate that the number must be greater than 100–1000 transmitters/photon.

A transmitter would have to be bound to the plasma membrane, and binding must affect its Na^+ conductance. Experimental manipulation of the transmitter concentration in the dark should mimic the effect of light.

In the dark, the transmitter concentration must return to its initial level rapidly enough to explain the restoration rate of plasma membrane voltage. The restoration rate has a half-time of 0.15–0.5 s with dim stimuli, and gets much longer with bright stimuli.

Rhodopsin in the disk membrane must somehow control the level of the transmitter in the light, possibly by changes in (1) the surface binding of the transmitter, (2) enzyme activity that controls the level of the transmitter, (3) its properties as a transmembrane carrier so that it effects the transport rate of a transmitter from inside the disks, (4) its properties as a transmembrane pore so that it allows the release of the transmitter from inside the disks.

Since the photoreceptor cell has such a high sensitivity, it is most likely that the dark-adapted level of intracellular transmitter is very low, and that photoexcited rhodopsin causes an increase in the level rather than a decrease. In this way a relatively small increase in the absolute amount of transmitter could cause a relatively large fractional increase in the transmitter. A large gain is required, because one photon absorbed in a frog rod blocks the entry of about 10^7 Na^+ ions per rod.[47]

10.5.3. Evidence for the Role of Calcium

Calcium appears to play a role in photoexcitation, and it is likely, although not proven, that the internal transmitter in the ROS is calcium. It was mentioned in Section 10.4. that high levels of extracellular calcium, of the order of 10 mM, mimic the effect of light in shutting off the Na^+ current through the plasma membrane. However, micromolar levels of extracellular calcium are sufficient to mimic the effect of light in slices of rat retina, if a transmembrane ion carrier (ionophore) called X537A is added.[52] This observation is a boost to the calcium-coupling hypothesis, but it has not yet been conclusively shown to operate in the simple way envisioned. The ionophore X537A has a rather low ion specificity; it transports many other ions as well as Ca^{2+}. X537A tends to carry other cations out as Ca^{2+} is carried in. There is some evidence that protons may be the initial trigger for visual excitation, and it is possible that changes in the Ca^{2+} levels are a secondary result of light excitation. It would be expected that the addition of X537A in the presence of very low levels of external Ca^{2+} would remove the Ca^{2+} from the disk, and abolish the light sensitivity of the rod. The light sensitivity of the rod does not decay in low calcium in the presence of X537A for at least 20 min.[52] The ionophore A23187 is more specific to divalent ions, but results with this ionophore have not been reported.

A large number of attempts have been made by various investigators to demonstrate Ca^{2+} release from the disks in the light, and uptake in the dark. Several groups have demonstrated Ca^{2+} uptake by ROS fragment preparations in the dark, but have not been able to show light-induced release in the same preparations. Mitochondrial contamination may be a problem here, because mitochondria take up calcium voraciously, but, of course, would not release it in response to light. Several groups have demonstrated a light-induced Ca^{2+} release, but most of these have so far been with a very low quantum efficiency (≤ 1 Ca^{2+}/photon).

Recent reviews go into the possible role of Ca^{2+} in more detail.[5,49,53] More work is clearly needed to identify the transmitter firmly and to develop an assay for transmitter release. Researchers in the field appear close to determining what rhodopsin does, and of course, once this is established, it should be possible to determine how rhodopsin does it.

10.5.4. Enzymatic Correlates of Photostimulation

Several enzyme activities have been found to change upon light excitation of rod photoreceptors. Most of these changes appear to be too slow to account for visual excitation, but may be involved in sensitivity control (light and dark adaptation); some may be fast enough, however, to be involved in excitation.

Most internal transmitter mechanisms require an active, energy-dependent removal of light-released transmitter in order to reset the system after the stimulus. Therefore, energy-dependent processes in the ROS are of interest. A Ca^{2+} or Mg^{2+} ATPase has been detected in frog ROS fragment preparations.[54] A small but highly reproducible fraction ($\sim 15\%$) of the activity was inhibited by illumination, which argues for the localization of at least this portion of the activity in the ROS. Up to 80% of the total activity was eliminated by inhibitors of mitochondrial Ca^{2+} ATPase, but the light sensitivity of the remaining portion was not tested. The energy substrate for this Ca^{2+} or Mg^{2+} ATPase seemed to be MgATP; it is stimulated by low levels of Ca^{2+} (a few micromolar) and inhibited by higher levels of Ca^{2+}. A Ca^{2+} pump in the disk membrane is required by most forms of the Ca^{2+} coupling hypothesis. The high sensitivity of the rod to light implies that the normal dark-adapted concentration of transmitter (e.g., Ca^{2+}) outside the disks would be very low. If this ATPase is the disk membrane Ca^{2+}

ion pump, it would be expected to work well at low levels of Ca^{2+}. The light sensitivity of the ATPase might reflect inhibition by high levels of Ca^{2+}, released by the bright stimuli used. Bownds and co-workers[55] showed that the ATPase activity of ROS is lowered upon extensive purification and found no activity in their purest fractions. The occurrence of ATPase in the ROS is still an unsettled point, since it is, of course, possible to inactivate enzymes by purification procedures.

Rhodopsin is phosphorylated in the light in intact animals or in membrane preparations (with added ATP present). The light sensitivity is in the rhodopsin substrate and not in the phosphorylating enzyme, because the enzyme phosphorylates exogenous substrates equally well in the light or in the dark. Kühn[56] has suggested that the phosphorylation may have a gain control function, since the dephosphorylation in intact animals has similar kinetics to dark adaptation. It has been reported that rhodopsin is phosphorylated more efficiently by GTP than ATP.

High activities of guanylate cyclase[57] and cyclic phosphodiesterase[58] have been found in purified vertebrate rod outer segment preparations; both of these enzyme activities appear to have a light sensitivity. The activity of the phosphodiesterase increases after illumination, and can be fully stimulated in dark-adapted preparations by the addition of small amounts of fully bleached ROS membranes.[59] In contrast, a light dependent inhibition of guanylate cyclase has been reported.[57] The cyclic guanosine 5′-monophosphate (cGMP) levels drop dramatically in intact retinas after light stimulation.[60] The role of cGMP in the photoreceptor is not yet clear. The guanylate cyclase has a calcium optimum, and the cyclase activity drops substantially at high or low calcium levels.[57] The kinetics of calcium activation or inactivation were not measured, but they could plausibly be quite rapid. It is possible that a chain of transmitters are operating in the ROS with the same gain at each stage. One chain for which there is some support is: $h\nu \rightarrow H^+ \rightarrow Ca^{2+} \rightarrow$ phosphodiesterase, where the cyclase would restore the system.

The early events in the conversion of light into an electrical signal that the brain can process are a fascinating problem. The time seems to be ripe for these events to be understood, as well as the mechanism responsible for light and dark adaptation.

10.6. CONCLUDING REMARKS

The visual system is a fertile ground for future work. There are more questions than answers at this time. Yet, enough is known to guide investigators to ask good questions. The mechanisms that cause the wide variations in the colors of the visual pigments are not established. Our understanding of the structure of the disk membrane is still rather sketchy, and little or nothing is known about the structure of the photoreceptor plasma membrane. We do not yet know what rhodopsin does, and so we have no firm idea of its mechanism of action. Most of the composition and structural information is restricted to rods. The cones (the color receptors) have scarcely been touched by biochemists.

The mechanisms of sensitivity control in light and dark adaptation have not been worked out, although some suggestive observations have been made. There are fascinating questions concerning the mechanisms of membrane synthesis, turnover, and control, which were only mentioned here, but which are being actively investigated. Little has been said about the exciting studies of visual signal processing in the retina and in the brain.[61,62]

As basic knowledge of the visual system is collected, we may hope to gain insight into medical problems of the retina. For example, many types of retinal photoreceptor degeneration are known. These are quite common in older people, but some types such as retinitis pigmentosa all too often afflict young and old alike. One surprising property of normal photoreceptors is that they can be severely damaged by exposure to light.[63,64] In a large number of animal species the light levels for severe damage are only slightly above the levels commonly encountered by humans. Damaging effects are marked in some animals exposed to the light levels equivalent to a bright beach for 8 h or to normal ambient illumination for 2–3 days.[65,66] In experimental animals the light-induced damage is largely reversible, but studies have not yet been done to see if there is a cumulative effect of repeated light damage.

The highly unsaturated membrane fatty acids in isolated, purified rod photoreceptor membranes are extremely sensitive to damage by atmospheric oxygen in the dark.[67] The regenerability of opsin is irreversibly lost in the presence of oxygen in the dark. Light might be expected to facilitate oxygen damage to the photoreceptors by a photodynamic mechanism. A group of Russian workers have recently reported that this is the case.[68]

High levels of vitamin E in the isolated membrane are helpful in retarding this oxygen damage. However, vitamin E is consumed in air; and other potent mechanisms must be present in the intact system to oppose oxygen damage.[67] The photoreceptors are highly metabolically active, containing unusually large numbers of mitochondria, and must function by absorbing large quantities of oxygen from the surrounding medium.

Even potent protective mechanisms against oxygen and light damage must have limits. This may be the reason that light levels slightly above those encountered during evolution are damaging to animal photoreceptors. Potentially damaging light levels are commonly encountered by some people, and might be expected to produce cumulative damage that is evidenced in advanced age. Nutritional factors (e.g., vitamin E) would also be expected to have a role in maintaining the long-term health of the photoreceptors. More drastic forms of photoreceptor degeneration may be due to genetic or disease-related deficiencies in the normally potent protective mechanism against lipid oxidation. The above hypotheses have potential medical importance that can be tested with the present knowledge of photoreceptors.

10.7. REFERENCES

1. R. A. Eakin, Structure of invertebrate photoreceptors, *Handbook Sensory Physiol.* 7(1), 625–684 (1972).

2. T. H. Goldsmith, The natural history of invertebrate visual pigments, *Handbook Sensory Physiol.* **7**(1), 685–719 (1972).
3. M. F. Moody, Photoreceptor organelles in animals, *Biol. Rev.* **39**, 43–86 (1964).
4. M. G. F. Fuortes (ed.), Physiology of photoreceptor organs, *Handbook Sensory Physiol.* **7**(2) (1972).
5. W. A. Hagins, The visual process: Excitatory mechanisms in the primary receptor cells, *Annu. Rev. Biophys. Bioeng.* **1**, 131–158 (1972).
6. A. E. Blaurock and M. H. Wilkins, Structure of frog photoreceptor membranes, *Nature* **223**, 906–909 (1969).
7. R. W. Young, Biogenesis and renewal of visual cell outer segment membranes, *Exp. Eye Res.* **18**, 215–223 (1974).
8. M. J. Hogan, I. S. Wood, and R. H. Steinberg, Phagocytosis by pigment epithelium of human retinal cones, *Nature* **252**, 305–307 (1974).
9. W. Stoeckenius, The purple membrane of *Halobacterium halobium,* in: *The Photosynthetic Bacteria* (R. K. Clayton and W. R. Sistrom, eds.), Plenum Press, New York, in press.
10. R. E. Anderson and L. Sperling, Lipids of ocular tissues. VII. Positional distribution of the fatty acids in the phospholipids of bovine retina rod outer segments, *Arch. Biochem. Biophys.* **144**, 673–677 (1971).
11. S. Futterman, J. L. Downer, and A. Hendrickson, Effect of essential fatty acid deficiency on the fatty acid composition, morphology and electroretinographic response of the retina, *Invest. Ophthalmol.* **10**, 151–156 (1971).
12. R. M. Benolken, R. E. Anderson, and T. G. Wheeler, Membrane fatty acids associated with the electrical response in visual excitation, *Science* **182**, 1253–1254 (1973).
13. S. Schwartz, J. E. Cain, E. A. Dratz, and J. K. Blaisie, An analysis of lamellar x-ray diffraction from disordered membrane multilayers with application to data from retinal rod outer segments, *Biophys. J.* **15**, 1201–1233 (1975).
14. R. A. Raubach, P. P. Nemes, and E. A. Dratz, Chemical labelling and freeze-fracture studies on the localization of rhodopsin in the rod outer segment disk membrane, *Exp. Eye Res.* **18**, 1–11 (1974).
15. B. J. Litman and H. G. Smith, The determination of molecular asymmetry in mixed phospholipid vesicles and bovine retinal rod outer segment disk membranes, *Fed. Proc.* **33**, 1575 (1974).
16. G. G. Santillan, and J. K. Blaisie, A direct analysis of lamellar X-ray diffraction from lattice disordered retinal receptor disk membrane multilayers at 8A resolution, *Biophys. J.* **15**, 109a (1975); G. G. Santillan and J. K. Blaisie, Comparison of the electron density profile for isolated, water-washed photoreceptor disk membranes in intact retina, *Biophys. J.* **16**, 35a, (1976); S. Schwartz and E. A. Dratz, The localization of rhodopsin in the photoreceptor membranes of frog retina, *Biophys. J.* **16**, 36a (1976).
17. B. Honig and M. Karplus, Implication of torsional potential of retinal isomers for visual excitation, *Nature* **229**, 558–560 (1971); B. Honig, A. Warshel, and M. Karplus, Theoretical studies of the visual chromophore, *Accts. Chem. Res.* **8**, 92–100 (1975).
18. C. D. B. Bridges, Absorption properties, interconversion and environmental adaptation of pigments from fish photoreceptors, *Cold Spring Harbor Symp. Quant. Biol.* **30**, 317–334 (1965); C. D. B. Bridges, The rhodopsin-porphyropsin visual system, *Handbook Sensory Physiol.* **7**(1), 417–480 (1972).
19. W. A. H. Rushton, The Ferrier Lecture: Visual adaptation, *Proc. Roy. Soc. B* **162**, 20–46 (1965).
20. P. A. Leibman, Microspectrophotometry of photoreceptors, *Handbook Sensory Physiol.* **7**, 481–528 (1972).
21. R. A. Cone, Rotational diffusion of rhodopsin in the visual receptor membrane, *Nature* [*New Biol.*] **236**, 39–43 (1972).
22. M. Poo and R. A. Cone, Lateral diffusion of rhodopsin in the photoreceptor membrane, *Nature* **247**, 438–441 (1974).
23. P. A. Liebman and G. Entine, Lateral diffusion of visual pigment in photoreceptor disk membranes, *Science* **185**, 457–459 (1974).
24. H. J. A. Dartnall and J. N. Lythgoe, The spectral clustering of visual pigments, *Vision Res.* **5**, 81–100 (1966).
25. G. Wald, Life and light, *Sci. Am.* **201**, 92–108 (1959).

26. B. Honig, A. D. Greenberg, U. Dinir, and T. G. Ebrey, Visual pigment spectra: Implications of the protonation of the retinal Schiff base, *Biochemistry* **15**, 4593–4599 (1976); D. S. Kliger, S. J. Milder, and E. A. Dratz, Solvent effects on the spectra of retinal Schiff bases: I. Models for the bathochromic shift of the chromophore spectrum in visual pigments, *Photochem. Photobiol.* **25**, 277–286 (1977).

27. A. S. Waggoner and L. Stryer, Induced optical activity of the metarhodopsins, *Biochemistry* **10**, 3250–3253 (1971).

28. B. Honig and T. G. Ebrey, The structure and spectra of the chromophore of the visual pigments, *Annu. Rev. Biophys. Bioeng.* **3**, 151–177 (1974).

29. A. Kropf, B. P. Whittenberger, S. P. Goff, and A. S. Waggoner, The spectral properties of some visual pigment analogs, *Exp. Eye Res.* **17**, 591–606 (1973).

30. E. L. Menger and D. S. Kliger, Photoisomerization kinetics of 11-*cis* retinal, its Schiff base, and its protonated Schiff base, *J. Am. Chem. Soc.* **98**, 3975–3979 (1976).

31. T. P. Williams, An isochromic change in the bleaching of rhodopsin, *Vision Res.* **10**, 525–533 (1970).

32. A. A. Lamola, T. Yamane, and A. Zipp, Effects of detergents and high pressures upon the Metarhodopsin I ⇌ Metarhodopsin II equilibrium, *Biochemistry* **13**, 738–745 (1974).

33. W. J. De Grip, G. L. M. Van der Laar, F. J. M. Daemen, and S. L. Bonting, Biochemical aspects of the visual process. XXIII. Sulfhydryl groups and rhodopsin photolysis, *Biochim. Biophys. Acta* **325**, 315–322 (1973).

34. S. L. Bonting and A. D. Bangham, On the biochemical mechanisms of the visual process, *Exp. Eye Res.* **6**, 400–413 (1967).

35. W. Zimmerman, M. Yost, and F. J. M. Daemen, Dynamics and function of vitamin A compounds in the rat retina after a small bleach of rhodopsin, *Nature* **250**, 66–67 (1974).

36. R. A. Cone and P. K. Brown, Spontaneous regeneration of rhodopsin in the isolated rat retina, *Nature* **221**, 818–820 (1969).

37. T. Tomita, Electrical activity of vertebrate photoreceptors, *Q. Rev. Biophys.* **3**, 179–222 (1970).

38. K. T. Brown and M. Murakami, Biphasic form of the early receptor potential of the monkey retina, *Nature* **204**, 739–740 (1964).

39. R. A. Cone and W. L. Pak, The early receptor potential, *Handbook Sensory Physiol.* **1**, 345–365 (1971).

40. R. A. Cone, Early receptor potential: photoreversible charge displacement in rhodopsin, *Science* **155**, 1128–1131 (1967).

41. W. A. Hagins and H. Rüppel, Fast photoelectric effects and the properties of vertebrate photoreceptors as electric cables, *Fed. Proc.* **30**, 64–78 (1971).

42. F. S. Werblin and J. E. Dowling, Organization of retina of the mudpuppy, *Necturus maculosus*. II. Intracellular recording, *J. Neurophysiol.* **32**, 339–355 (1969).

43. D. Baylor and A. Hodkin, Changes in time scale and sensitivity in turtle photoreceptors, *J. Physiol. (Lond.)* **242**, 729–758 (1974); D. Baylor, A. Hodgkin, and T. Lamb, The electrical response of turtle cones to flashes and steps of light, *J. Physiol. (Lond.)* **242**, 685–727 (1974).

44. W. A. Hagins, R. D. Penn, and S. Yoshikami, Dark current and photocurrent in retinal rods, *Biophys. J.* **10**, 380–412 (1970).

45. W. A. Hagins, H. V. Zona, and R. G. Adams, Local membrane current in the outer segments of squid photoreceptors, *Nature* **194**, 844–847 (1962).

46. S. Yoshikami and W. A. Hagins, Control of the dark current in vertebrate rods and cones, *Biochemistry and Physiology of Visual Pigments* (H. Langer, ed.), pp. 245–255, Springer-Verlag, Berlin (1973).

47. J. E. Korenbrot and R. A. Cone, Dark ionic flux and the effects of light in isolated rod outer segments, *J. Gen. Physiol.* **60**, 20–45 (1972).

48. M. Montal and J. I. Korenbrot, Rhodopsin in cell membranes and the process of phototransduction, in: *The Enzymes of Biological Membranes* (A. Martinosi, ed.), Vol. 4, pp. 365–405, Plenum Press, New York (1976).

49. K. Hong and W. L. Hubbell, Preparation and properties of phospholipid bilayers containing rhodopsin, *Proc. Natl. Acad. Sci. USA* **69**, 2617–2621 (1972).

50. Y. S. Chen and W. L. Hubbell, Temperature- and light-dependent structural changes in rhodopsin-lipid membranes. *Exp. Eye Res.* **17**, 517–532 (1973).

51. M. L. Applebury, D. M. Zuckerman, A. A. Lamola, and T. M. Jovin, Rhodopsin purification and recombination with phospholipids assayed by the Metarhodopsin I → Metarhodopsin II transition, *Biochemistry* **13**, 3448–3458 (1974).

52. W. A. Hagins and S. Yoshikami, A role for Ca^{2+} in excitation of retinal rods and cones, *Exp. Eye Res.* **18**, 299–305 (1974).

53. T. G. Ebrey and B. Honig, Molecular aspects of photoreceptor function, *Q. Rev. Biophys.* **8**, 129–184 (1975).

54. T. J. Ostwald and J. Heller, Properties of magnesium- or calcium-dependent adenosine triphosphatase from frog rod photoreceptor outer segment disks and its inhibition by illumination, *Biochemistry* **11**, 4679–4686 (1972).

55. D. Bownds, A. Brodie, W. E. Robinson, D. Palmer, D. J. Miller, and A. Shedlovsky, Physiology and enzymology of frog photoreceptor membranes, *Exp. Eye Res.* **18**, 253–269 (1974).

56. H. Kühn, Light-dependent phosphorylation of rhodopsin in living frogs, *Nature* **250**, 588–590 (1973).

57. R. G. Pannbacker, Control of guanylate cyclase activity in the rod outer segment, *Science* **182**, 1138–1140 (1973).

58. R. G. Pannbacker, D. E. Fleischman, and D. W. Reed. Cyclic nucleotide phosphodiesterase: High activity in mammalian photoreceptor, *Science* **175**, 757–758 (1972).

59. N. Miki, J. J. Keirns, F. R. Marcus, J. Freeman, and M. W. Bitensky, Regulation of cyclic nucleotide concentration in photoreceptors: An ATP-dependent stimulation of cyclic nucleotide phosphodiesterase by light, *Proc. Natl. Acad. Sci. USA* **70**, 3820–3824 (1973).

60. C. Goridis, N. Virmaux, H. L. Cailla, and M. A. Lelaage, Rapid, light-induced changes of retinal cyclic GMP levels, *FEBS Lett.* **49**, 167–169 (1974).

61. R. Rodieck, *The Vertebrate Retina,* Freeman, San Francisco, (Calif.) (1973).

62. T. Cornsweet, *Visual Perception,* Academic Press, New York (1970).

63. W. K. Noell, U. S. Walker, B. S. Kang, and S. Berman, Retinal damage by light in rats, *Invest. Ophthalmol.* **5**, 450–472 (1966).

64. T. Kuwabara and R. A. Gorn, Retinal damage by visible light, *Arch. Ophthalmol.* **79**, 69–78 (1968).

65. K. V. Anderson, F. P. Coyle, and W. K. O'Steen, Retinal degeneration produced by low intensity colored light, *Exp. Neurol.* **35**, 233–238 (1972).

66. T. Lawwill, Effects of prolonged exposure of rabbit retina to low-intensity light, *Invest Ophthalmol.* **12**, 45–51 (1973); J. Marshall, J. Mellerio, and D. A. Palmer, Damage to pigeon retina by moderate illumination from fluorescent lamps, *Exp. Eye Res.* **14**, 164–169 (1972).

67. C. C. Farnsworth and E. A. Dratz, Oxidative damage of the rod outer segment (ROS) disk membrane and the role of vitamin E, *Biochim. Biophys. Acta* **443**, 556–570 (1976).

68. V. E. Kagan, A. A. Shvedova, K. N. Novikov, and Yu. P. Kozlov, Light induced free radical oxidation of membrane lipids in photoreceptors of frog retinas, *Biochim. Biophys. Acta* **330**, 76–79 (1973).

11

Photomorphogenesis

11.1. INTRODUCTION

Photomorphogenesis is a collective term that describes the responses of organisms to light signals that regulate changes in structure and form, both at the gross macroscopic level, and at the subcellular and molecular level. If an organism develops in the dark, the rate of development and the size and shape of specific structures are controlled by its genome and the limits imposed upon the genome by the substrate or stored food materials. However, if light strikes the organism, the rate of development is altered, and new molecular products are produced that determine the morphological structure and form of the organism; e.g., the leaves of dark-grown seedlings are small and compact, whereas the leaves of light-grown seedlings are much larger and have expanded surfaces to capture an optimum amount of light for photosynthesis; in some fungi, dark-grown cells are

Walter Shropshire, Jr.   •   Radiation Biology Laboratory, Smithsonian Institution, Rockville, Maryland

nearly colorless but light-grown cells are brightly colored. In the case of leaves, the shape has been altered by both increased cell expansion and cell division, whereas in the fungus, the synthesis of pigment molecules has been greatly accelerated. Both processes are photomorphogenic.

These photomorphogenic responses evolved because they confer an enormous survival advantage on organisms living in environments that are periodically filled with light. Both morphological and physiological development of plants and animals are regulated by light; not only is the formation of structures for extracting optimal amounts of energy from the environment controlled, but also the rate of flow of energy within the organism.

In general, photomorphogenic responses are processes of growth, development, and differentiation that are regulated by light. Several parameters of light are important in this control: they are spectral quality, irradiance, frequency and duration of exposure, and spatial symmetry or asymmetry of the light. For convenience of discussion, photomorphogenesis is usually limited to those light responses in which the light exposure is nonperiodic and nondirectional. Thus photomovements, such as phototropism, and photoperiodic responses, such as flowering, are often excluded from a discussion of photomorphogenesis. (See Chapter 12 on photomovement, and Chapter 8 on chronobiology.) However, photoperiodic and phototropic responses are expressions of growth and differentiation under the special conditions of asymmetry in time and space of light exposure. They have the same pigment systems as photomorphogenesis, and the responses are mediated through growth and development. Therefore, I have used them, where suitable, as sources of information about photomorphogenic responses.

Photomorphogenic responses have several distinguishing characteristics. First, the energy required for the observable change in form and structure is not obtained from the light signal itself. This energy is derived from the metabolism of food, and the photomorphogenic light signal merely turns metabolic reactions on or off in order to regulate the overall rate of metabolism. Thus, photomorphogenesis can be thought of as a switching mechanism.

In photosynthesis, large numbers of light quanta are absorbed because the desired result is the capture and storage of energy as chemical potential energy. In photomorphogenesis, however, only small numbers of quanta are absorbed. It is a low-irradiance photoregulating process, as opposed to a high-irradiance photosynthetic process. Thus, a second characteristic is that, after the light signal is received, there is a large amplification of the signal that finally results in an observable change in form or the accumulation of a product.

Third, photomorphogenic responses have been subdivided arbitrarily into classes based on the spectral regions that are most effective. For example, leaf expansion is primarily a red light response, while the induction of pigment formation in some fungi is primarily a blue and near-ultraviolet light response. Typically, photomorphogenic responses are described as either red light or blue light responses. There also exist a few photomorphogenic responses with a maximum effectiveness in other spectral regions.

Finally, the photomorphogenic light signal interacts with a number of factors within the organism that influence the magnitude of the responses. For example, there are endogenous oscillations that appear to measure time. If a light signal is received by a plant in the middle of the night in the naturally occurring day–night cycle of the environment, it may be very effective in regulating the morphogenic response of flower formation. However, if the same light signal is given near the beginning or the end of the dark period, it may be completely without effect.

Similarly, as the development of the organism progresses to different stages, there are marked differences in its sensitivity to the light signals. This sensitivity change may be attributable to many factors, including variations in the content of photomorphogenically active pigment or the availability of structures that can respond rapidly.

In general, a quantitative relationship is observed between the size of the light stimulus (total number of quanta absorbed), and the magnitude of the response. However, great care must be taken experimentally in measuring photomorphogenic responses to evaluate the extent of interaction with endogenous rhythms and developmental patterns.

Such photoregulatory processes were recognized in the last century, but only since the 1950s have instrumentation and techniques for unraveling these processes at the molecular level been developed. One of the most intensively studied of these photomorphogenic responses has been the phytochrome system (named for the pigment involved in plants), which is the one that I will describe in some detail. A wide range of physiological responses is known to be regulated by phytochrome; the number of described responses[1,2] now exceeds 100, and includes such diverse processes as stem elongation, seed germination, leaf expansion, and the synthesis of pigments such as chlorophyll, carotenoids, and anthocyanins. It is a fascinating system with a large literature,[3] yet the field still has many unsolved problems that will tax the abilities of research scientists for some time to come.

11.2. THE PHYTOCHROME SYSTEM

In the early part of this century a number of seemingly disparate observations coalesced in an unexpected way. Plant physiologists had known for some time that plants produce flowers at different times of the year. Flowering is largely controlled by day length.[4] Thus, short-day plants, such as chrysanthemums or poinsettias, flower in the fall. Actually, these plants are sensitive to the length of night and must have a long, uninterrupted dark period for flowering to occur. If chrysanthemums are grown under completely controlled conditions with 8 h of light and 16 h of darkness they will flower. But if the dark period is interrupted by short (order of seconds) flashes of red light during the middle of the dark period, they will not flower.

Seed-bearing plants have a mechanism by which they can measure the length of the night and produce both flowers and seed at advantageous times. After a winter-long dormancy the seed germinate in the following spring when climatic conditions are favorable. One of the favorable conditions required is often the day length. This photoperiodic response is obviously of enormous advantage in the perpetuation of the species. Time measurement must be very precise in order for seed to be produced before the first killing frosts and yet allow the photosynthetic processes to accumulate an optimum of stored reserves to support seedling growth in the spring, until the seedling can manufacture its own food by photosynthesis.

After the discovery that night length is the critical factor and that light interruption of the dark period inhibits flowering, it was possible to learn about the nature of the pigment system that responds to light signals. Thus an action spectrum could be determined (Sections 1.5.4. and 3.7.).

An action spectrum is a measurement of the efficiency of light quanta at

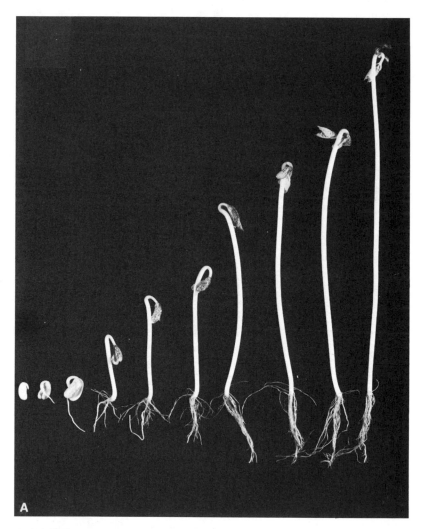

Fig. 11-1. (A) Dark-grown Black Valentine bean seedlings from 1 to 9 days of age. Note hook and length of stem (hypocotyl) between cotyledons and roots. (B) Light-grown Black Valentine bean seedlings from 3 to 9 days old. Note hook opening and inhibition of hypocotyl length between

different wavelengths in producing a response. The first law of photochemistry states that light must be absorbed to produce a chemical change; therefore, an action spectrum is related to the absorption properties of the absorber (i.e., pigment). Thus, if red light produced the response most efficiently, there must be a red-absorbing pigment present. Such a pigment probably would be blue, and its absorption spectrum could be experimentally predicted by action spectrum measurements.[3]

Another response that has been known by plant physiologists for a long time is the difference in growth pattern that occurs in seedlings of essentially identical genetic composition that are grown under identical growth conditions except for the light environment. A familiar example is the growth of shoots from potatoes

cotyledons and roots, as well as leaf expansion. Seedlings were irradiated continuously with a low irradiance of red light (630–700 nm, 100 mW m^{-2}). Seedlings were grown at 25°C in gravel beds subirrigated with tap water.

stored in a dark basement. These shoots are elongated, white, and have small leaves. They are referred to as *etiolated* shoots. Potato shoots that develop in the light are shorter, green, and have large, well-expanded leaves.

Similar responses occur in dark- and light-grown seedlings (Fig. 11-1). Several morphological differences are seen when a comparison is made of seedlings of exactly the same chronological age from the time of first imbibition of water. One of the most noticeable differences is in the length of the hypocotyl (that portion of the stem between the storage organs of the seed, the cotyledons, and the roots), which is regulated by light, and is extremely sensitive to red light.

When bean seedlings first push their way up through the soil, the stem forms a hook to protect the fragile growing tip from mechanical damage by soil particles (Fig. 11-1). This fragile growing point is enclosed between the cotyledons and is pulled rather than pushed up through the soil by the elongating stem. After the hook region erupts through the soil surface and is exposed to light, the hook rapidly unfolds so that the growing tip is at the top of the plant; the cotyledons open, and the young leaves rapidly expand, become green, and begin photosynthesis. With the exception of photosynthesis, all of these developmental processes—de-etiolation, stem elongation, hook opening, leaf expansion, and chlorophyll synthesis—are photomorphogenic responses. It was discovered by action spectrum measurements that red light was the most effective in producing them.

If the soil covering dormant seed is disturbed by plowing, the uncovered seed will sprout because of exposure to light. If the seed are not exposed to light they may remain dormant for many years. An action spectrum (Fig.11-2) determined in 1937 demonstrated that a measured amount of red light induced germination in 50% of the seed. However, if the seed were exposed to light at various wavelengths, blue and far-red light inhibited germination, but red light increased the germination percentage. Thus, under certain conditions, light can either promote or inhibit the germination of seed.

11.2.1. Phytochrome, a Photochromic Molecule

This promotive and inhibitory effect of light on seed germination was not appreciated until the 1950s, when it was suggested that both processes were under the control of a single pigment.[3] The hypothesis was tested on many red light-sensitive systems, and the concept emerged that all of these red light morphogenic responses were controlled by a single pigment that had the unexpected property of being photochromic. A *photochromic* pigment is one that changes its absorption properties after exposure to light, and if left in the dark will revert slowly to the original form. This reversion to the original form (dark reversion) may, in some cases, be enhanced by light (photoreversion). For example, an exposure to red light (660 nm) converts the pigment to a form that promotes the biological response, and its absorption maximum is shifted to the far-red (730 nm) portion of the spectrum. If the far-red-absorbing form of the pigment is exposed to far-red light, the pigment is converted to a form that is no longer promotive, and its absorption spectrum maximum is again in the red light region. If left in the dark, the far-red-absorbing form slowly reverts to the red-

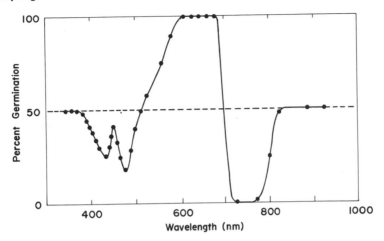

Fig. 11-2. Action spectrum for the promotion and inhibition of germination of light-sensitive lettuce seed. Percentage of germination is plotted as a function of wavelength. A control level of 50% germination was established by exposing the seed to a constant amount of red light prior to the experiment. Subsequent exposures were given at the wavelength indicated and the effectiveness for promotion or inhibition was measured. (From reference 5.)

absorbing form. The pigment acts truly as a switch that can be turned either on or off by exposure to red or far-red light, and this process can be reversed many times by sequential exposures to red and far-red light.

As soon as it was realized that only one photochromic pigment was involved, an intensive search was begun by several laboratories to isolate the active pigment. However, the pigment is present in very low concentrations (less than 10^{-7} M), and could not be detected *in vivo* in the 1950s by direct absorption measurements using conventional spectrophotometers. Advantage was taken of the photochromic absorption shifts of the pigment that occur after red and far-red light exposures. Photoreversible differences in absorption were detected at 660 nm (red) and 730 nm (far-red) using a sensitive dual wavelength spectrophotometer, and these differences provided a sensitive and quantitative physical assay for the presence and form of the pigment. This technique of measuring the absorption changes at 660 and 730 nm opened the way to the isolation and purification of the pigment, which was named phytochrome.

Phytochrome has been isolated and purified by several laboratories.[6] It is a protein that is readily soluble in aqueous buffers, and has an open-chain tetrapyrrole as the chromophoric (light-absorbing) group. By the usual techniques of protein chemistry, it has been purified manyfold, and it is generally accepted to have a molecular weight of about 120,000 daltons and to form dimers of about 250,000 daltons. The absorption spectra (Fig. 11-3) of highly purified material are photoreversible after exposure to red or far-red light, and match well the detailed action spectra for most of the physiological responses (Fig. 11-4). The ultrastructure of the molecule remains unknown, and previous electron microscope pictures of tetrameric, hexameric, or octameric double dumbbell shapes have now been identified as extraneous protein carried along with phytochrome in the

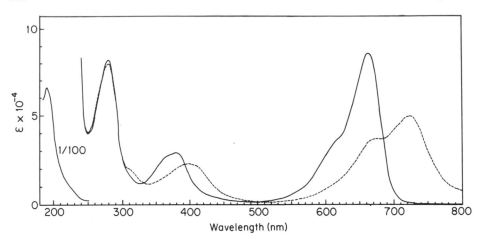

Fig. 11-3. Absorption spectra of highly purified phytochrome extracted from dark-grown oat seedlings. Molar extinction coefficient (liters $mol^{-1}cm^{-1}$) of P_r (red-absorbing form of phytochrome) (—), and P_{fr} (far-red absorbing form of phytochrome) (----). Spectra were determined using a Cary Model 15 spectrophotometer at 4.5°C using a 1.4×10^{-5} M (calculated on a molecular weight of 60,000) aqueous solution (pH 7.2), and a 1-cm path length. (From reference 7.)

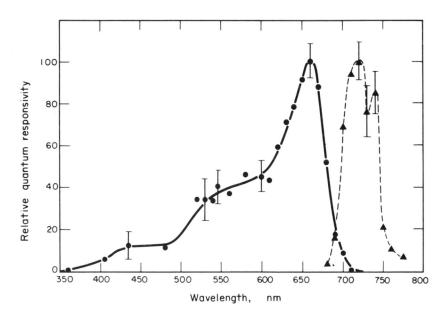

Fig. 11-4. Typical phytochrome mediated action spectra; photoinduction and photoinactivation of seed germination of *Arabidopsis thaliana*. The solid curve is induction; the dashed curve is inactivation. For the inactivation curve the seed were irradiated with 660-nm red light (0.15 J m^{-2}) immediately prior to irradiation with monochromatic far-red light. Standard errors of the mean are indicated at the inflection points. (From reference 8.)

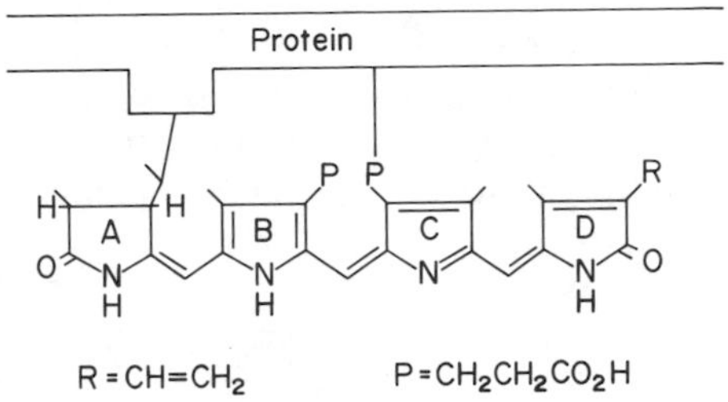

Fig. 11-5. Postulated molecular structure of phytochrome chromophore. The chromophores for both P_r and P_{fr} have identical ring structures. However, the chromophoric system of P_{fr} is shorter than the P_r chromophoric system by the double bond of the methine bridge between rings A and B. (From reference 12.)

R = CH=CH₂   P = CH₂CH₂CO₂H

isolation procedure.[9] Phytochrome behaves hydrodynamically as an ellipsoidal molecule with an axial ratio of about 10:1[10] and is reported to be a glycoprotein with about 4% of its weight as carbohydrate.[11]

The best data on the structure of the chromophoric group indicate that it is an open-chained tetrapyrrole (Fig. 11-5). However, the number of chromophores per protein moiety, the nature of the chemical attachment between chromophore and protein, or the changes that occur in the molecule immediately after absorption of a quantum are not known.[13]

11.2.2. The Reactions of Phytochrome

At the same time that isolation and purification techniques were being developed, much was learned about the *in vivo* reactions of phytochrome[14] by means of a sensitive dual wavelength spectrophotometer[6] to measure the difference in absorption occurring at two wavelengths in dark-grown tissue. The two wavelengths most frequently selected are 660 and 730 nm, to take advantage of the maximum photochromic shift. However, because of the problem of chlorophyll synthesis in dark-grown tissue after red light exposure, measurements often are made only at 730 nm, and compared to a stable reference wavelength of 800 nm where no absorption changes occur after exposure to the actinic red or far-red light. The photochromic nature of the pigment may be easily observed *in vivo*, using this spectrophotometer, and is represented formally by Eq. (11-1):

$$P_r \underset{730 \text{ nm}}{\overset{660 \text{ nm}}{\rightleftharpoons}} P_{fr} \qquad (11\text{-}1)$$

P_r denotes the red-absorbing form, and P_{fr} denotes the far-red-absorbing form of phytochrome. When attempts were made to correlate the *in vivo* measurements of the amounts of P_{fr} present with the physiological responses observed for given exposures to red light, it was soon realized that the correlations were not straightforward, and that the end response was influenced by other factors in addition to the apparent P_{fr} concentration.

The first complication was that the absorption of the two forms of the pigment overlap strongly in some spectral regions, and it is not possible to drive the pigment totally from one absorbing form to the other by irradiation. This fact is evident from the absorption spectra given in Fig. 11-3. For the usual exposure to 660-nm red light there is a mixture of the two absorbing forms present. The quantitative value for such a photoequilibrium obtained may be calculated by Eq. (11-2):

$$\frac{[P_{fr}]_\infty}{[P_r]_\infty} = \frac{I_\lambda \epsilon_{r\lambda} \Phi_r}{I_\lambda \epsilon_{fr\lambda} \Phi_{fr}} = \frac{\epsilon_{r\lambda} \Phi_r}{\epsilon_{fr\lambda} \Phi_{fr}} \tag{11-2}$$

The ratio $[P_{fr}]_\infty/[P_{tot}]$ has a value near 0.80, where $[P_{tot}] = [P_r] + [P_{fr}]$. $[P_r]_\infty, [P_{fr}]_\infty$ = concentrations of the two absorbing forms at photoequilibrium; I_λ = irradiance at wavelength λ; $\epsilon_{r\lambda}, \epsilon_{fr\lambda}$ = molar absorptivity for red- and far-red-absorbing forms at wavelength λ; Φ_r, Φ_{fr} = quantum yield for red- and far-red-absorbing forms.

Thus, in 660-nm light, a photoequilibrium between the two forms is produced such that about 80% of the pigment is in the P_{fr} form. For exposure to 730 nm light, only about 3% of the pigment is in the P_{fr} form. The importance of these values will be examined in a subsequent discussion of responses to high irradiances and the state of the pigment under white light conditions in the natural environment.

A second complication is that relatively slow dark reactions occur after irradiation. If the total amount of photoreversible phytochrome, $[P_{tot}]$, is assayed as a function of time after brief red irradiation, the total amount of reversible pigment decreases rapidly (Fig. 11-6). This process is known as *dark destruction;* P_{fr} appears to be converted enzymatically to an inactive form that is no longer capable of photoreversibility, since the process is temperature-dependent and is sensitive to metabolic inhibitors, such as sodium azide or cyanide. However, the physiological response does appear to be quantitatively correlated to the level of P_{fr} initially produced by brief (less than 5 min) exposures to red light.

Another process that occurs is *dark reversion,* in which the newly formed P_{fr} spontaneously reverts to P_r in the dark. This process is difficult to observe in some tissue, and is complicated by the fact that the initial photoequilibrium established by red light is about 80% P_{fr} and 20% P_r. As dark destruction occurs, the total amount of reversible pigment decreases, and the percentage of P_r appears to increase as a function of time simply because the original 20% P_r becomes available to subsequent exposures of red light as the P_{fr} content decreases. For example, after an hour, a second exposure to red light will establish the same photoequilibrium of 80% P_{fr} and 20% P_r, but since by this time more than half the total P_{fr} has been destroyed, an additional amount of P_r is converted to P_{fr}. These results might be incorrectly interpreted to suggest an apparent synthesis of P_r, when in fact a much lower rate of synthesis, if any, had occurred (Fig. 11-6).

P_{fr} may exist[16] in a protonated form, $P_{fr}H$, which can revert to P_r. P_r, of course, may be synthesized *de novo* from precursors. The increase in reversible phytochrome *in vivo* may be measured spectrophotometrically. This increase as a function of time roughly parallels the increase in fresh weight of the dark-grown

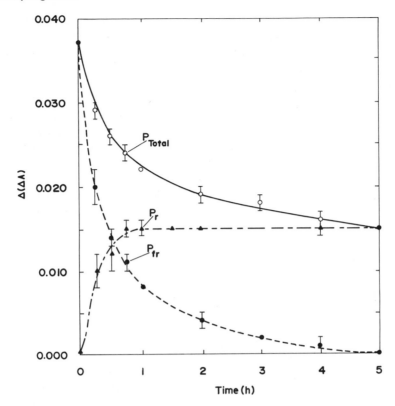

Fig. 11-6. Dark destruction. The time course of changes in the dark in the amount of spectrophoto-metrically detectable phytochrome in bean seedlings after a brief saturating exposure to 660-nm red light (1 kJ m^{-2}). The newly formed P_{fr} disappears rapidly. (o) represents values obtained by measuring the total reversible phytochrome present. (▲) represents the increase in absorbance change after a second exposure to 660-nm red light. (●) represents direct measured values of P_{fr}. In all cases, the absorbance differences were measured against a reference wavelength of 800 nm. The error bars denote standard errors of the mean for 20 replicate determinations. (From reference 15.)

seedlings. However, this increase in photoreversible phytochrome may not be due entirely to synthesis, e.g., as seed imbibe water and begin to germinate, the level of photoreversible phytochrome increases. All that is occurring is probably a change in the microenvironment of phytochrome molecules, which allows photoreversion. The incorporation of radioactive amino acids into extractable phytochrome from rye seedlings grown from seed or from excised embryos was unsuccessful,[14] even though other proteins in the tissues were labeled. There-fore, nearly all the phytochrome in these shoots had been present in the dry seed or embryos. It is obvious that phytochrome synthesis must occur in light-grown plants, but there are almost no data on the rate of synthesis or where synthesis occurs.

The synthesis of phytochrome has been observed in cultures of carrot cells.[14] The increase in the total amount of phytochrome parallels the increase in number of cells that are formed. One problem with the carrot cell culture system is that the levels of phytochrome produced vary by a factor of 2 in different diploid clones originally of common origin. In theory, however, such a tissue culture procedure offers many advantages for following the rates of synthesis of phytochrome. It would be of particular interest to know if there are different turnover rates for the protein portion of phytochrome compared to the chromophore.

A general scheme that summarizes all these processes is given in Fig. 11-7. The conversion of P_r to P_{fr} in the dark has been observed in lettuce seed, even though this conversion is thermodynamically very unlikely. However, such *inverse dark reversion* has never been observed *in vitro* or in seedlings, and it is probably an artifact having to do with the inbibition of seed. An intermediate form of phytochrome, P650, may be responsible for this observation.[17] P650 (formed in the extreme molecular environment of dehydration) may revert in the dark to P_{fr} and may account for the apparent synthesis of P_{fr} from P_r after exposure to far-red light.

11.2.3. The Primary Site of Action

The P_{fr} form is generally accepted to be the physiologically active form of phytochrome, although not all researchers are in agreement. The idea is based principally on two facts: (1) The predominant form of phytochrome in dark-grown seedlings is P_r. If the seedlings remain in the dark, most of the developmental processes, i.e., de-etiolation, do not occur except after very long periods of time (order of days or weeks). P_{fr} is assumed to be the active moiety that brings about the response. (2) P_r in dark-grown seedlings is relatively stable, but as soon as P_{fr} is formed it begins to disappear from the tissue, presumably because of its use to trigger developmental processes. It could just as easily be argued that P_r is an inhibitor of the developmental processes, and once it is removed the responses occur. This question will not be finally resolved until the precise molecular mode of action of P_{fr} is known.

In the summary scheme given in Fig. 11-7, P_{fr} is shown to combine with some reactant, X, which then brings about the observable physiological responses. The nature of X is unknown, but it is probably involved in amplifying the effect of the P_{fr} molecules. Although the effect is amplified, the response size is usually proportional to the amount of P_{fr} produced, and the *reciprocity* law is obeyed, i.e., if the two variables irradiance and time are adjusted so that their product is a constant, the response produced is the same over a fairly wide range of irradiances and times of exposure. Obviously, if the irradiance is extremely low, and the time required is of the same order of magnitude as the time required to observe the physiological response, then reciprocity fails.

11.2.4. The High-Irradiance Response System

Early in phytochrome research it was recognized that for responses produced by continuous irradiation, and particularly for high values of irradiance,[18] the response magnitude is not proportional simply to the amount of P_{fr} initially present, but rather to both the level of P_{fr} maintained over a period of time and

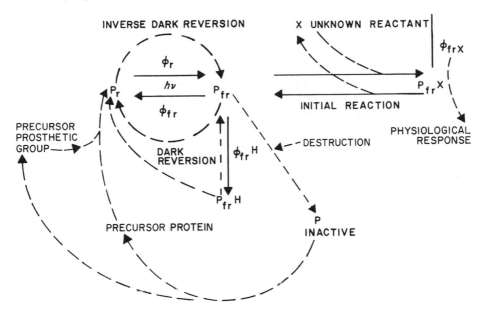

Fig. 11-7. Postulated relationships between the various forms of phytochrome. P_r is the red absorbing form, and P_{fr} is the far-red absorbing form. The solid arrows denote phototransformations. Φ_r and Φ_{fr} are the quantum efficiencies of the red and far-red forms, respectively. The dashed lines denote dark reactions. Combination of P_{fr} with the unknown reactant, X, leads to physiological responses. $P_{fr}H$ is the protonated form discussed in the text. (From reference 14.)

the irradiance value. These responses are irradiance-dependent, i.e., reciprocity is not obeyed, and it is often difficult to demonstrate that they are photoreversible. However, under correct experimental conditions, and for brief light exposures, the responses usually can be shown to be fully photoreversible by red and far-red light exposures, thus demonstrating that they are mediated by phytochrome.

An example of such a system is shown in Fig. 11-8 for anthocyanin synthesis in mustard seedlings. The accumulation of this red pigment product during exposure to either continuous red or far-red light is clearly irradiance-dependent. The far-red light maintains a nearly constant level of P_{fr} over the 12-h irradiation, but increasing the irradiance 1000-fold increases the response size 4-fold. However, for brief exposures, reciprocity can also be shown to hold.[19] If action spectra are determined for such high irradiance responses, they do not coincide with the absorption spectrum of either P_r or P_{fr} (Fig. 11-9; cf. Fig. 11-3). These action spectra typically have a maximum near 720 nm and multiple absorption maxima in the blue portion of the spectrum. In addition, simultaneous irradiation given at two wavelengths that are ineffective alone (i.e., 658 and 766 nm), is very potent in combination; in effect, they are synergistic. If an effective wavelength such as 717 nm is mixed with 658 nm, the effectiveness of 717 nm is greatly decreased, indicating an antagonistic interaction.[20]

From these kinds of data, the concept of a cycling process arose in which the system seems to "count" the number of times a molecule cycles between the P_r and P_{fr} form. For any given mixture of irradiances at two or more wavelengths, once photoequilibrium is reached the relative proportions of P_r/P_{fr} remain constant if the dark-reaction rate constants are assumed to be very much smaller than for the light reactions indicated in Fig. 11-7. The greater the irradiance, the more rapid the interconversion between the two forms, even though the equilibrium value might remain the same. If the response is proportional to the interconversion rate rather than simply proportional to the concentration of P_{fr} molecules, the magnitude of the response will be irradiance-dependent. The synergistic effects observed for two wavelength mixtures would also be explained, since the simultaneous exposure to two wavelengths that are ineffective alone (658 and 766 nm) would increase the rate of interconversion.

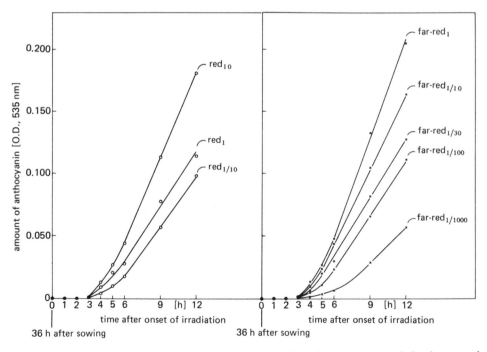

Fig. 11-8. The high-irradiance response. The time course for anthocyanin accumulation in mustard seedlings under continuous red (675 mW m^{-2}) or far-red (3.5 W m^{-2}) light as a function of irradiance. These irradiances are indicated by the subscript 1, and the other subscripts indicate multiples of these irradiances. The mustard seedlings were grown in the dark for 36 h, and then exposed continuously to the irradiances indicated. For far-red light, the level of P_{fr} remains nearly constant over 12 h, so it is clear that the magnitude of the anthocyanin accumulation is a function of irradiance. For brief exposures (< 5 min) of red or far-red light, this response obeys reciprocity and is fully photoreversible.

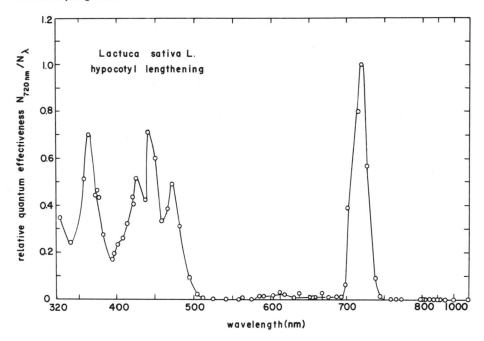

Fig. 11-9. High irradiance action spectrum for inhibition of hypocotyl growth in lettuce. The relative quantum effectiveness was normalized at each wavelength to the effectiveness at 720 nm. Grand Rapids lettuce seedlings were exposed continuously between 54 and 72 h after sowing, and the increase in hypocotyl length during this period was measured. A 50% increase in length compared to the dark control was observed for an irradiance of 145 pmol quanta $cm^{-2}s^{-1}$ at 720 nm at 25°C. (From reference 20.)

More recently, these high-irradiance responses have been explained in terms of the steady-state level of total phytochrome under continuous far-red irradiation.[21,22] If all the light and dark processes occurring for phytochrome (Fig. 11-7), including postulated intermediate forms, are considered mathematically by constructing a model to show the interdependence of these processes, the equilibrium value for total phytochrome is irradiance-dependent. The P_{fr} steady-state level is found to be proportional to the reciprocal of the irradiance of the far-red light. Using this same mathematical description,[21,22] and substituting known or estimated values for the rate constants of each process, theoretical action spectra can be constructed that coincide with a number of experimentally determined action spectra. The only assumptions made are: (1) a threshold value exists below which phytochrome no longer decays out of the system, and (2) the rate constants vary only slightly from one biological system to another. The mathematical description is very appealing in that it can predict quantitatively the results observed for a diverse range of responses having very different action spectra maxima. However, the general validity of this mathematical model awaits additional experimental verification.

11.2.5. Differential Gene Activation

The mathematical models and the calculations of the concentration of P_{fr} do not attack the problem of what the P_{fr} molecule does at the molecular level to bring about the physiological response. In other words, what is compound X with which P_{fr} reacts? Two general mechanisms that are not mutually exclusive have been proposed. One is that the primary mode of action of phytochrome is on differential gene activation (Fig. 11-10). As P_{fr} is produced it differentially exerts its effect upon potentially active and repressible genes, with the concomitant production or inhibition of enzymes that regulate product formation. The clearest examples in support of such differential gene activation are the control by P_{fr} of the induction of phenylalanine ammonia lyase, and the repression of

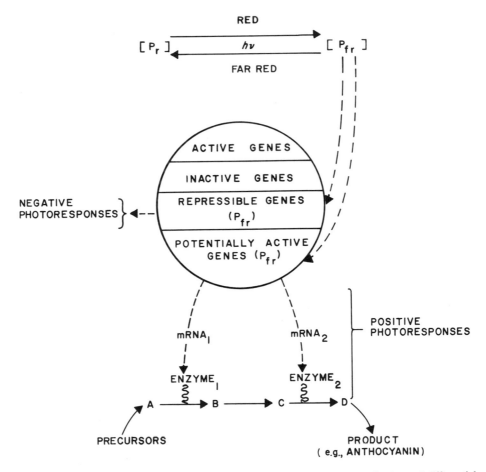

Fig. 11-10. Proposed scheme to illustrate the hypothesis of differential gene activation and differential gene repression controlled by P_{fr} for phytochrome mediated responses. The genome of the plant is represented in the circle. As the active P_{fr} is produced by light ($h\nu$) it can either repress genes to turn off the production of enzymes (negative photoresponses) or it can induce *de novo* enzyme biosynthesis (positive photoresponses). (From reference 14.)

lipoxygenase. A list has been tabulated of 35 enzyme systems that are controlled by the P_{fr} level, and 8 that are unaffected by P_{fr} (see reference 3, p. 488). However, density-labeling data using ^2H_2O indicate that phytochrome probably does not in all cases exert control directly upon the *de novo* synthesis of enzymes.[23]

11.2.6. Interaction with Membranes

Some responses mediated by the phytochrome system occur rapidly within a few minutes, suggesting that phytochrome exerts its control at the level of cellular membranes by altering their permeability. One of the most striking examples of such an effect is the Tanada effect.[14] If excised root tips of barley or mung bean are irradiated with red light, a positive charge is induced on the surface of the root tip. When the root tips are swirled in a liquid medium in a beaker in which the glass surface has previously been charged with phosphate ions, they will adhere to the negatively charged glass surface. If the tips are then irradiated with far-red light, the tips no longer adhere. This red, far-red photoreversibility is indicative of a phytochrome-controlled response. Direct measurements of changes in bioelectric potentials induced by light on the coleoptiles of grass seedlings indicate that these potential changes of membranes are under phytochrome control.[24] The rapid transmission of phytochrome-mediated signals between different organs of bean seedlings also implicates membranes.[25]

The isolation of phytochrome bound to particulate cellular material has been successful,[26] and characterization of this presumed phytochrome–membrane system is being intensively pursued. After exposure to red light, 40% of the phytochrome extractable from squash seedlings is pelletable as bound P_{fr}. If the red light exposure is followed by far-red light, only 4% of the phytochrome is pelletable. P_{fr} appears to bind selectively to particulate material. In addition, the bound P_{fr} is destroyed, since it gradually loses photoreversibility. If a 2-h dark period intervenes before extraction, little phytochrome is found in the bound condition. Thus, it is concluded that P_{fr} destruction occurs while it is bound on a membrane. However, it is clear that the demonstration of pelletable phytochrome does not prove that phytochrome is membrane-associated.[27]

From purified phytochrome, specific immunological stains have been prepared. This procedure has been used to localize phytochrome intracellularly, and the data are in support of a membrane-bound fraction after exposure to red light. Prior to red light exposure phytochrome is homogeneously distributed, but in the P_{fr} form it appears to migrate to specific receptor sites.[28]

These data are in agreement with the changes in physiological fluence–response curves for chlorophyll accumulation in dark-grown seedlings, and the change in sensitivity for seedlings given a prior exposure to any light.[29] If the seedlings have been given an exposure to red light, red followed by far-red, or far-red light alone, the fluence–response curves are shifted along the energy axis so that a greater light exposure is now required to bring about the accumulation of chlorophyll. However, the slopes of the fluence–response curves are greater, so the absorbed quanta now appear to be more effective.

The hypothesis that emerges is that P_r molecules are more or less homogeneously distributed throughout the cell, and, after any light exposure, P_{fr} molecules are concentrated in active sites on a membrane. The probability for capture of an incident light quantum is now diminished because of the phytochrome packing on the membrane, and a reduced area of the pigment absorber is presented to the incident energy. However, once the quantum is absorbed, it is more effective in stimulating the photomorphogenic response. After far-red light alone, or a photoreversion cycle of red followed by far-red, the total amount of P_{fr} remaining is small, but it becomes membrane-bound and affects the subsequent fluence–response curves. Perhaps this explains the fact that any light exposure that can be absorbed by the phytochrome molecule may result in de-etiolation. Whatever photosteady state of $[P_{fr}]/[P_r]$ is established, some P_{fr} becomes membrane-bound, and this bound material alters the subsequent light sensitivity of the system.

11.3. BLUE LIGHT RESPONSES

The second category of photomorphogenic responses are those conveniently referred to as blue-light-mediated. There are a large number of these types of photomorphogenic responses, and from action spectra data they appear to have a common photoreceptor. Some examples of these responses are photomovement (Chapter 12), light-induced growth responses of fungi, protoplasmic streaming, cytoplasmic viscosity changes, and many responses in which growth and differentiation are altered.[30,31]

11.3.1. Photoinduction of Carotenoid Synthesis

A typical blue light response system is the induction of carotenoid synthesis in the mycelium of the fungus *Neurospora*. The light requirement is nearly an absolute one with only trace amounts of carotenoids being synthesized in the dark. If dark-grown mycelial pads are exposed to brief, low total number of quanta, and are then kept at 6°C for 48 h, large amounts of carotenoids are synthesized. These are yellow, orange, and red carotenoids, but the major one is *neurosporoxanthin,* an orange-red, 35-C, acidic carotenoid. A detailed action spectrum is given in Fig. 11-11 and is typical of such blue-light responses. Multiple absorption maxima occur in the region between 400 and 500 nm; there is no activity beyond 520 nm, but there is activity in the near-UV region. The efficiency of near-UV radiation in these responses is variable, and has caused much dispute about the nature of the photoreceptor. Some blue light response action spectra have a clear maximum at about 365 nm, and this argues for a flavin-type photoreceptor. Other action spectra have only a small peak at 365 nm, and arguments are made for carotenoid photoreceptors.[31]

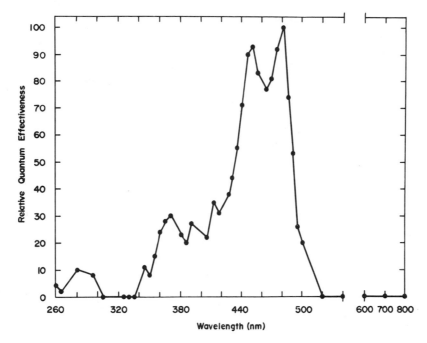

Fig. 11-11. Action spectrum for the photoinduction of carotenoid biosynthesis in mycelial pads of the fungus *Neurospora crassa* a blue-light photomorphogenic response. (From reference 32.)

11.3.2. Nature of Photoreceptor

Most of the convincing data argues for a flavin-type photoreceptor, but the issue has not been finally resolved.[33] The best proof, of course, is to isolate the photoreceptor system and observe it functioning *in vitro*.

In vivo transmission measurements show absorption shifts in *Neurospora,* and in the slime mold *Dictyostelium* following exposures to blue light. Extracted materials[34,35] also show these same absorbance changes, and a flavin-mediated reduction of a cytochrome has been demonstrated. However, it is not yet clear whether or not this system is the physiologically relevant system responsible for the blue-light photomorphogenic responses. In support of a flavin photoreceptor, direct optical excitation of the lowest triplet state of the photoreceptor has been achieved.[36] An action spectrum maximum was observed at 595 nm for both phototropic and light growth responses in sporangiophores of the fungus *Phycomyces,* using tunable laser stimulation. The maximum at 595 nm is 10^{-9} the height of the maximum at 455 nm and agrees with calculations for the absorption cross section for riboflavin in that spectral region.

Although it appears that the blue-light responses are mediated by a common photoreceptor system, this may not be true for all cases. In all probability the evolution of photoresponsive systems has occurred over a number of pathways and at a number of junctures in time. For example, in *Neurospora* and *Cephalosporium diospyri* carotenoid photoinduction occurs only for wavelengths shorter than 520 nm.[32,37] Some bacteria and mycobacteria respond also to red light, and the photoreceptor is believed to be a porphyrin.[38] In addition, carotenoid synthesis in higher plants can be regulated by the phytochrome system.[1]

11.4. ALGAL PHOTOMORPHOGENESIS

There are other photomorphogenic responses that occur in the algae. One of these is chromatic adaptation.[39] Depending on the wavelength distribution of the light energy in the environment in which algae are grown, they may alter their pigment composition. These alterations in pigment composition are usually due to differences in the relative rates of synthesis and total amounts of biliprotein pigments, phycocyanin, phycoerythrin, and allophycocyanin.

The biliproteins, as a whole, serve as accessory pigments in that quanta absorbed are transferred to photosynthetically active sites. Thus, if the algae are in a region where there is little photosynthetically active light, chromatic adaptation occurs, and pigments that will absorb strongly in this spectral region are synthesized. The organisms that have chromatic adaptation are able to harvest most of the available energy and have a survival advantage.

Coupled with the changes at the subcellular level in pigment composition, gross macroscopic photomorphogenic responses also occur, such as those described for the alga *Fremyella diplosiphon*.[40] When filaments are grown under fluorescent white light, the mean length is 460 μm, and they average 40 to 60 cells per filament. If they are grown under red light, the mean filament length is 50 μm, and they consist of only 10–15 cells. These effects of light of different wavelengths on morphology (Fig. 11-12) are fully reversible by sequential light treatments.

Action spectra for these responses (Fig. 11-13) indicate that green light stimulates long filaments *(aseriate)* and red light generates cultures with short filaments. A search for pigments matching the action spectra maxima has not been successful. Most authors feel that allophycocyanin is the receptor for the red maximum, but that the green maximum suggests an aggregate of pigments. Also, by analogy with phytochrome, a photoreversible pigment system has been suggested, but to date the active photoreceptor has not been isolated. Comparable reversible pigments have been reported for other algae with absorption maxima in the blue-green and yellow spectral regions.[41]

From an evolutionary standpoint, a dual role has been suggested for the phycobiliproteins.[39] They evolved not only as accessory pigments to aid photosynthesis under adverse conditions, but, at the same time, similar or identical pigments were used for controlling metabolism. It is completely reasonable that the processes of inhibition and promotion of filament length are under biliprotein control, and that it was only later that a single photochromic biliprotein evolved, as in the case of phytochrome for higher plants.

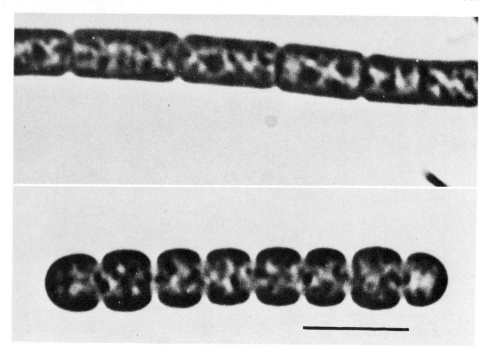

Fig. 11-12. Morphological differences between filaments of the filamentous blue-green alga *Fremyella diplosiphon* adapted to fluorescent (upper) and red light (lower) irradiation. The fluorescent irradiated (250 footcandles) cultures have a mean filament length of about 460 μm and 45–60 cells. The red irradiated (~0.1 W m^{-2}) filaments have a mean length of 50 μm and contain 10–15 cells. These light-induced changes are fully photoreversible. The bar length is 10 μm × 1950. (From reference 40.)

It is also of interest that such chromatic adaptation may serve a self-regulating morphogenic function, coupled with natural periodic fluctuations in the light environment. In the aquatic environment, the red portion of sunlight is attenuated faster by depth of water than is blue or green light. In addition, the phytoplankton themselves act as spectral filters. With an increase in dissolved nutrients, and an increase in phytoplankton population, there a relative increase in the green portion of the spectrum at lower depths. The development of the alga *Nostoc* exhibits delayed formation of filaments at increased depths and under competition for red light. Thus, when *Nostoc* populations develop, the aseriate stage develops in response to the penetration of green light. Such growth cycles become synchronized and may be the way in which planktonic blue-green algal blooms are regulated.[39]

These changes in morphology *(pleiomorphism)* are regulated by the spectral distribution of the light to which the organism is exposed. The morphology is environmentally controlled, and such alterations in form have caused a great deal of trouble for the systematists attempting to classify organisms on the basis of form and structure. Many of these algae have been classified as distinct species when in fact they are merely pleiomorphic variations of the same species.

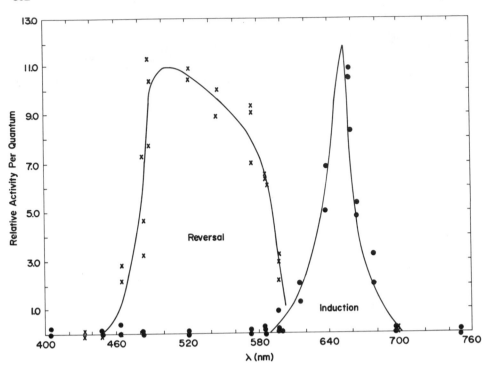

Fig. 11-13. Action spectra for photoinduction and photoreversal of filament formation from aseriate cultures of the blue-green alga *Nostoc muscorum* A. Dark-grown cells that had developed into aseriate microcolonies on agar (14–18 days old) were induced with fluences at each wavelength of 150 J m^{-2}. For reversal, the cells were given a saturating red light exposure followed by exposure at each wavelength to a fluence of 288 J m^{-2}. The development into filaments was scored after 4 days in the dark. (From reference 39.)

Another general principle for photomorphogenic responses is that they are rarely processes that are absolutely dependent on light. In many cases the developmental process evolved, and then a light-sensitive photoreceptor was adapted as the environmental sensor. For example, in *Nostoc* the rate of synthesis of the biliprotein phycoerythrin is regulated by light. If *Nostoc* is cultured in complete darkness, when an external energy source is added, phy-coerythrin is eventually synthesized after several weeks. Analogously, for the flowering process of higher plants, if peas or potatoes are left in complete darkness for periods of 6 or more weeks, abundant flowers develop. In American Wonder peas, both flowers and fruit develop in complete darkness.[42] Most light-sensitive seed will germinate in the dark if they remain there long enough or are stimulated by other physical manipulations, such as removal of the seed coat or osmotic shock. The conclusion is that light appears to exert a true regulatory function in photomorphogenic responses, and there is not an absolute require-ment for light to initiate the response.

11.5. IMPORTANCE OF PHOTOMORPHOGENESIS

Most of the responses that I have described have had their molecular mechanisms unraveled in the laboratory under well-defined and controlled conditions. Because of the complexity of interaction with the environment, this has been the only reasonable approach. However, we are now at the state of knowledge where we would like to be assured that what works in the laboratory actually represents what goes on in the natural environment.

To date, no one has been able to assay the $[P_{fr}]/[P_r]$ ratio in green plant materials grown under natural conditions. In fact, no one has successfully isolated phytochrome from fully green plants; however, we are reasonably confident that it is there because of the observed responses. For example, plant physiologists have long observed the differences in leaf size and shape between so-called sun- and shade-grown plants. Shade-grown leaves are usually larger in area than the same species grown in open fields, but sun-grown leaves are thicker and the leaves near the middle portion of the stem tend to be the largest.[43]

Confusion in the literature with respect to leaf size and shape controlled by light results from several factors.[44] (1) Not only is the irradiance important, but (2) the spectral quality of the irradiance reaching the leaves is crucial. Like the *Nostoc* system described in Section 11.4., the leaves of higher plants are excellent far-red transmitting filters. If the spectral quality of light passing through a canopy of leaves is examined as a function of the distance from the apex of a plant, the far-red portion of the spectrum increases. Leaves at the top of the plant may be in a red light-rich environment, but the leaves in the middle of the plant may be in a far-red light-rich environment. Stem elongation as well as leaf expansion, therefore, must be influenced strongly by natural shading. (3) Under laboratory conditions, the mineral and water relations of the roots are critical for the development observed, and in poplar leaves the photosynthetic capacity is a function of not only light intensity, but also of water and ion balance.[44]

A number of factors interact, but those that have been too long ignored are the subtle changes occurring in the spectral quality of the microenvironment where growth takes place. We know too little about this parameter and its effects. As a further example, much of the literature discusses *compensation points* (the value of light intensity needed to just balance the energy needs of the plant) for sun and shade plants.[45] The irradiances vary by a factor of 10-fold for compensation values, but rarely is consideration given to the spectral quality of the light environment. Regulatory photomorphogenic processes are probably involved in determining these compensation values.

The invasion of geographical areas by new species and the competition of plants are complex problems. The prevention of germination of weed species is probably regulated to a large extent by the enhanced far-red light environment beneath a green, leafy canopy.[46] The germination sensitivity of seed produced by a plant is regulated by the spectral quality of the light received by the parent plant just before seed maturation occurs; at least, this phenomenon can be demonstrated in the laboratory.[47] However, the survival value of such processes in nature is not known.

11.5.1. Importance of Spectral Quality in the Natural Environment

The importance of spectral quality in the environment is poorly understood. Calculations have frequently been made from theoretical models of the sun and sky, but rarely—because of experimental difficulties and cost—have measurements been made *in situ*. Model calculations[1] indicate minor changes in the red/far-red ratios, but measurements of direct solar radiation near sunrise and sunset,[47] as well as broad-band measurements of total sun and sky light,[48] indicate that changes in the ratios do occur (Fig. 11-14).

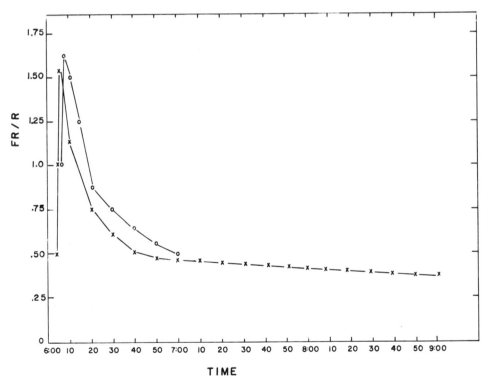

Fig. 11.14. Measured ratios of the irradiances of far-red (730 ± 4 nm, FR) and red (660 ± 4 nm, R) sunlight for 1-min intervals shortly after sunrise for two typical clear sky days in Washington, D.C. Measurements were made with a collimating tube that accepted a solid angle 2° larger than required to admit all the direct light from the solar disk, positioned over a red-sensitive photodetector. The instrument was manually aligned so that the center of the tube was aimed at the center of the sun. The flux transmitted through interference filters was measured by alternately positioning the filters in the light path. The time for successive measurements at the two wavelengths was no more than 5 s. The upper limb of the solar disk appeared at the horizon at 5:47 A.M. for the upper curve, and 5:45 A.M. for the lower curve. The measurements were made from the roof of the Smithsonian North Flag Tower on May 31 and June 1, 1963. (From Reference 47.)

The total sky light (direct, scattered, and reflected) is important because it is to this that an extended receiver such as a leaf is exposed—not just the direct solar radiation. An example of the importance of the total sky light is in the measurements for determining the short-wavelength minimum of UV radiation that reaches the earth's surface. If a measurement is made of the solar radiation by directing a collimating tube so that only the disk of the sun and the area immediately surrounding the sun are measured, the cutoff value on the short-wavelength side is near 290 nm. However, if a quartz diffuser is placed over the same measuring instrument so that a sample is taken from the entire hemisphere, including both sun and sky radiation, 285-nm radiation is recorded. This result is consistent with the hypothesis that the 285-nm radiation is apparently scattered out of the direct beam, but by multiple scattering and reflection is a significant component from the sky.[49] There is no substitute for direct measurements of the spectral distribution actually occurring in the light environment to ascertain how biological systems may be responding to light stimuli.

11.5.2. Commercial and Economic Importance

If the onset of flowering and fruit-set of commercial crops could be regulated by spraying the crops with a chemical that would shift the phytochrome molecule to the P_{fr} form, it would be of enormous economic importance. Unfortunately, we do not yet know enough about the mode or site of action of phytochrome to be able to design and synthesize a chemical that might produce this change. However, because light regulates this photochromic change, it has been suggested, and successfully demonstrated, that plant growth over relatively large areas can be controlled using a red laser beam.[50] Unfortunately, the area that can be controlled is not large enough to be commercially inviting, because large expensive lasers are required. In theory, such control might be economically feasible if the power requirements and size of the light sources were not prohibitive. However, from a photobiologist's point of view, there is a more important reason to understand photomorphogenesis and the processes occurring in the natural environment that are under photomorphogenic control. For example, it has been suggested that areas of the world might be kept light at night by the use of highly reflective surfaces on large satellites in stationary orbit. Technologically such satellites can be constructed and were seriously proposed with project Abel in 1966[51] for military purposes. Fortunately, this use never occurred because of the objections of the astronomers who bemoaned the loss of dark observational skies, as well as the concern of photobiologists interested in photomorphogenesis.

If such a light-reflective satellite accidentally developed a wobble such that a brief flash of white light at an irradiance of only three times full moonlight swept across the northern hemisphere, it is conceivable that many commercially important crops would be prevented from flowering or forming fruit. Imagine the consequences of enormous fields of vegetative plants unable to produce either fruit or grain before winter, not to mention the unknown effects of such light interruptions upon photoperiodically sensitive birds and mammals.[52]

In relation to proposed satellites, one of the questions posed to photobiologists was what is known about photoperiod requirements for tropical plants? The day length near the equator varies only slightly, and it has been assumed for a long time that photoperiod in the tropics plays a negligible role. Observations of woody species in Panama, however, indicate that this may not be so.[53] The times of flowering for some temperate zone species are as precise as ±15 min in a 12-h day, and tropical plants may be even more precisely measuring the day length. We simply do not know at present the extent of control for most tropical plants.

Photomorphogenic considerations arise also with the development of new light sources. Sodium vapor lamps, which are economical and conserve electrical power in their operation, have played havoc with woody species in our cities.[54] The sodium lamps are relatively rich in the red portion of the spectrum. Since they are allowed to burn all night, they interfere, probably through phytochrome, with both growth rates and the natural light–dark cycles required by trees to become dormant before the first killing frosts occur. Trees such as red maples or elms obviously cannot be planted in city areas where sodium vapor lamps are routinely used, or if they are, adequate shading of the trees must be provided.

Finally, there is an ever-increasing number of commercially important problems in our use of artificial light sources. Commercial greenhouse growers easily regulate the production of floral crops such as Easter lilies, poinsettias for Christmas, or chrysanthemums year round by regulating the photoperiod. For many areas of the world the availability of leafy vegetables or fruits such as tomatoes are a luxury in the winter. Commercial growers either supplement or grow such crops in controlled greenhouses and use prodigious amounts of electrical power for the production of these plants. The photomorphogenic responses may be important in regulating the efficiency of photosynthesis. If the spectral quality as well as the intensities and the time rate of application of light could be optimized, it should require far less electrical energy per unit of fresh weight of crop produced. At present we cannot predict the best combination of photomorphogenic lighting for producing the desired crop efficiently at minimum cost, because we simply do not know enough about the photobiology involved in the interactions between the photosynthetic and regulatory photomorphogenic processes.

11.6. CONCLUSIONS

Nature has evolved a number of pigments that enable biological systems to respond to the fluctuations of the natural light environment. Some of these pigment systems operate independently, but many seem to operate in coordination to enable individual species to survive. Little is known about the molecular events in most of these complex processes. Even for the system that we know the most about, phytochrome, there are a number of unresolved questions. For example: (1) What is the absorbing state of phytochrome in a green plant during the day as a function of both irradiance and spectral distribution? (2) Why do dark-grown seedlings produce so much phytochrome, or does this additional phytochrome have an action as yet undiscovered? (2) Why is it advantageous to a species growing at the edge of a forest in an environment rich in far-red light to have its seed require greater light intensity before germination occurs? (4) What advantage is it that these seed are less likely to germinate in the dark?

And for photomorphogenic responses in general: (1) What are the natural fluctuations of spectral distribution of sun and sky light, and how do they interact with the self-filtering action by vegetation to regulate growth and form? (2) What is the initial reaction product for any of these photomorphogenic processes after the photoreceptor pigment is excited by light? (3) Are there ways in which we can chemically circumvent the requirements of light signals for developmental processes that are commercially significant? (4) As we develop new artificial light sources, what biological processes are being affected by *light pollution?* (5) How many other regulatory processes that occur under the control of light are yet to be discovered? To unravel and discover the answers to these photomorphogenic questions is not only exciting and challenging to scientists, but also significant and important to mankind.

11.7. EXPERIMENTS

11.7.1. Experimental Procedures

Experiment 1. The simplest demonstration of photomorphogenic responses is to compare young seedlings of beans or peas grown in complete darkness with those grown in the light. A number of differences, such as pigmentation, stem length, and leaf area, can be observed after several-days growth.

Fill small flower pots three-quarters full with water-soaked vermiculite, and allow to drain. On the surface, sow 6 to 12 seeds of bean or pea, cover with 1 cm of moist vermiculite, and maintain at ~25°C. Maintain a relative humidity greater than 80% by covering the pots with large inverted beakers. Place one set covered with black cloth in a darkroom to ensure no light exposure, and the other set in the laboratory under white fluorescent lamps (intensity > 100 footcandles) for 6 to 7 days.

Most of the differences observed are photomorphogenic responses. If dry weights of seedlings grown under the two conditions are compared, the differences are slight because seedlings have relatively large food reserves in the seed, and thus are not dependent on photosynthesis during this time period.

Experiment 2. The photomorphogenic responses may be measured quantitatively by using physiological responses such as hook opening or leaf disk expansion in bean seedlings.

If bean seedlings are grown in the dark, as outlined above, after 6 days a hypocotyl hook may be excised by cutting the stem portion just below the cotyledons into 2.5-cm-long sections under a dim green safelight. These sections are placed in Petri dishes on moist filter paper where they may be exposed to low irradiance red light (0.1 W m^{-2}) and the angle of hook opening determined by measuring with a protractor. Dark grown hooks at the beginning of the experiment have the short arm and long arm of the hook nearly parallel. After red light exposure, the cells on the inner region of the hook expand and the angle between the short arm and long arm is proportional to the fluence. For exposure to 1.0 mW m^{-2} for 20 h an opening of about 65° is observed.[55] If brief red exposures (1 kJ m^{-2} up to 30 min) are followed by exposure (0.3 kJ m^{-2}) to far-red energy, the hook opening is reduced.

If 5-mm disks of leaf tissue are cut with a cork borer from dark-grown leaves and placed on filter paper wetted with 5.8×10^{-2} M sucrose and 1×10^{-2} M KNO$_3$, they will expand. The diameters of the disks may be measured with a millimeter rule, or with a calibrated eyepiece in a dissecting microscope. The disks should be cut from both sides of the midrib of leaves, and after placing them in Petri dishes, exposed to red or far-red radiant energy for 10 min. They are then placed in metal cans (coffee cans work well), and kept in the dark at ~25°C for 3 days. Typical values obtained by students[56] are

Treatment	Diameter of disks (mm)
Far-red	6.47 ± 0.31
Far-red, red	8.01 ± 0.37
Far-red, red, far-red	6.77 ± 0.45

Both of these experimental materials demonstrate the red, far-red reversibility of the phytochrome system in a quantitative manner. Similarly, both are amenable to adding exogenous materials, such as auxins, nucleotides, or light-mimicking chemicals (e.g., cobalt ions), to study the interaction with light exposures.

Experiment 3. Seed germination also affords a striking system for demonstrating the control of the phytochrome pigment. Lettuce *(Lactuca)* and mustard *(Arabidopsis)* seed have most often been used, but other seed may be tested in the same manner, and many demonstrate similar responses. (*Arabidopsis* seed may be obtained from Prof. A. R. Kranz, Fachbereich Biologie (Botanik) der Universität, Siesmeyerstr. 70, D-6000 Frankfurt/Main, West Germany.)

Seed are imbibed by placing them on filter paper moistened with 10^{-3} M KNO$_3$. (Typically three sheets of Whatman #1 filter paper, 10 ml 10^{-3} M KNO$_3$, and 50 *Arabidopsis* seed are distributed over the surface of 10-cm diameter Petri dishes.)[8] Wrap the dishes in black cloth and place them in a refrigerator (2–4°C) for 48 h. The dishes should be handled in complete darkness and exposed to red or far-red radiant energy either alone or in various sequences. Select your own fluence values or use comparable fluences to those used in Experiment 2. Seed are particularly good for measuring fluence-response curves by exposing for times ranging from 1 to 10,000 s. After exposure, keep the dishes in the dark for 4 days, in the case of *Arabidopsis,* and 2 days for lettuce at 25°C, and count the number of germinated seeds.

Experiment 4. Pigment production under the control of the phytochrome system may be demonstrated by measuring the amount of the red pigment anthocyanin produced in young mustard seedlings.[19] Mustard seed (Sinapis)

are placed on water-moistened filter paper in glass storage jars (5.0 cm deep) or in small transparent plastic refrigerator storage boxes. Thirty-six hours after sowing and growing in the dark, the seedlings are exposed to radiant energy. An effective red radiance is 0.7 W m^{-2} and far-red is 3.5 W m^{-2}. Five-minute exposures give good responses when assayed 3 to 24 h after exposure. Select 25 seedlings and extract their hypocotyl and cotyledons in 30 ml of propanol–HCl–H$_2$O (18:1:81 v/v/v) by immersing extraction vials (50-ml glass scintillation vials with plastic snap caps work well) with seedlings in boiling water for 1.5 min. The capped vials are clamped in a metal rack to prevent the tops from blowing off and allowing excessive loss of solvent during boiling. The vials are allowed to cool at room temperature for 24 h to allow the extractions to proceed to completion, are centrifuged for 40 min at about 5000g, and the absorbance measured at 535 and 650 nm. A value for corrected absorbance is calculated as

$$A_{535} - 2.2A_{650} = \text{corrected } A_{535} \qquad (11\text{-}13)$$

This system is very good for demonstrating the high-irradiance response. The time course of anthocyanin produced under continuous irradiation for both red and far-red exposures measured over a 12-h period is clearly a function of the irradiance levels used.[19]

This experiment may be modified for rapid qualitative results both by shortening the extraction times and filtering the extracts before measuring absorbances. Similarly, this system is a good one to demonstrate the validity of the reciprocity law (Bunsen–Roscoe) by determining the amount of anthocyanin produced by varying both time and irradiation fluence [(intensity) × (time) = k (a constant)].

Experiment 5. Photomorphogenesis under blue radiant energy may be demonstrated by measuring the photoinduction of the yellow carotenoid pigments in the fungus *Neurospora*. (Strains of *Neurospora crassa* can be obtained from the Fungal Genetics Stock Center, Humboldt State College, Arcata, California.) Wild-type cultures are maintained on 2% agar slants on Vogel's minimal medium.[57] A few drops of an aqueous suspension of conidia of the wild-type strain Em 5297a are added to each of several Erlenmeyer flasks (125 ml) containing 20 ml of Vogel's minimal medium supplemented with 0.8% Tween-80. The flasks are placed in the dark for 4 days at 25°C, or 6 days at 18°C. Under a red safelight (GE-BCJ 60-W incandescent lamp) the mycelial pads, which have grown from conidia, are poured out of the flasks, spread out on filter paper, filtered with mild suction on a Büchner funnel, and placed in 15-cm diameter Petri dishes with three pads per dish. The pads are floated on fresh medium (2–8 ml per pad), and equilibrated for 2 h in the dark at the temperature that is to be used during the subsequent light treatment. The pads are exposed to blue light (0.1 W m^{-2}), and allowed to grow at 25° or 6°C in the dark for 24 h. They are then extracted twice in 4 vol methanol per unit of fresh weight for 15 to 20 min. The methanol extracts are pooled. Then extract several times with four vol acetone until all pigment is removed from the mycelium. Combine all extracts, and add an equal volume of aqueous 5% NaCl. In a separatory funnel, extract the pigment by adding hexane (0.1 of the total volume). Pool the hexane extracts and dry over sodium sulfate. Decant off the hexane and measure the absorbance at 475 nm. The appearance of pigment occurs more rapidly at 25°C than at 6°C, but more total pigment is produced at the lower temperature during the 24-h development period.[32]

11.7.2. Source Materials and General Comments

For any photomorphogenic experiments, the sources of radiant energy are very important. For precise quantitative experiments, instrumentation for measuring the irradiance precisely is needed (radiometers, calibrated against standard lamps), as well as careful control of the spectral bandwidth employed (interference filters or optical dispersion systems, such as grating or prism monochromators). However, much information may be obtained by using broadband filters that are inexpensive and easily constructed. For example:

Red source: cool-white fluorescent lamps wrapped with several thicknesses of commercial red cellophane.[56]

Far-red source: incandescent reflector flood lamp mounted 15 cm away from several thicknesses of commercial red and blue cellophane. Heat can be removed from the system by passing the radiant energy through 10 cm of water in a large beaker before it passes through the cellophane filters. A simple wooden or cardboard box can be constructed to enclose the light source.

If a spectrophotometer is available, transmission of the filters should be measured. Good-quality red, far-red plastic filters may be obtained from Rohm and Haas, Chemische Fabrik, 61 Darmstadt, West Germany (U.S. Agent, B. E. Franklin, 421 Pershing Drive, Silver Spring, MD 20910) or if precise experiments are planned, interference filters should be purchased (see Section 1.3.2.).

11.8. REFERENCES

1. H. Smith, *Phytochrome and Photomorphogenesis:* An Introduction to the Photocontrol of Plant Development, McGraw–Hill, London (1975).
2. R. L. Satter and A. W. Galston, The physiological functions of phytochrome, in: *Chemistry and Biochemistry of Plant Pigments,* 2nd ed. (T. W. Goodwin, ed.), Vol. 1, pp. 680–735 Academic Press, New York (1976).
3. K. Mitrakos and W. Shropshire, Jr. (eds.), *Phytochrome,* Academic Press, New York (1972).
4. William S. Hillman, *The Physiology of Flowering,* Holt, Rinehart, and Winston, New York (1962); Daphne Vince-Prue, *Photoperiodism in Plants,* McGraw–Hill, London (1975).
5. L. H. Flint and E. D. McAlister, Wavelengths of radiation in the visible spectrum promoting the germination of light-sensitive lettuce seed, *Smithsonian Inst. Misc. Collections* **96,** 1–8 (1937).
6. R. E. Kendrick and H. Smith, Assay and isolation of phytochrome, in: *Chemistry and Biochemistry of Plant Pigments,* 2nd ed. (T. W. Goodwin, ed.), Vol. 2, pp. 334–364 Academic Press, New York (1976).
7. G. R. Anderson, E. L. Jenner, and F. E. Mumford, Optical rotatory dispersion and circular dichroism spectra of phytochrome, *Biochim. Biophys. Acta* **221,** 69–73 (1970).
8. W. Shropshire, Jr., W. H. Klein, and V. B. Elstad, Action spectra of photomorphogenic

induction and photoinactivation of germination in *Arabidopsis thaliana*, *Plant Cell Physiol.* **2**, 63–69 (1961).

9. W. O. Smith, Jr. and D. L. Correll, Phytochrome: A re-examination of the quaternary structure, *Plant Physiol.* **56**, 340–343 (1975).

10. W. O. Smith, Jr., Purification and Physiochemical Studies of Phytochrome, Ph.D. dissertation, University of Kentucky, Lexington, Ky., 90 pp. (1975).

11. S. J. Roux, S. G. Lisansky, and B. M. Stoker, Purification and partial carbohydrate analysis of phytochrome from *Avena sativa, Physiol. Plantarum* **35**, 85–90 (1975).

12. S. Grombein, W. Rüdiger, and H. Zimmermann, The structures of the phytochrome chromophore in both photoreversible forms, *Hoppe Seylers Z. Physiol. Chem.* **356**, 1709–1714 (1975).

13. T. Sugimoto, K. Ishikawa, and H. Suzuki, On the models for phytochrome chromophore. III, *J. Phys. Soc. Japan* **40**, 258–266 (1976).

14. W. Shropshire, Jr., Phytochrome, a photochromic sensor, *Photophysiology* **7**, 33–72 (1972).

15. W. H. Klein, J. L. Edwards, and W. Shropshire, Jr., Spectrophotometric measurements of phytochrome *in vivo* and their correlation with photomorphogenic responses of *Phaseolus, Plant Physiol.* **42**, 264–270 (1967).

16. F. E. Mumford and E. L. Jenner, Catalysis of the phytochrome dark reaction by reducing agents, *Biochemistry* **10**, 98–101 (1971).

17. R. E. Kendrick and C. J. P. Spruit, Inverse dark reversion of phytochrome: An explanation, *Planta* **120**, 265–272 (1974).

18. K. M. Hartmann, A general hypothesis to interpret high energy phenomena of photomorphogenesis on the basis of phytochrome, *Photochem. Photobiol.* **5**, 349-366 (1966).

19. H. Lange, W. Shropshire, Jr., and H. Mohr, An analysis of phytochrome-mediated anthocyanin synthesis, *Plant Physiol.* **47**, 649–655 (1971).

20. K. M. Hartmann, Ein Wirkungsspektrum der Photomorphogenese unter Hochenergiebedingungen und seine Interpretation auf der Basis des Phytochroms (Hypokotylwachstumshemmung bei *Lactuca sativa* L.), *Z. Naturforsch.* **22b**, 1172–1175 (1967).

21. E. Schäfer, A new approach to explain the "high-irradiance responses" of photomorphogenesis on the basis of phytochrome, *J. Math. Biol.* **2**, 41–56 (1975).

22. A. Y. Gammerman and L. Y. Fukshanskii, Mathematical model of phytochrome, receptor of photomorphogenic processes in plants, *Ontogenez* **5**, 122–129 (1974).

23. C. J. Lamb and P. H. Rubery, Interpretation of the rate of density labelling of enzymes with ^2H_2O, possible implications for the mode of action of phytochrome, *Biochim. Biophys. Acta* **421**, 308–318 (1976).

24. I. A. Newman, Electric responses of oats to phytochrome transformation, in: *Mechanisms of Regulation of Plant Growth* (R. L. Bieleski, A. R. Ferguson, and M. M. Gresswell, eds.), pp. 355–360, The Royal Society of New Zealand, Wellington Bulletin 12 (1974).

25. R. Caubergs and J. A. De Greef, Studies on hook-opening in *Phaseolus vulgaris* L. by selective R/FR pretreatments of embryonic axis and primary leaves, *Photochem. Photobiol.* **22**, 139–144 (1975).

26. B. Rubinstein, K. S. Drury and R. B. Park, Evidence for bound phytochrome in oat seedlings, *Plant Physiol.* **44**, 105–109 (1969).

27. P. H. Quail, Interaction of phytochrome with other cellular components, *Photochem. Photobiol.* **22**, 299–301 (1975).

28. L. H. Pratt, R. A. Coleman and J. M. MacKenzie, Jr., Immunological visualization of phytochrome, in: *Light and Plant Development,* Proceedings of the 22nd Nottingham Easter School in Agricultural Sciences, (H. Smith, ed.) Butterworth, London, 75–94 (1976).

29. C. W. Raven and W. Shropshire, Jr., Photoregulation of logarithmic fluence-response curves for phytochrome control of chlorophyll formation in *Pisum sativum* L., *Photochem. Photobiol.* **21**, 423–429 (1975).

30. P. Halldal, *Photobiology of Microorganisms,* Wiley, New York (1970).

31. K. Bergman, P. V. Burke, E. Cerda-Olmedo, C. N. David, M. Delbrück, K. W. Foster, E. W. Goodell, M. Heisenberg, G. Meissner, M. Zalokar, D. S. Dennison, and W. Shropshire, Jr., *Phycomyces, Bacteriol. Rev.* **33**, 99–157 (1969).

32. E. C. DeFabo, R. W. Harding, and W. Shropshire, Jr., Action spectrum between 260 and 800

nanometers for the photoinduction of carotenoid synthesis in *Neurospora crassa, Plant Physiol.* **57.** 440–445 (1976).

33. P.-S. Song and T. A. Moore, On the photoreceptor pigment for phototropism and phototaxis: Is a carotenoid the most likely candidate? *Photochem. Photobiol.* **19,** 435–441 (1974).

34. V. Munoz and W. L. Butler, Photoreceptor pigment for blue light in *Neurospora crassa, Plant Physiol.* **55,** 421–426 (1975).

35. K. L. Poff and W. L. Butler, Spectral characterization of the photoreducible b-type cytochrome of *Dictyostelium discoideum, Plant Physiol.* **55,** 427–429 (1975).

36. M. Delbrück, A. Katzir, and D. Presti, Responses of *Phycomyces* indicating optical excitation of the lowest triplet state of riboflavin, *Proc. Natl. Acad. Sci. U.S.A.* **73,** 1969–1973 (1976).

37. R. J. Seviour and R. C. Codner, Effect of light on carotenoid and riboflavin production by the fungus, *Cephalosporium diospyri, J. Gen. Microbiol.* **77,** 403–415 (1973).

38. P. P. Batra, Mechanism of light-induced carotenoid synthesis in nonphotosynthetic plants, *Photophysiology* **6,** 47–76 (1971).

39. N. Lazaroff, Photomorphogenesis and nostocacean development, in: *The Biology of Blue-Green Algae,* pp. 270–319 (N. G. Carr and B. A. Whitton, eds.), Blackwell, Oxford (1973).

40. A. Bennett and L. Bogorad, Complementary chromatic adaptation in a filamentous blue-green alga, *J. Cell Biol.* **58,** 419–435 (1973).

41. J. Scheibe, Photoreversible pigment: Occurence in a blue-green alga, *Science* **176,** 1037–1039 (1972).

42. A. C. Leopold, Flower initiation in total darkness, *Plant Physiol.* **24,** 530–533 (1949).

43. E. C. Humphries and A. W. Wheeler, The physiology of leaf growth, *Annu. Rev. Plant Physiol.* **14,** 385–410 (1963).

44. G. A. Pieters, The growth of sun and shade leaves of *Populus euramericana* "Robusta" in relation to age, light intensity and temperature, *Mededel. Landbouwhogeschool Wageningen* **11,** 1–106 (1974).

45. W. Starzecki, The roles of the palisade and spongy parenchymas of leaves in photosynthesis, *Acta Soc. Bot. Polon,* **31,** 419–436 (1962).

46. T. Gorski, Germination of seeds in the shadow of plants, *Physiol. Plantarum* **34,** 342–346 (1975).

47. W. Shropshire, Jr., Photoinduced parental control of seed germination and the spectral quality of solar radiation, *Solar Energy* **15,** 99–105 (1973).

48. B. Goldberg and W. H. Klein, Variations in the spectral distribution of daylight at various geographical locations on the earth's surface, *Solar Energy* **19,** 3–13 (1976).

49. W. H. Klein, personal communication (1974).

50. L. G. Paleg and D. Aspinall, Field control of plant growth and development through the laser activation of phytochrome, *Nature* **228,** 970–973 (1970).

51. Anonymous, "But who needs sun at night?", *Sky Telescope* **32,** 183, 210 (1966).

52. L. E. Scheving, Chronobiology, in: *Chronobiology* (L. E. Scheving, F. Holberg, and J. E. Pauly, eds.), pp. 221–223, Igaku Shoin, Tokyo (1974).

53. S. Rand, personal communication (1975).

54. H. Cathey and L. E. Campbell, Effectiveness of five vision-lighting sources on photo-regulation of 22 species of ornamental plants, *J. Am. Hort. Sci.* **100,** 65–71 (9175).

55. W. H. Klein, Some responses of the bean hypocotyl, *Am. Biol. Teach.* **25,** 104–106 (1963).

56. R. D. Powell, A simple experiment for studying the effect of red and far red light on growth of leaf disks, *Am. Biol. Teach.* **25,** 107–109 (1963).

57. H. J. Vogel, A convenient growth medium for *Neurospora* (medium N), *Microb. Genet. Bull.* **13,** 42 (1956).

12

Photomovement

12.1. INTRODUCTION

Photomovement may be described as any light-mediated behavioral act involving the spatial displacement of all or part of an organism. In order to understand better what these responses are and how they may be studied, we must first classify, or at least define, their basic characteristics. The scheme described here is the result of the better part of two centuries of work and is best summarized in the classic study of Fraenkel and Gunn.[1]

    A light response involving the differential development of an organism with respect to a directional light stimulus is classified as *phototropism*. These responses have been extensively examined in higher plants. Motile organisms may respond to a light stimulus in a number of ways. *Phototaxis* involves the directed movement with respect to a single source of light in which the long axis (anterior–posterior) of the organism is oriented in line with the stimulus. Locomotion then is either directed toward or away from the source of stimulation. Classically, as the definition implies, such reactions are thought to result from the unequal stimulation of symmetrically arranged sensors located on a bilaterally symmetrical organism. This arrangement makes possible the simultaneous comparison of intensities of stimulation on the creature's two sides. This description of phototaxis is easily applicable to metazoans, but must be modified for protistans, for many (such as *Amoeba*) are not bilaterally symmetrical. For these

William G. Hand • Department of Biology, Occidental College, Los Angeles, California

organisms, paired receptors may not exist, so that directed reactions may be effected via a single receptor complex capable of making successive comparisons of stimulus intensities in time. In either example, the critical point to keep in mind is that phototaxis is a *directed* response to light demanding that comparison be made, either between a source and its background, between two sources of differing intensity, or between successive intensities of a single source in time. In order to denote the direction of the response with respect to the stimulus origin, one calls the responses *positive* when movement is toward the stimulus, and *negative* when the movement is away from the stimulus.

In addition to directed movements, there are undirected movements that may be influenced by light stimuli resulting in an accumulation of organisms within a light or dark region of the environment. These responses, known as *photokineses,* are described by Gunn *et al.*[2] as "variations in generalized, undirected, random locomotory activity" due to variations in stimulus intensity. These variations in locomotion may be of two basic types: a change in linear velocity caused by a light stimulus *(orthokinesis)* or a change in the rate of change of direction *(klinokinesis)*. It is important to recognize that these responses bear no relation to the direction of stimulation, but only to changes in ambient intensity. Therefore, in order to avoid the directional connotations of the adjectives "positive" and "negative" in describing kineses, one refers to responses in which activity is directly proportional to stimulus intensity as *direct* kineses, while those in which activity is inversely proportional to stimulus intensity are known as *inverse* kineses. For example, if an organism increases its speed of swimming with a rise in ambient light intensity, it is demonstrating a direct orthokinetic response to light. If an organism decreases its rate of randomly directed turning (i.e., goes straighter) under these conditions, it shows an inverse klinokinetic response to light. A summary of these responses is given in Fig. 12-1.

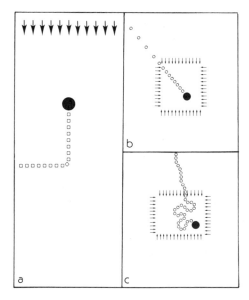

Fig. 12-1. The three types of motile responses that are the result of light stimulation. (a) Phototaxis. Note the directional nature of the stimulus (arrows). (b), (c) Photokinesis: (b) illustrates an inverse orthokinesis, and (c) shows a direct klinokinesis. Note that the stimulus in (b) and (c) is nondirectional.

There is one more type of light-related movement we must consider. These movements involve the use of a light source as a reference point by the organism with respect to its geographical or spatial navigation. Such phenomena are referred to as *light-compass* reactions, and they have been described for a number of metazoan forms, including crustaceans, insects, and birds. In many of these responses, the plane of polarization of the light source (usually the sun) is used as a navigational aid in migration or the maintenance of group, flock, or school formation.

12.2. PHOTOTROPISM

The study of tropisms has been largely a study of the hormonal control of the growth of higher plants. Charles Darwin, working with his son Francis, conducted experiments with canary grass seedlings.[3] The Darwins observed that the seedlings would bend toward a lighted window. In order to determine which part of the plant was responsible for the reception of the light stimulus, they established three groups of plants. One group was covered to the tip, one was left uncovered, and the last was covered on the tip only. The seedlings were then placed near the window, and the phototropism of the plants noted. The following day the Darwins observed that the seedlings that had been totally exposed as well as those that had been covered to the tip, were strongly bent toward the window, while those that had the tips covered exhibited no phototropism. The receptive surface was then obviously associated with the tip.

Building upon the experiments of Darwin, and later those of Paal and Boysen-Jensen, Frits Went conducted a series of experiments that were to form a basis for all future plant hormone study.[3] As a graduate student (1926) working in his father's laboratory at Utrecht, Holland, Went removed the tips of oat *(Avena sativa)* coleoptiles, placing the tips in darkness on a block of agar. After a time, he then placed this treated agar block asymmetrically on a properly prepared decapitated coleoptile. This coleoptile grew faster on the side with the agar block causing bending in the opposite direction (Fig. 12-2). This experiment established the presence of a chemical growth substance associated with the tip of the plant. In continued experiments, Went showed that in laterally lighted coleoptiles, more of this substance would diffuse into an agar block placed below the dark side of the tip. He named the agar block substance *auxin*. Auxin was then described chemically in the early 1930s by Kogl of Holland and Thimann of the United States, and given the chemical name indoleacetic acid (IAA).[4]

Since these pioneering studies, the role of auxin in phototropism has been elucidated. Briggs, and later Leopold,[5] demonstrated the lateral transmission of auxin from the lighted to the darkened side of the coleoptile. This results in the asymmetric elongation and growth of cells on the dark side of the seedling, bending the seedling toward the light.

A determination of the nature and mode of action of a receptor responsible for the growth response remains an unsolved but widely attacked problem. While some higher plants (notably *Avena*) have been studied, considerable effort has been directed toward the fungus *Phycomyces* (Fig. 12-3) by a host of investiga-

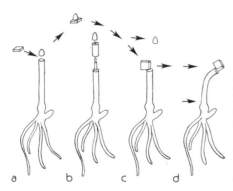

Fig. 12-2. The experiment of Frits Went determining the chemical nature of the phototropic response of *Avena* coleoptiles. The coleoptile tip is removed (a), and placed on an agar block (b) and allowed to remain. The block is then transferred to a decapitated coleoptile (c), resulting in shoot bending (d). [After F. B. Salisbury and C. Ross, *Plant Physiology,* p. 446, Wadsworth, Belmont, Calif. (1969).]

tors.[6] The action spectrum for phototropism exhibited by the sporangiophores of this fungus have three major maxima at 385, 455, and 485 nm. This action spectrum suggests carotinoids or possibly flavins as possible receptor pigments. *Phycomyces* does contain large quantities of carotene. A better prospect, however, may be flavoprotein. These proteins occur, in some cases, as crystals in *Phycomyces.* Wolkin suggests that such crystals may be the unit photoreceptor. He argues that similar flavinoid crystals are associated with the retinal systems of

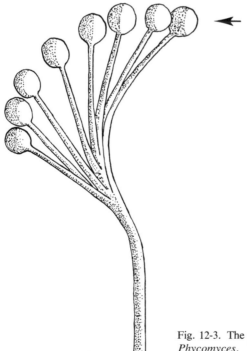

Fig. 12-3. The phototropic response of the sporangiophore of *Phycomyces.* Positional shifts represent 5-min intervals. Arrow indicates direction of stimulus origin. (Adapted from reference 6.)

vertebrates.[7] Needless to say, until the problem of what constitutes a receptor is solved, the mode of action of a receptor cannot be adequately described.

12.3. PHOTOTAXIS

With the added dimension of rapid movement of the whole organism, many technical problems in quantifying phototaxis supervene, including agreement as to what constitutes a phototactic response. Historically, the foundations of phototaxis research are based on the direct observations of motile organisms, most notably the protists. Throughout the latter half of the last century, investigators observed the migration of motile and algal cells to a directional light stimulus.[8,9]

In recent years, a variety of techniques have been adopted to examine and quantify the phototactic response.[10] These techniques may be divided into two categories: the mass movement method and the individual cell (organism) method. All mass movement methods utilize the principle of measuring the change in optical density of cell suspensions. The basic components are a flask containing a suspension of cells, a test light, and a photocell connected to a recording potentiometer. As the cells respond to the test light, they change the intensity of the light path going to the photocell, which in turn registers a change in potential on the recorder. This system accurately measures the *result* of light stimulation on a population of cells, but does not measure the process by which the cells achieve that result. Many investigators have failed to make this important distinction, and hence have concluded that they were quantifying phototaxis, when in effect they were observing the results of what might have been phototaxis, photokinesis, or a combination of the two.

The individual cell method consists of a microscope, recording device (film, videocamera), field illuminator (usually filtered to a phototactically "neutral" wavelength), and a stimulating light source. The cells being examined are placed in a cuvet or well slide on the stage of the microscope, stimulated in a plane parallel to the surface of the microscope stage, and the response is recorded. Recently, Davenport and co-workers[11] have successfully married the computer to a closed circuit television microscope technique. This system holds great promise in the close examination of the phototactic act, for it is capable of examining the minute movements of many responding organisms simultaneously. The only disadvantage of this system is that it can only be used effectively on organisms that move primarily on an x-y plane. A full three-dimensional analysis of movement has been achieved by Berg and co-workers[12] using a mechanized microscope stage. In this system the cell is kept stationary by moving the microscope stage in an x, y, or z direction. These movements can be transferred to a paper recorder, producing a "three-dimensional" tracing of the cell's movement. The major disadvantage of this system is that only one cell may be examined at a time, hence statistical information concerning the movements of a population becomes a laborious task. For a complete review of methods, see Hand and Davenport.[10]

One may ask first whether phototaxis occurs in like fashion in all organisms. In metazoan forms where bilaterally arranged, paired receptors are the rule, the response is reasonably uniform from one form to the next. It is in the unicellular forms that one observes differences. For instance, the phytoflagellate *Euglena,* which has been extensively studied,[13] exhibits a response that is much less precise than any metazoan. Early investigators, e.g., Jennings[8] and Mast,[9] observed that the *Euglena* cell orients with respect to a light stimulus in reasonably precise stepwise shifts in the body axis until the cell is aligned with its long axis parallel to the stimulus direction. The most widely accepted hypothesis for this movement assumes the existence of a single photoreceptor, which determines the direction of a light stimulus by measuring the intensity of the stimulus at two points in time through the effects of a second organelle, the stigma or "eyespot." The stigma, which is laterally positioned with respect to the receptor, periodically "shades" the receptor as *Euglena* rotates. This brief, intensity transient results in a momentary change in motor activity (flagellum) with a resulting long-axis shift in swimming direction (the term "phobic response" is used by many authors). When these periodic changes in receptor stimulation cease, the cell is oriented either toward or away from the light source. This response is summarized in Fig. 12-4. A cell oriented toward the stimulus source would have the receptor constantly illuminated, while a negatively phototactic *Euglena* would have the receptor continually shaded by the large number of light-absorbing organelles within the cell. A reaction sequence of

Fig. 12-4. The phototactic response of the algal cell *Euglena*. Intervals between cell positions are ~0.1 s. Arrows indicate region of directional stimulation. Note that the flagellum pushes the organism through the generation of successive helical waves. (A composite adapted from reference 13.)

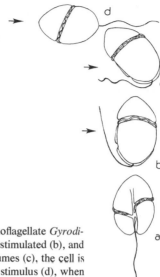

Fig. 12-5. The phototactic orientation response of the marine dinoflagellate *Gyrodinium*. Positional shifts represent ~0.07 s. A swimming cell (a) is stimulated (b), and exhibits a cessation of flagellar activity. As flagellar activity resumes (c), the cell is turned toward the stimulus until it is in a plane parallel with the stimulus (d), when normal swimming resumes. (Adapted from references 14 and 15.)

this nature is slow when compared, for instance, with the response of any bilaterally symmetrical metazoan.

What data beyond the early observations of Jennings and Mast support the shading hypothesis? Unfortunately, in the case of *Euglena* there are few studies that give real support. One reason for this is the elusive phototactic response of this cell. There are other cells that far better support the hypothesis in a modified form. One such series of studies involves the marine dinoflagellate *Gyrodinium dorsum.* Unlike *Euglena,* the phototactic response is easily observed, and is very predictable. The response of *Gyrodinium* differs from *Euglena* in degree rather than kind. Upon stimulation, *Gyrodinium* ceases swimming (this "stop" response being dependent on intensity), makes a single anteroposterior axial rotation until it faces the stimulus, and then resumes forward swimming (Fig. 12-5). This is a very rapid response when compared with *Euglena,* for it is completed within a few milliseconds to half a second, depending on stimulus intensity. Recent studies give some indication as to how orientation is accomplished by a *Gyrodinium* cell.[15] It can be argued that the "phobic response" observed with *Euglena* is similar to the "stop response" of *Gyrodinium.* Upon stimulation, *Gyrodinium* elicits a distinct and predictable flagellar response. This response can be directly related to the reorientation of the cell. How effector is related to receptor is conjectural. Orange stigma-like organelles appear in the cell, but their position and number are highly variable from cell to cell and are usually not closely related with the flagellar base as they are in *Euglena.* In addition, there are many red-brown lipid-containing chromoplasts, any of which could act as a shading device should the receptor area be located close to the cell surface.

And what of the receptor itself? The best evidence for proposed receptors comes from electron microscope studies. As in *Euglena, Gyrodinium* should have a receptor associated with the base of the flagellum. Indeed, in a closely related form, *Glenodinium foliacium,* the proposed receptor appears to be located at the junction of the two indentations in the cell (the *sulcus–girdle* junction).[16] Similarly placed receptors in *Gyrodinium* would enable the variety of surrounding red plastids to serve as a functional shading body. If one further implicates the sulcus and girdle, using them as functional light channels, we can predict the chain of events involved in a phototactic orientation by a *Gyrodinium* cell. As the cell is stimulated, it stops in a position so that the sulcus–girdle junction is facing the light source. This response involves a single, but predictable lateral repositioning of the longitudinal flagellum. These two predictions are supported by direct photographic evidence. As the longitudinal flagellum resumes its beat, it turns the cell in a predictable manner, for as it begins to beat it simultaneously repositions itself in the normal trailing position behind the cell. In doing this, it rotates the cell so that it now faces the light stimulus. The anterior of the cell, with its associated chromoplasts, effectively shades the receptor from continued stimulation, and the cell resumes its forward swimming in a light-oriented fashion. Should the cell deviate from a course parallel to the stimulus, light will enter the sulcus channel and partially stimulate the receptor, causing a flagellar response, and a subsequent realignment with the stimulus. Such responses have been observed and recorded.[6]

In conclusion, one may say that phototactic orientation (at least for the forms studied to date) involves two steps: (1) a change in the level of stimulation on the receptor that leads to (2) a predictable flagellar response that realigns the cell with respect to the stimulus origin. As I have pointed out, these cell responses require continued study if the hypotheses set forth are to become established doctrine.

What constitutes a receptor? With respect to metazoans, refer to Chapter 10 on vision, for visual systems are well documented. With unicellular organisms, electron microscope studies offer the best current prospect for determining the nature of the receptor. Direct comparisons can be made between the rod outer segment of vertebrate retinas and the primitive receptors of flagellated protists. Both have a lamellar structure, a flagellar or ciliary origin, and a carotenoid pigment implicated as the chemical agent involved in light reception (see Chapter 10). Circumstantial evidence links carotenoids to the receptor of unicellular forms, i.e., action spectra analyses correlated with absorption spectra data support this. The action spectra for most forms studied are in the blue region of the spectrum. This correlates positively with the absorption spectra of carotenoids found in these and other similar cells. In *Euglena,* a flavinoid, probably riboflavin, is implicated as a receptor pigment.[17]

Some secondary pigments are also implicated in the receptive state of the cell, most notably phytochrome.[18] In this instance, the pigment governs the sign as well as the sensitivity of the response. One also must assume that pigments such as chlorophyll play an indirect role in the response potential of the algal flagellates.

The next major question is: How is a light stimulus transduced into a behavioral act? Metazoan reception and transduction involves the isomerization of a carotenoid pigment, which in turn is somehow translated into an electrical impulse. Many studies are currently being conducted on the mechanisms underlying information changes from a chemical to an electrical state in this system. The nature of information transfer in unicellular organisms is very speculative. As in metazoans, electrical events may be involved. Membrane potentials have been recorded in phototactic dinoflagellate cells, and have been implicated in the swimming reversal response of the ciliate *Paramecium*. [19] These small, but measurable, potentials have been directly linked to the reversal of ciliary beat in *Paramecium,* and therefore may possibly be intrinsic to the change of flagellar beat associated with light orientation. One might expect the initial response of a pigment stimulated by light to "trigger" a chain of events similar to a poised "redox" state. [13] This would result in the flow of electrons, thereby altering adjacent membrane flow states. Since the flagellum is a "miniature muscle" biochemically, altering the state of the boundary membrane might elicit a change in the contractile response of the flagellum. [20] In addition, the possible piezoelectric properties of the flagellum should be considered. As Jahn and Bovee [13] have discussed, the flagellum exists in a quasi-crystalline state, hence it should be piezoelectric. If so, the voltage change generated by the initial photochemical event could cause the flagellum to bend locally and also alter its entire beat profile. There are many uncertainties here, but some of these hypotheses are testable using standard physiological dose–response methods. The entire complex problem requires a more thorough knowledge of membrane systems and their interaction with other subcellular components.

12.4. PHOTOKINESIS

Unlike phototaxis, photokinesis is not concerned with the direction of the light stimulus, but with the intensity of stimulation. The flatworm *Planaria gonocephala* moves at an average velocity of 57 mm/s in total darkness. When placed in dim light, this flatworm increases its velocity approximately 25%. [21] The functional significance of such a response is effectively to "trap" the organism in darkness. This concept is consistent with the life habits of the worm, for it is most frequently found under rocks and in other dark places. The reaction is a direct orthokinetic response to light. Such responses are not difficult to document, for they may be photographed or traced, and the lengths of the paths measured and referenced in time. Klinokinetic responses on the other hand, are more difficult to document, for measuring angle changes in a continuously turning path becomes a difficult and often subjective task. Such a study was conducted by Ullyott [22] on the flatworm *Dendrocoelom lacteum*. He used the angle of turning as his measure of the rate of change of direction (RCD). With increasing intensity of illumination, the RCD values increased correspondingly. However, after a time, sensory adaptation occurred, and the RCD value for a given high intensity returned to a basal level. This led *Dendrocoelom* to congregate in a darkened area, because the RCD level was reduced as the light level decreased, thus compelling the worms to seek the darkest recesses of the region. Conversely, when the intensity was increased, the forward progress of the worms was impaired, because of the high RCD value. This explanation of the adaptive significance of a klinokinetic response has been tested mathematically by Rohlf and Davenport.[23] A hypothetical "bug" was placed on a grid and subjected to a nondirectional stimulus. The resulting klinokinetic response led the "bug" to some predicted location on the grid.

It is often difficult to separate the two types of kineses, for each component is often observed in the overall motile response. Surtees,[24] working with several species of beetles, devised a simple analysis for determining which of the two responses was the major contributor to the dispersion or aggregation observed. The relative irregularity of the pathway, or the intensity of turning, is computed as the ratio of the distance traveled (DT)/linear displacement (LD). By this method: "if the DT is significantly reduced but the DT/LD is not, the mechanism is primarily a modification of speed. If DT is reduced, or remains the same, but DT/LD is significantly larger, the mechanism is primarily a modification of irregularity of pathway." This simple comparison removes many of the problems in designing experiments to demonstrate kineses. However, one major problem in designing experiments remains, and that is the problem of assuring that the test stimulus gives no directional cues.

Photoreceptive surfaces involved in kinetic responses may be localized receptors, or they may be some generalized membrane or tissue response. For instance, the avoidance response of *Paramecium,* which may be classified as a klinokinesis, requires no specialized receptor, but is the result of a change in the state of the plasmalemma (cell membrane) at the anterior of the cell. As the *Paramecium* encounters an object, physical contact deforms the membrane, causing ion flux changes, resulting in ciliary reversal.[19] As the membrane reestablishes normal flux, cilia resume their normal beat direction. This process is repeated until the cell bypasses the obstacle.

12.5. CELL ORGANELLE MOVEMENTS

A final consideration must be given to the oriented movements of organelles within cells.[25] These responses may be either kinetic or tactic in nature. For example, the rotational streaming of chloroplasts in *Elodea* and *Valisneria* is the result of the movement of the cytoplasm around the inside periphery of the cell. This *cyclosis* can be altered by light intensity, with higher intensity increasing the rate of movement. The movement is purely an orthokinetic response, for no orientation is observed.

Oriented chloroplast movements have been documented for several algal species, and have the functional significance of orienting chloroplasts for maximum light use. In these responses, the chloroplasts may be locomotory, or they may be moved passively as in cyclosis. How these movements occur in either case, is yet undetermined. One of the best studied examples of such a cell is that of the alga *Mougeotia*. Each elongate *Mougeotia* cell contains a single flat chloroplast that, when illuminated with low-intensity light, turns from a profile position to one facing the stimulus. This response can be triggered by stimuli of very short duration. Once stimulated, the cell will respond in total darkness. The photoreceptor in this cell appears to be a phytochrome that is probably localized in the plasmalemma; hence it responds to two finite wavelengths of red light, 660 and 720 nm. The position of the chloroplast appears to be the result of two primary factors: first, the orientation of the phytochrome molecules, and second, the ratio of P660 to P730 in the system at any time.[25] A strong dichroic orientation of phytochrome molecules exists, with their main vector of absorption being oriented parallel to the surface, and obliquely to the cell's long axis. This orientation produces an absorption gradient, with the side facing toward (front) and away (rear) absorbing approximately twice the energy as the flanks. Irradiation with red light changes P660 to P730, establishing a gradient of P730. The chloroplast will orient with its edges closest to P660 (avoids close association with P730), and therefore orients with its flat surface facing the light. The response is pictured in Fig. 12-6. This example is rather unique with respect to receptor–effector systems governing oriented behavior, but it presents a new and exciting way of looking at such systems.

12.6. LIGHT-COMPASS REACTIONS

Many studies conducted between the late 1940s and the mid-1960s indicate that animals use the sun as a directional compass for migratory or food-gathering activities. Two studies are most often cited; the flight of the starling, and the "waggle" dance of the honeybee.

In 1953, Karl von Frisch[26] reported that bees communicated the direction of a food source to other bees by a dance. This dance consists of a circle and straight waggle, the latter communicating the food direction as a function of the relative position of the sun with respect to the hive and the food source. Von Frisch showed that this angle changed during the day relative to the sun's

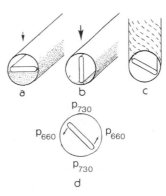

Fig. 12-6. The chloroplast rotational response of the alga *Mougeotia*. (a) A cell with the chloroplast facing a low-intensity stimulus; (b) in profile position in response to a high-intensity stimulus; (c) dichroic orientation of phytochrome molecules in the plasmalemma of this alga cell; (d) a cross section demonstrating the movement of the chloroplast as regulated by the position of P660 and P730 phytochrome. (Adapted from reference 25.)

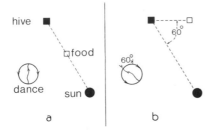

Fig. 12-7. The sun-compass orientation response of the honeybee, and its relation to the bee dance. (a) When the food source is placed in line with the sun and the hive, the straight run of the dance is vertical on the face of the hive. (b) when the food source is shifted, the dance angle shifts correspondingly. (After P. Marler and W. J. Hamilton, *Mechanism of Animal behavior,* p. 543, Wiley, N.Y. (1967).

position. (Fig. 12-7).[27] Similar relationships between the solar cycle and the orientation of bee dances were demonstrated by translocation experiments.[28] Bees were trained to fly to feeding stations in Long Island at a particular time of day. They were then moved during the night to Davis, California. The following morning the bees foraged normally, but their search direction had shifted, the degree of the shift corresponding to the change in the sun azimuth between Davis and Long Island. It should be mentioned, however, that the hypothesis of information content in the dance has been questioned in recent years. The simpler hypothesis of Wenner and Wells[27] suggests that scent recognition by recruit bees explains the ability to locate food sources.

The European starling *(Sturnus vulgaris)* has been shown to navigate using the sun as a point of reference.[29] Starlings were put in a circular enclosure where their view of horizon and landmarks was obscured, but the sun was visible. While so restricted, the bird (in a migratory condition) would flutter or attempt flight while perched. The role of the sun in migratory behavior was determined by the direction of fluttering. An oriented response was not observed if the sun was occluded by clouds, or not visible from inside the apparatus. When mirrors were introduced to change the apparent direction of the sun, the oriented response was predictably shifted. Such oriented "flight" was maintained for several hours.

These and other studies of vertebrates and invertebrates clearly indicate the sun as a reference point in movement. It is important to note that the movement is not a true taxis or kinesis, but may contain elements of both. For example, the crustacean *Daphnia longispina* migrates away from shore during the daylight hours, returning during the night.[30] When placed on a circular Plexiglas grid, *Daphnia* swim in a direction perpendicular to the shoreline. These movements are randomized in darkness. During daylight, a taxis is involved in the movement, but it is not directed *toward* or directly *away* from the sun, but at some angle to it, this angle being dependent on where the shoreline is with reference to the light source.

Minor attention has been given to lunar orientation as demonstrated, e.g., by the beach hopper *Talitrus saltator.*[31,32] When these crustaceans are placed in a cylindrical chamber, they immediately move toward the direction of the moon. This is clearly a positive phototactic response, but to a very low light level when compared to similar responses for sun-oriented movement. However, more sophisticated moon-reference orientation is exhibited when *Talitrus* is placed on the beach. The circadian mechanism characteristic of beach hoppers is apparently a compensation for the variations in relative positions of earth and moon.

12.7. CONCLUSIONS

How light-mediated behavior benefits a species is often obvious by the nature of the response. There are some situations, however, where the ecology of a species may be influenced in very subtle ways by the organism's behavior.

Phototaxis for the flagellate is an ecologically important response, for it places this photosynthetic cell in an optimal light environment. The vertical migrations of several marine dinoflagellates correspond with the daylight cycle; the organisms rise near the surface during the daylight hours and descend at night.[33] If these cells were not phototactically active, and remained at some fixed depth (this being determined by the physical topography), would they be functionally capable of reproducing themselves? Would they be able to carry on their photosynthetic processes efficiently enough to produce the energy "surplus" necessary for fission? This can be tested and requires further investigation before we can understand the role of phototaxis in the dynamics of population growth. One can ask other questions of this system as well. For example: What environmental factors can alter the "normal" photoresponse? How does temperature affect the photoresponse at the behavioral, physiological, or biochemical level? None of these questions has yet been approached adequately.

Marine larvae have been shown to change the sign of the light response as they develop from one stage to another.[33] The functional significance of such responses in not clear, and awaits proper investigation. The most obvious explanation for the behavior is that the response places the organism in: (1) a more favorable feeding position in the water mass, and/or (2) a position offering maximum protection from predation. It is nevertheless a fact that light-oriented responses are a common constituent in the overall behavioral profile of marine fish, crustacea, annelids, and other lesser taxa.

It is in the examination of these ecological problems that the universal importance of the light behavior of organisms will become known. It is an exciting prospect to consider the possible significance of a simple behavioral change that may have saved a species countless generations of anatomical adaptation. For example, if it were not for the presence of a phototactic response in marine dinoflagellates, and the possibility of behaviorally altering their light environment, would they be able to exist as they do today? Granted, they have undergone several anatomical adaptations to arrive at their present condition, but how much more difficult would it have been to cope with the problem some other way?

Let us think of the problem of the adaptive significance of photomovement from a practical viewpoint. I have discussed at length the phototactic response of the dinoflagellate *Gyrodinium*. This organism and its relatives are the major constituent species in what is commonly called the "red tide." The occurrence of red tide is world-wide, and, as yet, no method has been found to control it. It is possible that the phototaxis response of dinoflagellates may be a major factor in determining whether a population of dinoflagellates undergoes the explosive growth in numbers associated with a red tide. The presence of specific ionic materials in the water (NH_4^+, Ca^{2+}, Mg^{2+}) could contribute to the overall sensitivity of the cells, causing them to remain at the surface for a longer period. This, in turn, along with the existence of adequate nutrient levels, could produce the excess biochemical energy necessary to promote accelerated cell division. What I am stating is pure speculation, but one can easily see that it suggests many areas of future exploration (see Section 6.3.3.).

These questions and concepts are in need of experimental verification. Until such information is available, a clear understanding of the importance of photomovement will not be possible. Interdisciplinary cooperation will be necessary to integrate present knowledge and to uncover new phenomena.

12.8. EXPERIMENTS

12.8.1. Phototropism

Phototropism can be easily demonstrated using coleoptiles of *Avena,* the oat plant. Seeds may be procured from a biological supply house.

Germinate the seedlings in the dark for 24 h at 22–24°C prior to use. During the 12th hour of germination, subject the seedlings to a 5-min period of red illumination. After the 24-h period, transfer the seedlings to a beaker or tray lined with wet filter paper. Insert the seedlings through the filter paper, so that the paper holds the coleoptile erect, with the root down in the water. Allow the coleoptiles to continue to grow until they are approximately 30 mm high (3 to 5 days). It is important that this entire procedure be conducted in darkness or in a green safelight.

Subject the coleoptiles to differing periods (30 s, 1 min, 5 min, 15 min, 1h) of directional incandescent illumination (60-W lamp), and note the result. Devise a method of quantifying your observation, and plot the effect of illumination.

Further experimentation involving the causal agent (auxin) of phototropism may be conducted using the method devised by Went. For a detailed description of this experiment see A. Dunn, and J. Arditti, *Experimental Physiology,* Holt, Rinehart, Winston, New York (1968).

12.8.2. Phototaxis

A simple demonstration of photoaccumulation may be achieved using a young culture of the flagellate *Euglena gracilis.*

A unialgal suspension of *Euglena* may be prepared by obtaining a starter culture from a biological supply house and adding to 2–3 liters of Difco *Euglena* medium (broth). This suspension should be placed in a well-lit area for approximately 1 week until it turns a bright green. Pour some of the cell suspension into a large cuvet (20 × 20 cm) made from two glass plates cemented together with silicone cement. Illuminate the cuvet from one side with an image projected from a slide projector. Allow 5 min for the accumulation to take place, then turn off the slide projector. The image projected should be "outlined" in *Euglena*. The image will then fade as the cells randomize.

In order to determine the spectral sensitivity of the response, color slides may be introduced, these being made by photographing a spectral chart from a physics or chemistry textbook. Note where the greatest accumulation occurs, correlating it directly with the position of the spectral display on the cuvet.

In order to determine whether the accumulation is due to phototaxis or photokinesis, a direct examination of the cells with a microscope is advised. The basic procedures are discussed in Sections 12.3. and 12.4.

12.9. REFERENCES

1. G. S. Fraenkel and D. L. Gunn, *The Orientation of Animals: Kineses, Taxes and Compass Reactions,* Dover Press, New York (1961).
2. D. L. Gunn, J. S. Kennedy, and D. P. Pielou, Classification of taxes and kineses, *Nature* **140,** 1064–1067 (1937).
3. N. G. Ball, Plant tropisms, *Vistas Botany* **3,** 228–254 (1963).
4. L. J. Audus, *Plant Growth Substances,* Interscience, New York (1959).
5. R. K. De La Fuente and A. C. Leopold, Lateral movement of auxin in phototropism, *Plant Physiol.* **43,** 1031–1036 (1968).
6. K. Bergman, P. B. Burke, E. Cerda-Olmedo, C. N. David, M. Delbrück, K. W. Foster, E. W. Goodell, M. Heisenberg, G. Meissner, M. Zalokar, D. S. Dennison, and W. Shropshire, Jr., *Phycomyces, Bacteriol. Rev.* **33,** 99–157 (1969).
7. J. J. Wolken, *Invertebrate Photoreceptors: A Comparative Analysis,* Academic Press, New York (1971).
8. H. S. Jennings, *Behavior of the Lower Organisms,* Columbia University Press, New York (1906).
9. S. O. Mast, *Light and the Behavior of Organisms,* Wiley, New York (1911).
10. W. G. Hand and D. Davenport, The experimental analysis of phototaxis and photokineses in flagellates, in: *Photobiology of Microorganisms* (P. Halldal, ed.), pp. 253-282, Wiley, London (1970).
11. D. Davenport, G. J. Culler, J. O. B. Greaves, R. B. Forward, and W. G. Hand, The investigation of the behavior of microorganisms by computerized television, *IEEE Trans. Biomed. Eng.* **17,** 230–237 (1970).
12. H. C. Berg, How to track bacteria, *Rev. Sci. Instr.* **42,** 868-871 (1971).
13. T. L. Jahn and E. C. Bovee, Locomotive and motile response in *Euglena,* in: *The Biology of Euglena* (D. E. Beutow, ed.), Vol I, pp. 45–108, Academic Press, New York (1968).
14. W. G. Hand, Phototactic orientation by the marine dinoflagellate *Gyrodinium dorsum* Kofoid. I. A mechanism model, *J. Exp. Biol.* **174,** 33–38 (1970).
15. W. G. Hand and J. Schmidt, Phototactic orientation by the marine dinoflagellate *Gyrodinium dorsum* Kofoid. II. Flagellar activity and overall response mechanism, *J. Protozool.* **22,** 494–498 (1975).
16. J. D. Dodge and R. M. Crawford, Observations of the fine structure of the eyespot and associated organelles in the dinoflagellate *Glenodinium foliaceum, J. Cell Sci.* **5,** 479–493 (1969).
17. G. Tollin and M.I. Robinson, Phototaxis in *Euglena.* V. Photosuppression of phototactic activity by blue light, *Photochem. Photobiol.* **9,** 411–418 (1969).
18. R. B. Forward, Jr., Phototaxis in a dinoflagellate: Action spectra as evidence for a two-pigment system, *Planta* **111,** 167–178 (1973).
19. Y. Naitoh and R. Eckert, Ionic mechanisms controlling behavioral responses of *Paramecium* to mechanical stimulation, *Science* **164,** 963–965 (1969).
20. M. A. Sleigh, *Cilia and Flagella,* Academic Press, New York (1974).

21. H. E. Walter, The reactions of planarians to light, *J. Exp. Zool.* **5,** 35–162 (1907).
22. P. Ullyott, The behaviour of *Dendrocoelom lacteum.* II. Responses in non-directional gradients, *J. Exp. Biol.* **13,** 265–278 (1936).
23. J. F. Rohlf and D. Davenport, Simulation of simple models of animal behavior with a digital computer, *J. Theor. Biol.* **23,** 400–424 (1969).
24. G. Surtees, Laboratory studies on dispersion behaviour of adult beetles in grain. VIII. Spontaneous activity in three species and a new approach to analysis of kinesis mechanisms, *Anim. Behav.* **12,** 374–377 (1964).
25. W. Haupt, Perception of light direction in oriented displacements of cell organelles, *Acta Protozool.* **11,** 179–188 (1972).
26. K. von Frisch, *Bees: Their Chemical Senses, Vision and Language,* Cornell University Press, Ithaca, N.Y. (1950).
27. P. H. Wells and A. Wenner, Do honeybees have a language?, *Nature* **141,** 171–175 (1973).
28. M. Renner, The contribution of the honey bee to the study of time-sense and astronomical orientation, *Cold Spring Harbor Symp. Quant. Biol.* **25,** 361–367 (1960).
29. D. R. Griffin, Bird navigation, in: *Recent Studies in Avian Biology* (A Wolfson, ed.), pp. 154–197, University of Illinois Press, Urbana, Ill. (1955).
30. O. Siebeck, Researches on the behaviour of planktonic crustaceans in the littoral, *Int. Ver. Limnol.* **15,** 746–751 (1964).
31. F. Papi and L. Pardi, On lunar orientation of sandhoppers *(Amphipoda talitridae), Biol. Bull.* **124,** 97–105 (1963).
32. L. Pardi, Innate components in the solar orientation of littoral amphipods, *Cold Spring Harbor Symp. Quant. Biol.* **25,** 395–401 (1960).
33. R. B. Forward, Jr., Light and diurnal vertical migration: Photobehavior and photophysiology of plankton, in: *Photochemical and Photobiological Reviews* (K. C. Smith, ed.), Vol. 1, pp. 157–209, Plenum Press, New York (1976).

13

Photosynthesis

David C. Fork • Department of Plant Biology, Carnegie Institution of Washington, Stanford, California

13.1. INTRODUCTION

Photosynthesis, the conversion of light energy into stabilized chemical energy, involves the absorption of light by a pigment, energy transfer, energy trapping or stabilization by reaction centers, and the initiation of chemical reactions from donor to acceptor molecules. The process continues with a sequence of oxidation–reduction reactions that comprise the electron transport that leads to the formation of reduced nicotinamide-adenine dinucleotide phosphate (NADPH) and adenosine triphosphate (ATP). Reactions leading to the fixation of carbon are powered by the energy available in the molecules of NADPH and ATP. Photosynthesis can be studied from many points of view, including those of the physicist, physical chemist, biochemist, biologist, ecologist, and agronomist.

    This chapter will emphasize photosynthesis that occurs in green plants. Bacterial photosynthesis is covered more fully elsewhere.[1-3] A number of reviews and books have appeared recently and can be consulted for more information than can be included here.[4-14]

13.1.1. How Did Photosynthesis Evolve?[15,16]

    Conditions on the primeval earth were drastically different from those that exist today. The Russian biochemist Oparin suggested[17] that no oxygen existed in the atmosphere, but rather hydrogen and methane were the dominant components with lesser amounts of ammonia, nitrogen, hydrogen sulfide, and water being present. When such substances thought to be present in the primeval earth's atmosphere are exposed to ultraviolet (UV) radiation, radioactivity, heat or electric discharges, a wide variety of substances can be produced, such as HCN, sugars, amino acids, peptides, nucleosides and nucleotides, as well as polyphosphates and porphyrins.

    On primeval earth, since no living organisms would have been around to metabolize these substances, it is conceivable that substantial amounts could have accumulated over long periods of time. When amino acids are heated they form proteinlike molecules that can aggregate into small spheres about the size of bacteria.[18] Nobody knows just how living forms may have arisen from such aggregations, but they did not appear probably for more than 1.2 billion years after the earth was formed (4.0–4.6 billion years ago).

These primitive organisms probably could obtain all the energy they required simply by utilizing the substances that were found abundantly in the environment. With an increase in the number of such "living organisms," the supply of these compounds would eventually run out. Substances from which it is easy to obtain energy would be used (oxidized) first. Survival depended upon the primitive bacterialike cells being able to utilize the remaining compounds that were more difficult to oxidize, such as H_2S and water. In order to do this, these "cells" would be greatly assisted if a method could be developed by which the energy in light could be used. Thus, a primitive photosynthetic system probably evolved that could utilize light energy to oxidize an electron donor, and thereby synthesize an energy-rich substance for later metabolic use. Since water is one of the most difficult compounds to oxidize, the ability of an organism to accomplish the photosynthetic oxidation of water granted a great evolutionary advantage over other life forms, since it would no longer be necessary to depend upon the availability of energy-rich compounds.

Photosynthesis using water as the donor of electrons to reduce carbon to carbohydrate and releasing O_2 as a waste product appears to have evolved very early in the earth's history. The fossil record shows that the O_2-evolving blue-green algae appeared about 3 billion years ago. The ability of the blue-green algae to use water as their electron donor gave them great flexibility in the habitats that they could occupy, and provided them with a great evolutionary advantage over photosynthetic bacteria that were confined to areas where they could obtain simple organic compounds or H_2S. Oxygen-evolving photosynthesis became the dominant type of energy-converting system of plants, and gave rise to the accumulation of oxygen in the atmosphere, and eventually to the protective ozone (O_3) layer in the upper atmosphere via the photochemical alteration of O_2.

13.1.2. An Overview of Photosynthesis

The time scale for photosynthesis is large. The absorption of a photon produces the excited state within about 10^{-15} s, but the enzymatic reactions of CO_2 fixation and cellular synthesis take place within seconds or minutes.

Essentially, photosynthesis in green plants removes electrons (and hydrogen) from water and adds them to carbon dioxide, forming carbohydrate and leaving O_2 as a waste product. This is an "uphill" process and requires energy that is captured by chlorophyll (chl) molecules.

The very first step in photosynthesis may be considered to be the light-harvesting act. After a pigment molecule has captured a photon and has been raised to an excited state, it may either lose its energy by a number of different pathways in a nonuseful manner or may pass its energy to chl in a specialized environment called a reaction center where conversion of light to chemical energy takes place. Upon receiving excitation, chl in the reaction center expels an electron and becomes oxidized. The electrons expelled by reaction centers are received by other compounds called primary acceptors, and these become reduced. The raising of an electron to the higher energy level of the acceptor by absorption of a photon of light by the reaction center is analogous to the operation of a pump elevating water to a higher level. The primary acceptor can pass its electron to a secondary acceptor, and the secondary acceptor to a third, and so on in an orderly sequence of steps called electron transport. All of these secondary electron transport reactions are downhill, and proceed with the thermodynamic gradient much as water runs downhill. Bacterial photosynthesis uses one light conversion step that is connected to an electron transport system, while plants utilize two independently acting reaction centers, each pumping electrons uphill against the thermodynamic gradient.

These two reaction centers are connected together in series so that an electron expelled by one light reaction eventually reaches the other reaction center via a number of transfers of the electron from one carrier to another along the transport chain. During this process, electronic energy is stored as chemical energy with the formation of NADPH and ATP. These two products are the energy-rich end products of photosynthesis that are used in the dark reactions of CO_2 fixation to yield carbohydrates (CH_2O).

The overall reactions of photosynthesis are:

$$2H_2O \xrightarrow{h\nu} O_2 + 4e^- + 4H^+ \tag{13-1}$$

$$2NADP^+ + 4e^- + 2H^+ \rightarrow 2NADPH \tag{13-2}$$

$$2NADPH + 2H^+ + CO_2 \rightarrow 2NADP^+ + H_2O + (CH_2O) \tag{13-3}$$

$$\text{Net:} \quad CO_2 + H_2O \xrightarrow{h\nu} (CH_2O) + O_2 \quad \text{(by 8 quanta)} \tag{13-4}$$

The elaboration of sugars in the carbon cycle (see Section 13.9.), in addition, requires ATP that is formed during electron transport from water to NADP.

Photosynthesis is an oxidation–reduction process between an oxidant, CO_2, at a potential of about -0.4 V and a reductant, H_2O, at about $+0.8$ V that requires the transport of $4e^-$ uphill against a total gradient of about 1.2 V. Thus, to move $4e^-$ in photosynthesis from water to CH_2O produces a potential difference of 4.8 eV or 110 kcal mol^{-1} (1 eV = 23 kcal mol^{-1}).

13.2. THE PHOTOSYNTHETICALLY ACTIVE PIGMENTS IN PLANTS

The light reaching the earth has wavelengths extending from approximately 290 to 1100 nm. There are pigments in different photosynthetic organisms that can absorb light over this entire range of wavelengths (Fig. 13-1). The bacteriochlorophyll in bacteria absorbs the near-UV and infrared wavelengths not effectively absorbed by chl a of algae and higher plants (Fig. 13-1A). Chl a, by contrast, has its maximum absorption in the blue and red regions near 440 and 680 nm, respectively. Other photosynthetically active pigments such as chl b and c, the carotenoids, phycoerythrin and phycocyanin have absorption maxima that lie somewhere between 400 and 650 nm, and are discussed more fully in Section 13.2.2.

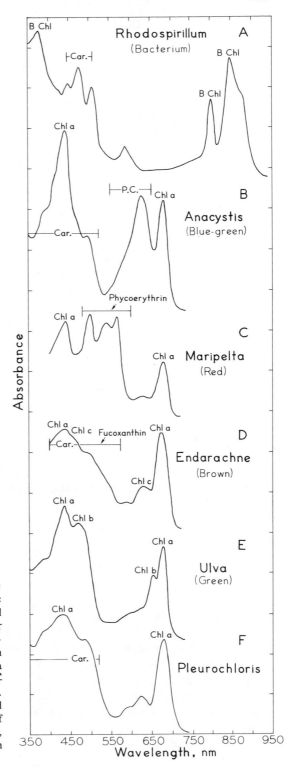

Fig. 13-1. Absorption spectra made *in vivo* for the major kinds of photosynthetic organisms. Spectrum A was measured using chromatophores of the nonsulfur purple bacterium *Rhodospirillum spheroides*. Other spectra were obtained with intact organisms. The deep-water red alga *Maripelta rotata* grows in submarine canyons off California at depths up to 60 m. The approximate absorption maxima, and in some cases, the range of absorption of photosynthetic pigments are shown. Car., carotenoid; P.C., phycocyanin. (From Fork, unpublished.)

13.2.1. The State of Chlorophyll *in Vivo*

Chemically, chl consists of four pyrrol rings joined to form a flat ring that is complexed with magnesium in its center (Fig. 13-2A). This tetrapyrrol ring has attached to it a long and almost completely saturated hydrocarbon phytol tail. The flat head part of the chl molecule is more soluble in water, while the phytol tail is fat-soluble. The single molecular species of chl *a* apparently exists in more than one state in the living plant. If one measures absorption spectra of plants very accurately, the absorption band of chl *a* is complex and appears to be composed of a number of chlorophylls, each absorbing at slightly different wavelengths. These so-called forms of chlorophyll *a* apparently result from the association of chlorophyll with different protein molecules in the plant. When it is extracted from plants, only a single molecular species of chl *a* is obtained. We can summarize the results of much work [19–21] by noting that there are probably four different kinds of chl *a in vivo* and that they have their absorption maxima at 662, 670, 677, and 684 nm, respectively.

The ability of plants to elaborate chl that absorbs at different wavelengths turns out to be a very important feature of the photochemical apparatus, for a photon will be absorbed and transferred from a chl having a short wavelength maximum to a chl having a longer wavelength maximum (see Section 13.3.4.). In addition, the so-called accessory pigments such as chl *b* and *c*, the red and blue phycobilin pigments of red and blue-green algae, and certain carotenoids can absorb short wavelengths of light and transfer them to a form of chl *a* absorbing at longer wavelengths. The different accessory pigments and different forms of chl *a* thus function somewhat like "antennas" to collect light quanta for the reaction center chl that absorbs at the longest wavelength and receives quanta that are absorbed and transferred to it from a large assemblage of other pigments absorbing at shorter wavelengths.

13.2.2. The Distribution of Pigments in Algae and Higher Plants[19,21]

13.2.2.1. Blue-Green Algae

These primitive plants probably evolved from the photosynthetic bacteria, and, like the bacteria, do not have their photosynthetic pigments localized in chloroplasts as do other kinds of algae and higher plants. Rather, they have flattened sacklike structures termed lamellae or thylakoids extending throughout the cytoplasm of the cell.

The blue-green algae contain chl *a* as their main photosynthetic pigment. The maximum absorption bands of chl *a* occur near 680 nm in the red and around 440 nm in the blue part of the spectrum. Figure 13-1B shows the absorption curve for cells of the blue-green alga *Anacystis*.

In addition to β-carotene (Fig. 13-2B), which has been found in all photosynthetic plants, the blue-green algae contain myxoxanthophyll, a carotenoid not found in other organisms.

The blue-green algae also contain phycocyanin, absorbing maximally near

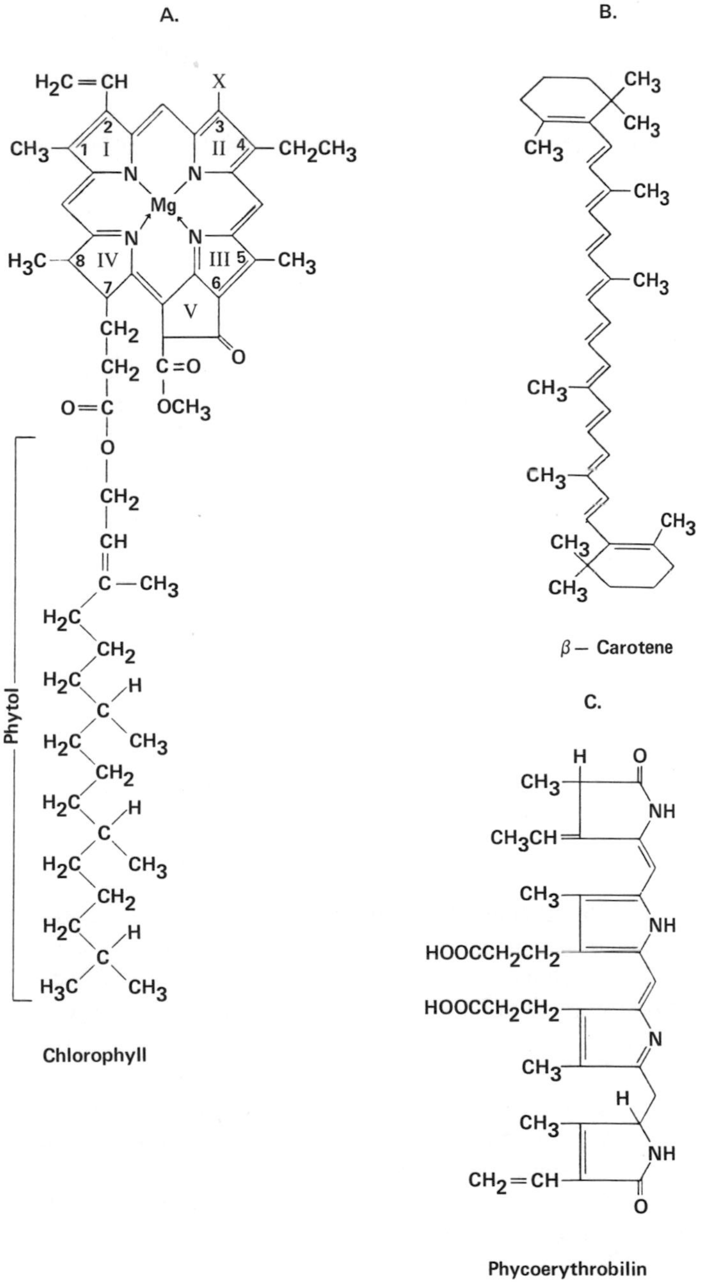

Fig. 13-2. Molecular structures of chlorophyll *a*, β-carotene, and phycoerythrobilin.

626 nm, as well as allophycocyanin, absorbing at 650 nm, that serve as second (or "accessory") pigments. Phycocyanin as well as the pigment phycoerythrin belong to a class of water-soluble proteins called the phycobilins (algal bilins). Phycocyanin, allophycocyanin, and phycoerythrin constitute the major accessory pigments of blue-green and red algae, and are present in other less well-known algal groups as well. The chromophores of phycobilins are related to chl, but here the four pyrrol rings are arranged linearly (Fig. 13-2C). This arrangement is the fundamental structure for the bilin pigments, so called because they were first isolated from animal bile. These pigments serve by absorbing green to orange light, wavelengths not effectively absorbed by chl.

13.2.2.2. Red Algae

These algae live largely in marine habitats and frequently occur in deep waters where red and blue wavelengths are largely filtered out by the upper water layers. Red algae containing the phycobilin pigment phycoerythrin can absorb the remaining green light in deep water, and use it effectively for photosynthesis (see Section 13.5.1 on action spectra). Red algae contain, in addition, varying amounts of phycocyanin as well as allophycocyanin. Red algae living in deep water have more phycoerythrin than phycocyanin. The combination of phycoerythrin with chl *a* permits red algae to absorb almost all the wavelengths in the visible part of the spectrum (Fig. 13-1C). The absorption bands of chl *a* at 440 and 678 nm can be seen, as well as the peaks at 500, 540, and 566 nm produced by phycoerythrin absorption. In addition to β-carotene, the red algae contain α-carotene that has been found in no other plants examined so far, except for certain kinds of green algae *(Siphonales)*. Lutein is the principal xanthophyll of red algae.

13.2.2.3. The Brown Algae, Diatoms, and Dinoflagellates

Diatoms and dinoflagellates living in the upper ocean layers account for about 33% of the world's photosynthesis. In addition to chl *a*, these algae contain chl *c* as well as the characteristic xanthophylls fucoxanthin (brown algae and diatoms) and peridinin (dinoflagellates). The absorption maxima of chl *c* occur near 460 and 640 nm, while fucoxanthin and peridinin absorption maxima are near 490 nm (Fig. 13-1D). When in its native state, probably attached to a protein, the absorption maxima of fucoxanthin and peridinin are shifted by about 40 nm to longer wavelengths, compared to the absorption maxima of these pigments in organic solvents. Together with chl *c*, these xanthophylls make available a large fraction of the green light that otherwise would be lost for photosynthetic purposes.

13.2.2.4. Green Algae and Higher Plants

These plants constitute the most commonly seen photosynthetic organisms in everyday life. In these plants, chl *a* is found in conjunction with chl *b*, the latter having blue and red maxima near 470 and 650 nm, respectively (Fig. 13-1E). Unlike the blue-green algae, diatoms, and dinoflagellates, the higher plants and green algae do not have special pigments to absorb light between about 500 and 600 nm, and so reflect these wavelengths and appear green to the eye. In addition to the universal β-carotene, the higher plants and green algae contain lutein as their major xanthophyll, and lesser amounts of violaxanthin and neoxanthin. Unlike xanthophylls, such as fucoxanthin and peridinin, the xanthophylls of higher plants and green algae are not particularly active in absorbing light used for photosynthesis.

There are certain algae that contain only chl a. An example is given in Fig. 13-1F for *Pleurochloris* (a yellow-green alga in the class Eustigmatophyceae). This spectrum clearly shows the contribution to absorption of only chl a and carotenoids.

The distribution of the different pigments is summarized in Table 13-1.

13.2.3. The Function of Carotenoids

Carotenoids are found in all photosynthetic organisms. As mentioned above, certain xanthophylls such as fucoxanthin and peridinin absorb and transfer light energy to the reaction center(s) of photosynthesis. In addition to their role as absorbers of photosynthetically active wavelengths of light, the carotenoids appear to function by protecting plants against lethal photooxidations. Plants that have lost their carotenoid pigments through mutation are quickly killed in the light.

It appears that carotenoids may also have other functions. The level of the xanthophyll violaxanthin in leaves is altered by light in a reaction sequence that involves the reversible interconversion of violaxanthin to zeaxanthin via antheraxanthin as an intermediate.[22, 23]

13.3. LIGHT HARVESTING AND PHOTOCHEMICAL CONVERSION[13,24]

13.3.1. Absorption of Light and Fluorescence of Pigments

In order for a light reaction to take place it is necessary for a pigment to absorb light. When a substance absorbs a photon of light, it acquires additional energy, and is described as being in the electronic excited state. According to the laws of quantum mechanics, the transition from the ground state to the excited state can take place only via certain discrete energy states. To be absorbed, the energy content of the photon must be exactly the same as the difference in energy between two electronic states. For this reason, only certain wavelengths of light can be absorbed by a particular substance (see Section 2.3.).

Usually the energy levels of the first excited singlet state and the ground state are not overlapped by sublevels, as are the first and higher excited states. Therefore, a transition from the ground to the first excited singlet state requires the absorption of a photon, and likewise, the transition from the first excited

TABLE 13-1.   Distribution of Photosynthetically Active Pigments in Different Plants

	Chl a	Chl b	Chl c	Phycoerythrin	Phycocyanin	β-Carotene	Major xanthophyll
Blue-green algae	+			+	+	+	Myxoxanthin
Red algae	+			+	+	+	Lutein
Brown algae, diatoms	+		+			+	Fucoxanthin
Dinoflagellates	+		+			+	Peridinin
Green algae, higher plants	+	+				+	Lutein
Yellow-green algae	+					+	Violaxanthin; vaucheriaxanthin

singlet state to the ground state results in the emission of a photon of light. This light emission is termed fluorescence. Fluorescence emission comes only from the lowest excited singlet state and not from any higher excited state, because internal conversions (transitions producing heat) are rapid and take place within $\sim 10^{-12}$ s, while fluorescence emission needs about 10^{-9} s. Also, small rapid internal conversions (energy losses) occur between sublevels of the lowest excited singlet state. For these reasons, fluorescence is seen at longer wavelengths than absorption. The photosynthesis sensitized by the absorption of high-energy blue quanta is no more of an advantage to the plant than is absorption of lower-energy red quanta, because the higher energy state produced by absorption of blue light is degraded rapidly to the lower energy state before the slower photochemical reactions have a chance to take place.

13.3.2. Delayed Fluorescence[25,26]

If a plant is transferred to the dark after a period of illumination, it can continue to emit a faint red glow. This glow is only detectable with sensitive light-sensing equipment, and was discovered unexpectedly in 1951 in an experiment intended to demonstrate light-induced ATP formation in photosynthesis by using an extract of firefly tails that glows when ATP is present (see Chapter 14). It was found that algae emitted a glow even though no firefly extract had been added. It appears that all photosynthetic organisms emit this "afterglow," that has been termed "delayed fluorescence" to distinguish it from "prompt fluorescence" that stops within about 10^{-9} s after the exciting light is turned off.

The two types of emission have the same spectral distribution, indicating that they both originate from chl (or bacteriochlorophyll). Delayed fluorescence appears to be produced by the back-reaction of primary (or secondary) electron acceptors with the oxidized chl that is formed by light at the reaction centers. This back-reduction of chl apparently returns it to its singlet excited state, from which light can be emitted, rather than to the ground state. Moreover, delayed fluorescence seems to be a phenomenon that is associated only with the reaction center of photosystem 2 (as is prompt fluorescence). This is the case because delayed fluorescence is still seen in algal mutants that lack the photosystem 1 reaction center P700 (see Sections 13.3.5.1. and 13.5. for a discussion of reaction centers and the two photochemical systems). Moreover, delayed fluorescence in dark-adapted plants show oscillations like those seen for O_2 evolution (described in Section 13.6.2.3.) when a sequence of flashes is provided.

13.3.3. Energy Transfer[27,28]

Before a quantum that is absorbed by a chl molecule can be trapped by a reaction center, it can be transferred from one molecule to another, again and again, much like one billiard ball hitting another until it eventually falls into a pocket. A photon that is absorbed by any of the 400 or so chl molecules that comprise the "antenna" molecules for the photochemical trapping center is rapidly transferred (in about 10^{-12} s) from one chl molecule to another, and is

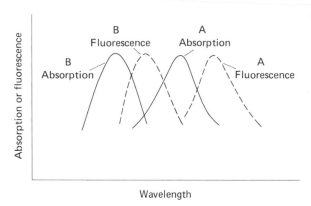

Fig. 13-3. Diagrammatic representation of pigments having absorption and fluorescence bands close together as is the case, e.g., with chl *b* and chl *a* in green algae and higher plants. Since the fluorescence of B overlaps the absorption of A, there is a good probability that a photon absorbed by B will be transferred to A. This can be demonstrated by exciting B, and observing only the fluorescence of A (see text).

eventually trapped by a reaction center; the whole process taking place within 10^{-9} s.

Energy transfer can be demonstrated experimentally in a green alga such as *Chlorella*, e.g., by illuminating it with a wavelength of light (i.e., 650 nm) that is absorbed in large measure by chl *b*. The overlap between the absorption of red bands of chl *b* and chl *a* is somewhat like the diagrammatic representation given in Fig. 13-3. The absorption maximum of chl *b* occurs *in vivo* at a shorter wavelength (650 nm) than the absorption maximum of chl *a* (678 nm). In this case, only the characteristic curve for chl *a* fluorescence is seen. No fluorescence can be measured that corresponds to the fluorescence of chl *b*, even though this substance is strongly fluorescent in solution. In fact, studies done *in vivo* with *Chlorella* demonstrate that the transfer of energy from chl *b* to chl *a* is nearly 100% efficient. Similarly the phycobilins can transfer their energy to chl with efficiencies ranging from 80 to 90%. By contrast, the efficiency of energy transfer from carotenoids to chl range from 20 to 50%, depending upon the particular carotenoid involved.

Förster[29,30] provided one of the theoretical foundations for understanding the mechanics of energy transfer between molecules. He suggested that substantial overlap is needed between the fluorescence of the donor molecule and the absorption of the acceptor molecule (Fig. 13-3), and that the transfer rate is proportional to the reciprocal of the sixth power of the distance between the two molecules. For a probability of 50% for energy transfer to occur from one molecule to another, the calculated distance between molecules should be of the order of 5 nm. In chloroplasts the distances are much smaller than this, so that the probability of transfer is very high.

13.3.4. The Reaction Center

Photochemical reactions take place in specialized places in the photosynthetic apparatus called reaction centers. The reaction center chl receives a photon directly by absorption or indirectly from other pigments by energy transfer. The electron in chl, elevated to a higher state after absorption of a photon, can be received by a suitable compound that acts as an acceptor. The transfer of an electron to an acceptor molecule causes it to become reduced, and chl to become oxidized. Chl can be reduced when it, in turn, accepts an electron in a downhill process from a suitable donor. If the acceptor molecule is at a lower electrochemical potential than the excited state of chl, this transfer will also be with the thermodynamic gradient (downhill), and proceed spontaneously. It is believed that something like this happens in photosynthesis. The chemical identities of the acceptors in photosynthesis are not yet known.

The reaction center of the photosynthetic bacteria is probably best understood, since it can be prepared in a "pure" form without antenna bacteriochlorophyll attached. Progress has also been made in the isolation of reaction centers from one of the photochemical systems (photosystem 1) of higher plants and algae. By contrast, almost nothing is known about the reaction center of photosystem 2 in these plants, except that a light-induced absorption change has been seen that may correspond to the oxidation of the reaction center chl.

In reaction centers prepared from photosynthetic bacteria, light-induced decreases of absorption in the region from 870 to 890 nm can be seen. Similarly, small particles can be prepared from higher plants and algae, by sonication and/ or detergent treatment followed by differential centrifugation, that show light-induced decreases of absorption with a maximum near 700 nm. This was discovered by Kok, who suggested that the unknown substance producing this change was a form of chl, and labeled it pigment 700 (or P700).

In the case of the bacterial reaction center as well as for P700, chemical agents of the right oxidation–reduction potential can produce the same effect as can the absorption of a photon of light. Titration of the reaction center in bacteria with ferricyanide/ferrocyanide mixtures shows that the midpoint potential E'_0 of this reaction is +0.5 V, and that oxidation produces a loss of one electron from the reaction center. A similar titration of P700 shows that it has a redox potential of about +0.43 V.

13.3.5. The Photosynthetic Unit

Plants contain large amounts of pigments, but only a small fraction of these pigments function in the crucial photochemical steps of photosynthesis. This has been known for over 40 years from the experiments of Emerson and Arnold, who measured O_2 evolution from the green alga *Chlorella*. They used very short flashes of red light that were effectively absorbed by chl. With sufficiently intense flashes, they were able to excite every chl molecule, i.e., to saturate the photochemical reactions. It was possible also to measure the O_2 given off as a function of dark time between the flashes. If the intervals between flashes were sufficiently long, there was enough time for the dark reactions to process the products made during the light reactions. If the flashes came too quickly, however, the dark reactions were unable to keep pace, and photosynthesis, as measured by O_2 evolution, slowed down. The dark time needed for the enzymes to react with the photoproducts was found to be of the order of 10^{-2} s.

Interestingly, in experiments with increasing flash intensities it was found that the maximum yield attainable was one O_2 molecule for about every 2500 chlorophylls that absorbed the light flash. Since 8 photons are needed to produce one O_2 molecule, the flash yield experiment showed that 2500/8 or about 300 chl molecules cooperate to ''process'' one photon.

Calculations show that, even in full sunlight, each chl molecule absorbs only a few photons per second. The experiments of Emerson and Arnold show how nicely plants have adapted to this situation with their assembly of 300 chl molecules (called a photosynthetic unit), all cooperating to capture and funnel energy to a single specialized chl that does the photochemical conversion. With this antenna arrangement to concentrate light energy, the dark reactions are thus not limited by the rate at which photochemical conversions take place.

13.4. SPECTROSCOPY IN PHOTOSYNTHESIS RESEARCH[31]

Many of the substances involved in the reactions of photosynthesis absorb in the near-UV and visible regions of the spectrum, and their absorption spectra are different depending upon whether they are in the oxidized or reduced state. Plotting the difference spectrum (i.e., the oxidized-minus-reduced spectrum) shows the changes in the absorption spectrum that occur on going from one form to the other. Figure 13-4 gives the difference spectrum for cytochrome f as well as for other photosynthetic intermediates that can be detected spectroscopically as light-induced absorbance changes, and plotted as light-minus-dark difference spectra.

The use of techniques to measure small and rapid light-induced changes of absorption provides an important aid for fitting together the pieces of the puzzle of photosynthesis. By this technique, a monochromatic ''measuring beam'' is passed through a sample of plant material, and the intensity of this beam is detected by a suitable device, usually a photomultiplier, and then recorded in a convenient way, such as with an oscilloscope or on a strip-chart recorder. Another light source provides the exciting or actinic illumination. Filters are placed in front of the detector so that only the measuring beam is ''seen'' by the photodetector. With appropriate light sources and amplifiers this type of apparatus can measure very rapid and small changes of absorption that take place in many of the substances involved in photosynthesis.

13.5. THE TWO PHOTOCHEMICAL REACTIONS IN ALGAE AND HIGHER PLANT PHOTOSYNTHESIS

13.5.1. Enhancement and Two Light Reactions[32]

An action spectrum is a measurement that can reveal what pigments are functional in sensitizing a certain photochemical reaction (see Sections 1.5.4. and 3.7.). In the case of photosynthetic organisms, e.g., the effectiveness of different

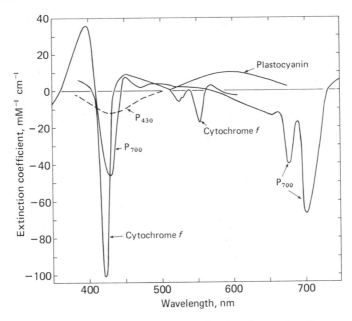

Fig. 13-4. Oxidized-minus-reduced difference spectra for the electron transfer components P700, P430, cytochrome f, and plastocyanin. The difference spectra for P700, cytochrome f, and plastocyanin are produced upon conversion by light from the reduced to the oxidized form. The negative P430 band results from the reduction of an unidentified compound (see text) (Data from Dr. Tetsuo Hiyama.)

wavelengths of light in producing O_2 evolution can be plotted.[32] Such a measurement was made in an elegant manner in 1882 by the Dutch scientist T. Engelmann, who used filamentous green and blue-green algae. He added a suspension of oxygen-requiring motile bacteria to a slide holding the algal filament, and sealed the slide from the outside air. After illuminating the algal filament with a spectrum projected onto the stage of the microscope, he was able to see that the motile bacteria congregated more in certain areas of the spectrum than others. In the case of the green alga, most bacteria congregated in the blue and red regions of the spectrum where chl absorbs maximally (Fig. 13-1E), while with blue-green algae higher concentrations of bacteria gathered in orange light, that is absorbed by phycocyanin (Fig. 13-1B), than in red light absorbed by chl a itself.

A very similar result, but in more detail, was obtained[33] using rapid polarographic techniques to monitor oxygen evolution from variously pigmented algae as a function of wavelength. The action spectrum for O_2 evolution of a green alga gave a result in agreement with Engelmann's, that blue and red light absorbed by chls were most effective in promoting O_2 evolution. In brown algae, not only was chl effective in sensitizing photosynthesis, but the accessory xanthophyll, fucoxanthin, was almost as active as well. An unexpected result was found in the red and blue-green algae. Here the activity of the accessory pigments phycoerythrin and phycocyanin were more active than chl a itself in promoting photosynthesis. The action spectrum dropped away rapidly at wavelengths where chlorophyll a absorption was still strong. A closer reexamination of the action spectrum for the green alga showed a similar ''red drop,'' but this effect was not nearly so pronounced as in red algae. A clear red drop was also seen with blue-green algae. A similar red drop in the action spectrum for fluorescence excitation can be seen in the unicellular red alga *Porphyridium*.

These action spectra measurements were difficult to reconcile with the fact that any photon absorbed by a pigment should be just as effective as any other photon absorbed. Emerson discovered that the red drop could be abolished if plants were illuminated with two wavelengths of light absorbed by different pigments. For example, in red algae, green light absorbed by phycobilins given simultaneously with red light absorbed by chl induced more O_2 evolution than the total amount of the O_2 evolution obtained when these irradiations were given separately. This synergistic effect of two wavelengths of light has been termed *Emerson enhancement,* or simply enhancement (for a review, see reference 34).

The discovery of the enhancement effect was a milestone in photosynthesis research. Prior to its discovery it was assumed that photosynthesis needed only one light reaction, and that the accessory pigments served only to close ''optical windows'' in the visible spectrum, and to transfer their absorbed energy to chl a where a single kind of photochemical reaction center fixed the quanta. The amount of enhancement that could be obtained depended upon how the value was expressed, but for red algae the ratio of O_2 produced by the irradiations given simultaneously over the sum of the O_2 production produced by each irradiation alone was about 1.8.

The red drop can be eliminated in all of the differentially pigmented algae if appropriate wavelengths are paired. In fact, enhancement spectra (i.e., the effectiveness of different wavelengths of supplementary light in eliminating the red drop) showed that photosynthesis occurred most efficiently only when accessory pigment excitation was provided along with chl excitation. The two wavelengths need not be presented simultaneously, however. Alternation of the appropriate wavelengths with periods as long as 6 s still produced enhancement.[35]

13.5.2. The Hill and Bendall Scheme

The enhancement results can best be explained if one were to assume that photosynthesis proceeds via two photochemical reactions that are connected by a series of dark enzymatic steps. Hill and Bendall[36] in 1960 provided a theoretical framework for a current concept of how photosynthesis works (Fig. 13-5). In the Hill and Bendall model (sometimes called the ''Z-scheme'') a cytochrome of the c type found in leaves, termed cytochrome f, was proposed to function as an intermediate between two different types of photochemical reaction centers; one reaction center (called photosystem 1) served to oxidize cytochrome f, and another photochemical reaction center (photosystem 2) reduced cytochrome f. The light-induced oxidation of the reaction center of photosystem 2 eventually would give rise to the production of a strong oxidant leading to the oxidation of water. The operation of photosystem 1 would produce a strong reductant to reduce NADP that is needed for CO_2 fixation.

The first experimental result that could best be explained by the Hill and Bendall model was the finding[37] that red light, absorbed predominantly by chl a of the red alga *Porphyridium*, produced oxidation of cytochrome f, but green light, absorbed largely by the accessory pigment phycoerythrin (cf. Fig. 13-1C), produced cytochrome f reduction. This antagonistic effect of different wavelengths of light was consistent with the Hill and Bendall model. P700 shows a similar "two-wavelength effect" since, in the blue-green alga *Anacystis*, red light absorbed by chl (Fig. 13-1B) brings about the oxidation of P700, while orange light, absorbed by the phycocyanin, reduces P700.[38] Similar antagonistic two-wavelength effects have been found for other photosynthetic intermediates, such as plastoquinone and plastocyanin. In fact, the observation of a push–pull effect caused by excitation of one pigment system or the other can be used to localize the site of action of a compound in the electron transport between photosystems 1 and 2.

13.5.3. Light Distribution between the Two Photosystems

Since the two light reactions in photosynthesis appear to operate in series, it is necessary that both be excited equally so that one of them does not become rate-limiting. There appears to be a mechanism in plants that can regulate the way in which light absorbed by a chl molecule may end up trapped either by photosystem 1 or by photosystem 2. Studies[39–41] have shown that if a plant is illuminated with light that is preferentially absorbed by photosystem 2 (e.g., with 650-nm light for the green alga *Ulva*; Fig. 13-1E) there is a kind of adaptation to the color of the light so that a change in light quantum distribution between the two photosystems occurs. The redistribution is such that more quanta are delivered to photosystem 1, thus increasing the photosystem 1 reactions, and preventing them from limiting the overall rate of electron transport. In this condition the plants are said to be in "state 2." The other way around, illumination with light absorbed largely by photosystem 1 (far-red light) causes the plants to go into "state 1," in which more absorbed quanta can be delivered to photosystem 2. This flexibility of quantum distribution is of importance to a plant, since it allows it to utilize light energy efficiently even if the wavelength distribution of the light is unfavorable for a balanced excitation of both photosystems.

The distribution of quanta between the two photosystems is under the control of divalent cations such as Mg^{2+}. Since it has been shown that illumination of chloroplasts results in ion movements across thylakoid membranes, it has been suggested that the cation-induced changes in transfer of excitation energy are related to the state changes induced by illumination.[42–44]

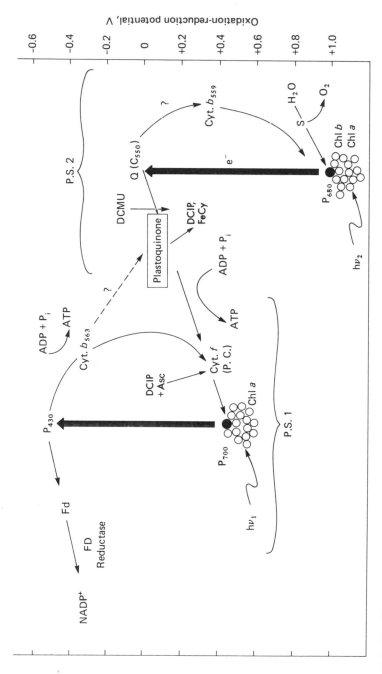

Fig. 13-5. The Hill and Bendall or "Z" scheme for photosynthesis plotted to show redox potentials of the reacting components. The upward arrow signifies the raising of an electron against the thermodynamic gradient to a higher energy level upon absorption of a photon by a reaction center. The solid circles represent P700 and P680, specialized chl molecules functioning as the reaction centers for photosystems 1 and 2, respectively. The circles around the reaction centers represent other chl forms and accessory pigments that absorb short-wavelength light and transfer the energy to the reaction centers. The different forms of chl a and the accessory pigments are both present in the pigment assemblages of photosystems 1 and 2, but more pigments absorbing at short wavelengths are associated with photosystem 2 than with photosystem 1. P430 and C550 (Q) are the primary acceptors for photosystems 1 and 2, respectively. Cyt represents cytochrome; PC, plastocyanin; FD, ferredoxin; NADP$^+$, nicotinamide adenine dinucleotide phosphate; S, the accumulator of charge in photosystem 2 (manganese enzyme?); Asc, ascorbate; FeCy, ferricyanide; DCIP, dichlorophenol indophenol. Two phosphorylation sites may be coupled to electron flow. One site, for noncylic photophosphorylation, probably occurs somewhere between plastoquinone and P700. A site for cyclic phosphorylation may be coupled to the electron pathway that includes P430 and cytochrome B_{563}.

13.6. ELECTRON TRANSPORT AND THE TWO PHOTOCHEMICAL SYSTEMS IN PHOTOSYNTHESIS

13.6.1. Photosystem 1

13.6.1.1. P700

Little is known about the physical or chemical conditions that make up the specialized environment that is thought to exist in the reaction center containing P700. As mentioned before, the redox potential (E') of P700 has been measured to be about +0.43 V. In green plants and algae the concentration of P700 ranges from one P700 to about 200 to 1000 bulk chl molecules. Very detailed flash-induced difference spectra from 250 to 850 nm have been measured[45] that show negative peaks for spinach at 430, 682, and 700 nm. Careful measurements of the differential absorption coefficient of P700 at 700 nm gave a value of 64 mM^{-1} cm^{-1} (see Fig. 13-4). Good absorption coefficients must be available before reliable quantitative measurements can be made.

Experiments with chloroplasts have shown that they can reduce viologen dyes, such as methyl viologen, having redox potentials that range from −0.5 to −0.7 V.

13.6.1.2. P430[46, 47]

Short flashes of light can produce absorbance decreases that are largest around 430 nm (Fig. 13-4). These changes can be distinguished from those produced by P700 on the basis of their different kinetic behavior. The difference spectrum of P430 has a much broader band in comparison to that produced by P700, and has no minima in the red region. Since the iron–sulfur compound ferredoxin is found in all photosynthetic organisms examined, and in early measurements was found to have a redox potential of about −0.42 V, it was proposed that this ubiquitous substance served as the primary electron acceptor of photosystem 1. The primary acceptor of photosystem 1 seems not to be the soluble form of ferredoxin, because P700 is able to reduce compounds with a lower redox potential than ferredoxin. Spinach chloroplasts washed free of soluble ferredoxin still contain a bound form that could be detected by its characteristic EPR spectrum.[48] The bound form of the ferredoxin could still be reduced at the temperature of liquid nitrogen, suggesting that this reduction is a primary photochemical process, since ordinary chemical reduction would not take place at this temperature.

The shape of the difference spectrum of P430 (Fig. 13-4) is different from that of the iron–sulfur protein ferredoxin. The primary electron acceptor may consist of a complex that includes bound ferredoxin and unknown substances, and this complex may have a lower redox potential than the soluble form of ferredoxin.

Light-induced reactions of bound ferredoxin and P700 can take place at low temperatures (below 150 K), and a stoichiometry exists between the P700 that can be oxidized and the ferredoxin that can be reduced under varying conditions.[49]

13.6.1.3. Ferredoxin and Flavoprotein (Ferredoxin-NADP-Oxidoreductase)

Ferredoxins are members of a group of nonheme iron–sulfur proteins (a protein with an iron and sulfur but not a heme moiety) that have been shown to occur in all organisms so far investigated, from obligate anaerobic bacteria to plants and animals. These proteins serve as electron transfer agents for a number of different types of reactions.[50]

Plant ferredoxin has a low molecular weight of 12,000, and consists of a single polypeptide chain containing a high percentage of acidic amino acids. There are many similarities in the amino acid sequences of plant and bacterial ferredoxin, suggesting that higher plants and photosynthetic bacteria had a common ancestor in the anaerobic bacteria, such as the present-day *Clostridium*.

The reduction of NADP, the last step of photosynthetic electron transport, requires reduced ferredoxin and a flavoprotein enzyme termed ferredoxin-NADP-oxidoreductase. This flavin protein has been isolated from spinach and crystallized.

13.6.1.4. Cytochrome f and Plastocyanin

These electron transport components can arbitrarily be considered in a discussion of photosystem 1, since they both appear to react rapidly with P700. They appear to function as intermediate electron carriers between photosystems 1 and 2, since they are oxidized by photosystem 1 and reduced by photosystem 2. As discussed in Section 13.5.3., the antagonistic effect of excitation of photosystems 1 and 2 on the redox state of cytochrome *f* served as the first experimental verification of the Hill and Bendall scheme of photosynthesis. Spectroscopic evidence for light-induced redox changes can readily be seen for cytochrome *f*, but light-induced redox changes of plastocyanin are difficult to detect, since the excitation coefficient of the oxidized-minus-reduced form of plastocyanin is low, about $10 \text{ m}M^{-1} \text{ cm}^{-1}$ at 597 nm (Fig. 13-4) only about one-half of that of the α-band of cytochrome *f*, thus making the detection of redox changes of plastocyanin *in vivo* more difficult than for cytochrome *f*.

Plastocyanin is an acidic protein with two atoms of copper per molecule. The copper in plastocyanin accounts for about 50% of the total copper in the chloroplast. This protein was discovered in the green alga *Chlorella*, and it has since been found in a wide variety of other plants, including the blue-green algae (although in a slightly different form). However, it apparently is lacking in certain yellow-green algae.

Studies based on the ability of added plastocyanin to restore partial electron transport in photosystem 1, and studies with mutants and with EPR spectroscopy indicate that plastocyanin serves as the primary reaction partner to P700. Other studies using gentle procedures to wash out plastocyanin, and leave cytochrome *f* in chloroplast particles, show that cytochrome *f* functions just as rapidly as in the intact chloroplasts, indicating that cytochrome *f* serves as the primary donor to P700. Cytochrome f and plastocyanin may function in parallel, both being reduced by photosystem 2, and both, in turn, reacting directly with P700. A definite resolution of these conflicting results awaits further research.

13.6.1.5. Cytochrome b_{563}

In addition to cytochrome f, cytochrome b_{563} is localized in the chloroplasts of higher plants. Light-induced oxidation and reduction of this cytochrome can be detected as changes in its α-band at 563 nm, and, under appropriate conditions, can be distinguished from the α-band of cytochrome f that occurs at 554 nm. The redox potential of this cytochrome is about -0.06 V, and it appears to be reduced by photosystem 1 and to mediate a cyclic flow of electrons around photosystem 1 (Fig. 13-5).

It is possible to obtain reduction of $NADP^+$ by photosystem 1 even though electron flow from photosystem 2 is blocked (DCMU addition; see Section 13.6.2.2.) or photosystem 2 is damaged. In this case, electron donors such as 2,6-dichlorophenolindophenol (DCIP), diaminodurene (DAD), or N-tetramethyl-p-phenylenediamine (TMPD) that are reduced by ascorbate can "feed" electrons in at various points along the transport sequence after photosystem 2 (Fig. 13-5).

13.6.2. Photosystem 2[51]

Although relatively little is known about the detailed functioning and interaction between the components of photosystem 1, even less is known about the functioning of photosystem 2.

Considerable effort has been expended looking for a special chl that functions as the reaction center chl analogous to the situation of P700 for photosystem 1. A reversible light-induced absorbance decrease near 680 nm ("P680") has been seen[52] that, like P700, may correspond to the bleaching (oxidation) of a special chl component. Since this change is especially large in subchloroplast particles enriched in photosystem 2, it is likely to reflect the functioning of the reaction center chl of photosystem 2. It has been suggested[53] that P680 functions like P700 as the photosynthetic reaction center of photosystem 2, because a 20 ns ruby laser flash given to a preparation maintained at $-196°C$ produced a decrease of absorbance at 680 nm, and the dark recovery of P680 (reduction) was coupled with the oxidation of a b-type cytochrome (termed cytochrome b_{559} for the minimum in the α-band of the oxidized-minus-reduced form). Absorbance changes are particularly difficult to measure in the region around 680 nm in particles enriched in photosystem 2, because the fluorescence of chl a is at its maximum in this region, and interferes with absorbance change measurements.

13.6.2.1. Q and C550[48]

The existence of a hypothetical substance termed "Q" for quencher was postulated[54] to be the primary electron acceptor for photosystem 2. Its existence was reflected by changes in intensity of fluorescence that can occur even at liquid nitrogen temperatures. When Q is oxidized, fluorescence is lower (quenched), because light absorbed by photosystem 2 can be used to reduce the primary acceptor and not be wasted as fluorescence. Accumulation of Q in the reduced form produces increased fluorescence, since it can no longer accept excitation energy from the reaction center. Q is present at a concentration of about one Q for every 400 chl molecules.

A new reversible light-induced absorbance decrease at 550 nm (C550) was discovered[55] in spinach chloroplasts, and was found to be produced by the excitation of photosytem 2. This change was associated with photoreduction and could still be seen at liquid nitrogen temperature (77 K). For these reasons, the change was proposed to be produced by the primary acceptor of photosystem 2.

A number of studies have been done to see whether Q and C550 were reflections of the activity of the same compound. Oxidation–reduction titrations showed[56] that both compounds had identical midpoint potentials and both followed the same titration curve. Mutants of algae lacking Q also lacked C550.[57] Recent results suggest that the C550 absorbance change is produced by a light-induced blue shift in the absorption maximum of a compound from 546 to 544 nm, rather than by a change in absorption caused by a redox reaction.[58] Extraction of lyophilized chloroplasts with hexane leads to a loss of the C550 absorbance change that can be seen at low temperatures. Addition of β-carotene to this extract restored the change. Under certain conditions, the C550 change can be inhibited, and yet, photochemical reactions driven by photosystem 2, such as DCIP reduction, can continue.[59] On the basis of these results it has been suggested that C550 is a carotenoid–protein complex, and not the actual primary acceptor for photosystem 2. It appears to be an "indicator" of another substance that is actually undergoing reversible oxidation–reduction reactions.

13.6.2.2. Plastoquinone

Plastoquinone, along with α-tocopherol-quinone and vitamin K_1 (a naphtho-quinone), are the most abundant electron carriers found in chloroplasts. Like lipids, these compounds are soluble in hydrocarbons and insoluble in water. Some of the quinones present in plants are in the form of globules, and in this state are probably not active in photosynthesis. There is about one plastoquinone molecule to about 40 chl molecules, making it about 10 times higher in concentration than Q, cytochrome f, or P700. Plastoquinone plays an important role in photosynthesis, since it apparently serves as a large reservoir for electrons generated upon the oxidation of water by photosystem 2. Its function can be monitored as a decrease of absorption near 254 nm, upon its reduction. Measurements of this kind have shown that plastoquinone can be reduced by photosystem 2, and oxidized by photosystem 1. The oxidation of reduced plastoquinone by photosystem 1 light is relatively slow, requiring about 10^{-2} s, about the same dark time needed between flashes to get maximum O_2 evolution, as found by Emerson and Arnold (Section 13.3.5.1.). It would thus appear that the bottleneck reaction of photosynthesis occurs at the oxidation of plastoquinone. The reduction of plastoquinone is blocked by the herbicide DCMU [3-(3,4-dichlorophenyl)-1,1-dimethylurea]. The fluorescence increase seen in the presence of DCMU is thought to represent the accumulation of the reduced form of Q, suggesting that DCMU acts on electron transport at the step(s) between the oxidation of Q and the reduction of plastoquinone.

If compounds such as DCIP or ferricyanide are added to chloroplasts, they can act as acceptors and become reduced by the electrons produced by the excitation of photosystem 2. This reduction has been called the *Hill reaction* after Robin Hill, its discoverer.

13.6.2.3. Oxygen Evolution

It is necessary to remove four electrons from two molecules of water in order to release one molecule of oxygen. Since the absorption of one quantum produces one free electron, it is necessary to put together somehow the results of four photoreactions before oxygen evolution can take place. It has been long known that a very short flash of light given to completely dark-adapted algae yields no O_2 evolution. Recently, Joliot *et al.*[60] and Kok *et al.*[61] have provided important new data about the mechanism of photosynthetic oxygen evolution. In their experiments, very sensitive polarographic techniques were used to study the O_2 evolved from dark-adapted algae illuminated with successive brief flashes of light strong enough to excite every reaction center. In algae kept in the dark for a sufficiently long period, the first flash yielded no O_2 evolution, and the second flash only a negligible amount. The third flash, however, yielded the highest amount, and the fourth flash somewhat less (Fig. 13-6). This sequence

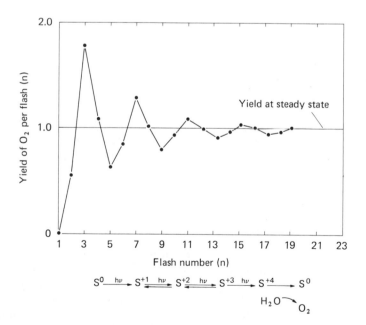

Fig. 13-6. Relative yield of O_2 per flash in dark adapted spinach chloroplasts in relation to the steady state value. (Adapted from reference 60.)

continued with the next two flashes showing low yields of oxygen evolution followed by two flashes having high yields. Interestingly, the peaks of highest O_2 yield occurred after every four flashes, except for the first peak of O_2 evolution that was seen after the third flash. Kok *et al.*[61] proposed a model by which an enzyme "S" successively accumulates four plus charges from the reaction center of photosystem 2. The enzyme, in the fully oxidized (S^{4+}) state, oxidizes water releasing a molecule of O_2, and returns the system to the discharged state (Fig. 13-6). The intermediate oxidation states, S^{2+} and S^{3+}, are unstable and decay in the dark to the stable S^+ state. Thus, in dark-adapted algae given three flashes, O_2 evolution is maximum, and continued flashes give maximum O_2 evolution with a periodicity of four; this periodicity continues as a damped oscillation with a period of four.

Kok's model provides a useful framework to explain the mechanism of O_2 evolution, but as yet we have no good idea about the biochemical nature of the substance that serves as the charge accumulator for photosystem 2. It is known, however, that the O_2-evolving system is very labile, and is rapidly destroyed by heating for a few minutes at 50°C, by washing with Tris buffer, and by UV irradiation. Cheniae[62] has shown that manganese plays an important role in O_2 evolution. There are six manganese atoms for every 400 chl. Four of these six manganese atoms can be easily extracted, leading to a decrease of O_2 evolution proportional to the loss of this easily removed manganese fraction. The four manganese atoms are bound to a protein that may be functioning as the enzyme mediating O_2 evolution.

If the O_2-evolving system is inactivated, e.g., by washing with Tris buffer, or by mild heating then the electrons that normally come from the oxidation of water can be provided by adding compounds such as diphenylcarbohydrazide, phenylenediamine, hydroxylamine, or benzidine.

13.6.2.4. Cytochrome b_{559}

Cytochrome b_{559} has been seen to undergo low temperature oxidation mediated by photosystem 2. This oxidation of cytochrome b_{559} is accompanied by the reduction of C550. But cytochrome b_{559} cannot be the primary electron donor to photosystem 2, because, prior to freezing, this cytochrome can be chemically oxidized, e.g., by the addition of potassium ferricyanide. Then at low temperature, the C550 change can still be seen to take place even though cytochrome b_{559} is oxidized.

Grana preparations (*i.e.*, lamellar stacks, see Section 13.8.2.) made from spinach chloroplasts by gentle methods show light-induced oxidation of cytochrome b_{559} that is mediated by photosystem 2. One of the functions of this cytochrome may be to mediate a cyclic electron flow around photosystem 2 (Fig. 13-5), which can act as a safety valve. If for some reason photosynthetic electron flow were stopped, then such a cyclic flow around photosystem 2 could protect the photochemical system from damage by preventing the accumulation of strong photooxidizing equivalents formed by photosystem 2.

13.7. PHOTOSYNTHETIC PHOSPHORYLATION[63,64]

13.7.1. Noncyclic and Cyclic Photophosphorylation

The job of the light reactions with their coupled electron transport is completed with the formation of NADPH and the production of ATP. These two substances are required in order to drive CO_2 reduction reactions to form carbohydrates. We have seen how electron transfer leads, as its end result, to the formation of NADPH, but this flow yields ATP as well. The coupling of ATP formation with the one-way transport of electrons to NADP was termed "noncyclic" photophosphorylation. It has been shown that ATP is the only product formed in the light when adenosine diphosphate (ADP) and inorganic phosphate (P_i) are added to the reaction mixture containing chloroplasts or chromatophores. This process is known as "cyclic photophosphorylation."

The light and dark phases of phosphorylation can be separated.[65] After first illuminating chloroplasts and then transferring them to a solution kept in the dark that contains ADP, P_i, and Mg^{2+}, it is still possible to obtain some ATP formation. It is clear from this type of experiment that something is formed in the light, with a lifetime of several seconds, that can lead to the dark formation of ATP.

13.7.2. The High-Energy Intermediate and the Chemiosmotic Hypothesis[66]

Two ideas have been developed to explain the conservation of energy in high-energy phosphate bonds of ATP when cyclic or noncyclic electron flow occurs. One idea supposes the formation during electron flow of a "high-energy state" of a hypothetical substance that undergoes both oxidation and reduction reactions as well as hydration and dehydration reactions. This concept has not been very successful so far, because no compound has been found that serves this dual role.

Mitchell[67,68] has developed a "chemiosmotic" hypothesis whereby the electrochemical potential generated across a photosynthetic (thylakoid) membrane as a result of the light reactions and electron transport gives rise to movements of H^+ from the outside to the inside of the membrane. The higher H^+ concentration inside the thylakoid membrane favors the enzymatic dehydration of ADP and P_i to form ATP and water. The separation of the light and dark phases of phosphorylation can be explained by the chemiosmotic model, as the generation against the thermodynamic gradient of a high internal H^+ concentration produced by the light reactions. In the dark, these H^+ ions leak out again, and, in doing so, generate ATP by a coupled process. Many other results favor the Mitchell hypothesis. For example, it is known that in order to get ATP formation it is necessary to have a membrane that maintains a separation between an inner and an outer space so that differences in ion concentration can be achieved. Water movement as a result of such ion transport gives rise to volume changes in chloroplasts that can be detected as large light scattering changes. Interestingly, illuminated chloroplasts cause the pH of their bathing medium to rise, presumably as a result of H^+ being taken up by the illuminated chloroplasts. This H^+ disappearance is reversible. The pH decreases again (H^+ goes out) after the chloroplasts are put in the dark. Even more interesting and perhaps providing more support for Mitchell's hypothesis are experiments[69,70] showing that the action of light could be substituted for by a dark soaking of chloroplasts in a medium at pH 4, so that the concentration of H^+ inside the chloroplasts was increased. Raising the pH to a value of about 8 in the presence of ADP and Pi gave ATP production.

Figure 13-7 represents diagrammatically what is thought to represent the sequence of events according to the chemiosmotic hypothesis. The water-evolving system is put on the inside surface of the thylakoid. One proton is released for each electron extracted from water by photosystem 2. The reduction by photosystem 2 of Q (C550) eventually leads to the reduction of plastoquinone. To be reduced, plastoquinone requires the addition of a proton that is taken up

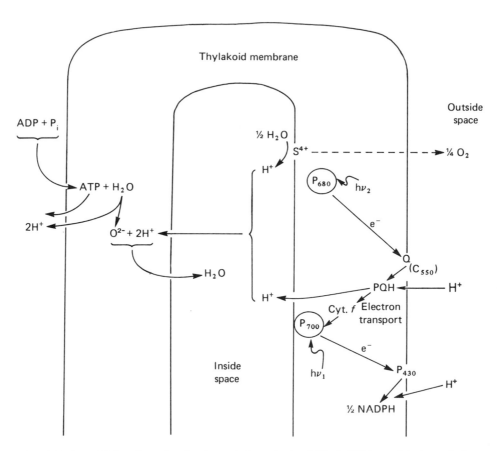

Fig. 13-7. A model based on the chemiosmotic hypothesis of Mitchell[67] to explain how electron transport can be linked with proton uptake and phosphorylation in the photosynthetic membrane. See text for details.

from the outside. Oxidation of plastoquinone is achieved by removal of only the electron by the reactions of electron transport. The proton released from plasto-quinone preferentially diffuses into the inside space. ATP formation results when two H^+ on the inside combine with the equivalent of O^{2-} (enzyme bound?) formed by the dehydration of ADP and P_i to form ATP that is mediated by an ATPase enzyme. The two H^+ released are discharged to the outside space. This polarity of O^{2-} to the inside and two H^+ to the outside is an arbitrary assumption of the chemiosmotic hypothesis.

The Mitchell idea explains the experimental observation that electron transport seems to be coupled to phosphorylation in that the rate of electron transport is controlled by the rate at which ATP formation can take place.

If phosphorylation is "uncoupled," then electron transport can proceed at a much faster rate. Uncoupling of phosphorylation from electron transport is somewhat like removing the load from an engine thereby allowing it to run more freely. Many different types of substances act as uncouplers. The effectiveness of all of these substances seems to lie in their ability to remove electrochemical potentials generated across membranes. The uncoupling action of ammonia or of an amine such as methylamine, apparently results from the ability of these compounds to diffuse freely through the membrane and combine with the H^+ on the inside space of the thylakoid to form NH_4^+, or the positive amine ion.

There are a number of antibiotic-like compounds, termed ionophores, that act as ion-translocating substances. Much research is currently being done on the effects of these substances on ion movements and membrane potentials in relation to ATP formation.[71,72]

The antibiotic nigericin allows the exchange of H^+ for K^+ and other alkali metal cations, or vice versa, thus affecting phosphorylation but not diminishing the electric charge across the membrane that results from an unequal ion concentration on the inside and outside spaces. Another antibiotic, gramicidin, apparently makes "holes" in the membrane allowing free diffusion of H^+ in and out, thus abolishing the charge across the membrane as well as phosphorylation. The antibiotic valinomycin causes membranes to be permeable to ions such as K^+ or NH_4^+, but not to H^+. Thus, a membrane potential will either be formed or dissipated depending upon the direction of electron flow.

While the Mitchell hypothesis is emphasized here, there are other ideas of how to relate ion transport to electron flow and photophosphorylation, as reviewed by Dilley.[71]

13.7.3. Electrochemical Changes and Absorption Shifts of Pigments

Light-induced changes in absorption discussed so far, except perhaps for C550, are produced by oxidation–reduction reactions of reaction-center chl molecules P700 or P680, or by redox carriers such as cytochromes, plastocyanin, or plastoquinone. There are, however, absorbance changes of photosynthetic pigments that are not correlated with redox changes, but rather with an energized state of the photosynthetic membrane.

The electric potentials that are thought to be formed across membranes as a result of an accumulation of excess H⁺ or other ions seem to produce light-induced shifts in the absorption maxima of bulk chl and carotenoids. Commonly what is seen is a small shift to longer wavelengths in the absorption maxima, and a return of the absorption maxima to their original positions in the dark. For example, a shift to a longer wavelength of the absorption spectrum of a carotenoid pigment with its three peaks each separated by about 30 nm will yield a

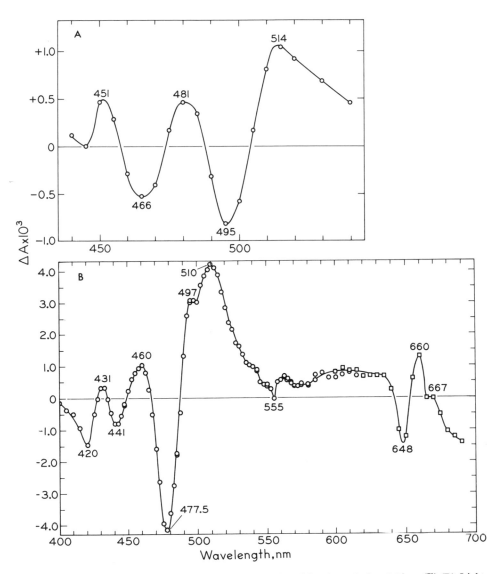

Fig. 13-8. (A) Light-induced absorbance changes produced by the red alga *Iridaea*.[73] (B) Light-induced absorbance changes produced in the green alga *Ulva* by red actinic light.[74] The minima at 420 and 555 nm are produced by cytochrome *f*, but the other peaks are probably produced by light-induced shifts in absorption of carotenoids and chl *a* and *b*.

characteristic difference spectrum, with three negative and three positive peaks each separated by about 30 nm. Difference spectra having characteristics very similar to these have been seen in many photosynthetic organisms. They are seen particularly clearly in certain photosynthetic bacteria, and in certain higher plants (Fig. 13-8). Bulk chl as well as carotenoids seem to show light-induced shifts of their absorption maxima upon illumination. In green algae and higher plants, a large light-induced absorbance increase can be seen in the region around 510 to 518 nm that is correlated kinetically with a negative peak near 475 nm (Fig. 13-8B). This change has been the subject of much research, particularly in the last several years. It apparently is produced both by carotenoid and by chl *b* absorption shifts, and apparently takes place within 20 ns. It has been proposed[75] that light produces an electric field across the thylakoid membrane, inducing an electrochromic shift in the positions of the absorption peaks of bulk chl and carotenoids. In continuous light, these shifts are sustained and show complicated kinetics, presumably caused by the interactions of both light reactions on the production of an electrical potential across membranes. The idea has been proposed[76] that the 515 nm absorption shift is an indicator of the driving force for photophosphorylation.

13.8. STRUCTURAL ORGANIZATION OF PHOTOSYNTHESIS[77]

13.8.1. The Chloroplast

Except for photosynthetic bacteria and blue-green algae, the higher plants and other algae have their photosynthetic apparatus localized in definite structures termed chloroplasts. Chloroplasts vary widely in shape, depending upon the plant species, from the cup-shaped structures in *Chlorella,* to spiral bodies in *Spirogyra,* to star-shaped structures in other algae. Spinach chloroplasts are flattened spheres with a diameter of about 5–7 μm.

The chloroplast is surrounded by a continuous double membrane. The outer membrane is connected to the endoplasmic reticulum. Apparently the inner part of the double membrane is not connected to other membranes of the cell, but is involved with the formation of the extensive lamellar system found inside the chloroplasts. This envelope acts as a selective barrier to the passage of metabolites in and out of the chloroplast.

13.8.2. Lamellae, Grana, and the Stroma

A characteristic internal feature of higher plant chloroplasts is the occurrence of a complex structural arrangement of lamellae (or thylakoids). Frequently the lamellae are closely stacked one upon the other to form grana that look somewhat like a stack of pancakes. Portions of the lamellae may extend out from the grana stacks into the surrounding space, called the stroma, which is composed of a proteinaceous matrix. The light reactions and coupled electron transport are localized in the lamellae and grana structures. The products of the light reactions, ATP and NADPH, diffuse out of these structures into the stroma where the enzymes of CO_2 fixation are located.

Chloroplasts can be prepared with and without their surrounding double membranes. The class I chloroplasts, with intact surrounding membranes, retain their soluble stroma components, and can fix CO_2 at high rates. However, the class II chloroplasts, which lack surrounding membranes, lose their soluble components from the stroma and require the readdition of these lost components in order to fix CO_2.

The stroma also contains ribosomes and DNA, which are involved in chloroplast regulation and replication. Starch grains are also commonly seen in the stroma region as are globules containing the lipophilic chloroplast quinones, such as plastoquinone, vitamin K_1, and both α-tocopherol and α-tocopherylquinones. These globules do not play a direct role in photosynthesis, but appear to serve as a storage pool for these substances.

Since grana stacks are found in almost all higher plant chloroplasts, it would seem that they play some significant role in photosynthesis. We still do not know what it is that leads to grana stack formation, and chloroplasts lacking grana show all the photochemical activities thus far measured in chloroplasts containing grana. For example, the red algae carry on complete photosynthesis, including O_2 evolution, yet have no grana stacks, but only unstacked lamellae extending throughout the chloroplasts. Studies comparing photochemical activities have shown that photosystem 1 appears early in the development of chloroplast lamellae. The appearance of photosystem 2 activity seems to be correlated with the later formation of grana.

The grana and stroma lamellae of chloroplasts of higher plants, such as spinach, can be separated from each other by mechanically disrupting the chloroplasts. Differential centrifugation yields a dense fraction containing almost pure grana, and a less dense fraction consisting of intergrana lamellae (or stroma lamellae). The intergrana lamellae have a high chl a/b ratio (about 5 to 6), very low fluorescence, and a high concentration of P700. This fraction does not evolve O_2 or show other reactions that are associated with photosystem 2.

In contrast to the less dense intergrana lamellae that contain only photosystem 1, the denser grana fraction seems to contain both photosystems 1 and 2. This fraction has a low chl a/b ratio (2.5), is highly fluorescent, evolves O_2, and has cytochrome b_{559} and P680.

13.8.3. Granal and Agranal Chloroplasts

Some plants that possess the so-called C_4-dicarboxylic acid pathway for CO_2 fixation (Section 13.9.2.) have different types of chloroplasts depending upon their location in the plant. In leaves of a C_4 plant, e.g., corn, the cells that surround the vascular bundles contain almost no grana but only stroma lamellae, while chloroplasts contained in the mesophyll regions have extensive grana as well as stroma lamellae. Initially it was suggested that the chloroplasts from vascular bundles that have no grana would have no photosystem 2 activity, by analogy to intergrana lamellae of spinach, and that the mesophyll chloroplasts of corn would contain both photosystems 1 and 2. A good deal of evidence now suggests that the agranal bundle sheath chloroplasts do have high photosystem 1 activity and are somewhat deficient in photosystem 2, having perhaps only about 40% the activity of that seen in mesophyll chloroplasts. These results suggest that a direct correlation cannot be made between the extent of grana stacking and the level of photosystem 2 activity.

Currently there is an extensive research effort[77,78] being devoted to the study of chloroplast structure and function. This work employs, in addition to light and electron microscopy, the use of polarization microscopy, X-ray diffraction, fragmentation using detergents or physical treatments such as sonication or the French pressure cell, site-specific labeling of macromolecules, and immunochemistry, to name a few.

13.9. CARBON DIOXIDE FIXATION

13.9.1. The Calvin–Benson or C_3 Carbon Cycle[79]

CO_2 fixation requires the NADPH and ATP that is formed by the light reactions of photosynthesis, and proceeds via a number of dark reactions. In general, there are two major types of CO_2 fixation—the 3-carbon and the 4-carbon pathways.

The C_3 pathway is also termed the Calvin–Benson cycle after its discoverers. Melvin Calvin was awarded the Nobel Prize for chemistry in 1961 for this work that elucidated the pathway of carbon in photosynthesis. The crucial reaction of the C_3 pathway is the addition of CO_2 to the 5-carbon sugar ribulose diphosphate (RuDP) to form two molecules of phosphoglyceric acid (PGA). The reducing power of NADPH along with ATP is used to change the acid group in PGA into the aldehyde group of the 3-carbon sugar phosphoglyceraldehyde (triose phosphate). Energy storage of photosynthesis is achieved with the formation of this 3-carbon sugar. The other steps of the Calvin–Benson cycle serve to regenerate the initial CO_2 acceptor, RuDP, via a complex series of reactions involving 3-, 4-, 5-, 6-, and 7-carbon sugar phosphates. Although sugars and carbohydrates are formed as end products of photosynthesis, other compounds can be formed as well. Substances such as fats, fatty acids, amino acids, and organic acids can also be synthesized in photosynthetic CO_2 fixation. The formation of these substances seems to be partially under the control of environmental factors such as light intensity and the concentrations of CO_2 and O_2. A better understanding of how these factors interact may lead to the ability to control growth conditions so as to produce plants that have the desired amounts of sugars, fats, or proteins.

13.9.2. The C_4 Cycle[80,81]

There are a number of plants scattered widely throughout the plant kingdom that seem to thrive in hot and arid environments. In these plants the initial fixation of CO_2 does not occur via condensation with RuDP, as in the Calvin–

Benson cycle, but via an intermediate cycle termed the C_4 cycle by which CO_2 first combines with the 3-carbon acid phosphoenolpyruvate (PEP) to form the 4-carbon acids, malic and aspartic (Fig. 13-9). These reactions proceed via the intermediate 4-carbon oxaloacetic acid, and are catalyzed by the enzyme PEP carboxylase. The 4-carbon acids, malic and aspartic, cannot be converted into sugars, however, until they are decarboxylated (lose the CO_2 fixed previously) to form the 3-carbon pyruvic acid. By reaction with ATP formed photosynthetically, the pyruvic acid is again converted into PEP that can again accept a CO_2 molecule. Essentially what this cycle accomplishes is to pump CO_2 from the outside and release it into the inside of the plant. Once inside the plant, the released CO_2 reacts in the usual manner with RuDP via the Calvin–Benson cycle. This may seem like a useless endeavor for the plant especially when it is considered that an extra ATP must be used to operate this "CO_2 pump." Plants having the C_3 cycle require 3 ATP and 2 NADPH molecules, but C_4 plants require 4 molecules of ATP and 2 of NADPH. As mentioned in Section 13.8.3., C_4 plants have a characteristic anatomy and two quite distinct types of chloroplasts. The vascular bundles in these plants are tightly surrounded by the so-called bundle sheath cells containing chloroplasts that can be distinguished from mesophyll cells located outside and around the bundle sheath cells (Fig. 13-9).

It appears that the CO_2 pump of C_4 plants is located in the mesophyll cells, and functions efficiently by trapping any CO_2 that diffuses through the stomatal pores on the epidermis of the leaf. This trapped CO_2 is then released into the interior bundle sheath cells where it is fixed via the Calvin–Benson cycle. The CO_2 pump operates efficiently to keep the concentration of free CO_2 in the

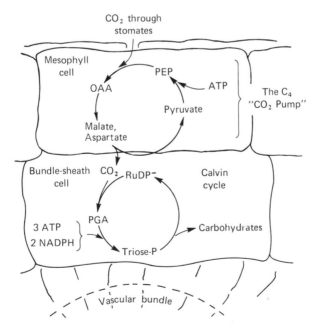

Fig. 13-9. The pathways of carbon fixation in a plant having C_4 photosynthesis. PEP represents phosphoenol pyruvic acid; OAA, oxaloacetic acid; PGA, phosphoglyceric acid; and RuDP, ribulose diphosphate.

mesophyll cells low so that CO_2 tends to diffuse from the atmosphere, where it is present in concentrations of about 300 ppm (0.03%), into the inside space of the leaf even though the stomates may be almost closed. The ability to obtain CO_2 even with stomates partially closed means that less water will be lost—a vital requirement for plant survival in climates that are both hot and dry. It thus appears that it is worth it for the plant to spend the extra ATP needed to operate the CO_2 pump, since it enables it to collect sufficient CO_2 but not to lose substantial quantities of water in the process. In any case, this extra ATP may be rather easily obtained by cyclic photophosphorylation, since plants living in arid regions usually receive an ample amount of light.

The ability of C_4 plants to photosynthesize and grow in arid regions has important practical consequences for agricultural production in these regions. Until now, most of the agricultural effort has been directed to developing crop plants that are suited to temperate regions. In the future, more effort will undoubtedly have to be devoted to developing crop plants that are well suited to grow in the earth's vast arid regions. Plants with the C_4 pathway may play an important role in this undertaking.

13.10. PHOTOSYNTHESIS AS A SOURCE OF ENERGY

There is currently considerable interest in the possibility of using photosynthetically converted light energy as a means of expanding our energy supply. It is possible, e.g., to use solar energy trapped by bacteria or hydrogen-adapted algae to make hydrogen, a nonpolluting and easily transportable fuel.

The anaerobic purple photosynthetic bacteria do not evolve oxygen, and since they cannot oxidize water they must depend upon a large variety of organic or inorganic substances for their metabolic needs. The photosynthetic production of H_2 requires that the appropriate bacteria be supplied with an organic (or inorganic) electron donor, the right amounts of amino acids (or limiting NH_4^+) and that they be in a condition where there is a high activity of energy conversion in relation to the cell's biosynthetic needs. Under these conditions, the bacteria regulate themselves to produce H_2 by energy-dependent means. Of course, any large-scale system based upon bacterial H_2 production will also require large amounts of inexpensive and easily available electron donors. This requirement may impose a severe limitation on the economic feasibility of this kind of system.

The most readily available electron donor is, of course, water, but only higher plants and algae can extract electrons from water with the operation of their two-pigment systems. We have seen that photosystem 1 can reduce compounds that have redox potentials as low as -0.7 V (Section 13.6.1.). Since at the physiological pH of 7 the E_0' of the hydrogen electrode is about -0.42 V, there is sufficient potential to produce molecular hydrogen. But algae do not normally produce hydrogen with the reductant generated by photosystem 1 unless they have been hydrogen adapted. Hydrogen adaptation is achieved by keeping an alga such as *Chlorella* in the dark for 10–12 h under nitrogen or hydrogen. Hydrogen production can apparently be seen after appropriate adaptation conditions in a large number of variously pigmented algae, including the marine red and brown algae, as well as green algae.

The adaptation procedure in algae apparently activates a hydrogenase enzyme; but this enzyme is inactivated if O_2 is present. The photoproduction of hydrogen is mediated only by photosystem 1, and the O_2-evolving system of photosystem 2 is not operative under anaerobic conditions. The substrate for H_2 production may be a product of anaerobic metabolism that accumulates during the adaptation period at a very slow rate in the algal cell. This slow formation of the substrate of H_2 production may be the cause of the low rate of H_2 evolution that usually amounts to only about 10% of the maximum rate of O_2 produced photosynthetically.

To utilize the ability of algae to oxidize water, release O_2, and at the same time produce hydrogen, requires considerable reengineering of the photosynthetic machinery. A separation of incompatable reactions has already been achieved by certain blue-green algae that fix nitrogen in their resting cells (or heterocysts). These special cells provide physical protection of the nitrogen-reducing enzymes against the destructive effects of O_2 by removing it quickly by means of a high respiratory activity in the heterocysts.

It may be possible to isolate from the living cell the various subsystems of photosynthesis such as the photochemical reaction centers, the O_2-evolving system, and the H_2-evolving system, and use them *in vitro*. For example, it might be possible to absorb photosynthetic components on appropriate supports so that the products of one system would be made close enough to be used by the next enzyme system, and so on until desired products could be removed leaving behind the intact absorbed system. In order to avoid inactivation of a hydrogenase enzyme by O_2 it may be possible to separate across opposite sides of a synthetic membrane the reactions yielding O_2 from those producing H_2. This may be accomplished, e.g., by a system employing microencapsulation, but at present our knowledge about the separation of components of the photosynthetic system is crude at best.

A laboratory-scale production of H_2 has been demonstrated by coupling the reducing power produced by chloroplasts with an algal or bacterial hydrogenase. A problem with this system is, again, O_2 production that inactivates the hydrogenase system. This problem may be alleviated by removing the O_2 from the reaction mixture, e.g., with the addition of glucose and glucose oxidase. Unfortunately, the yield of H_2 from these composite systems is disappointingly low— only a few percent of the average rate of normal photosynthesis. Moreover, these systems are very labile and expensive.

Only about 10% of agricultural production is used for food. "Waste" products from agriculture and forestry represent an enormous energy potential. Chaff from grain, stalks, sawdust, wood chips, paper, as well as table scraps all contain energy trapped by the photosynthetic process. Wood chips can be converted by pyrolysis into charcoal and numerous valuable hydrocarbon products. Plant material can be fermented into methane, converted into wood alcohol, or burned efficiently to release its energy.

It has been suggested[82] that the sugar (as well as the cellulose) produced photosynthetically by cane or sugar beet plants could be converted economically with almost no energy loss into ethyl alcohol. Alcohol can be added to gasoline to reduce the amount of petroleum needed. It may also be economically feasible to obtain hydrocarbons directly from plants rather than obtaining them indirectly, such as by the conversion of carbohydrates into ethanol. The rubber tree (*Hevea*), e.g., might serve as a direct source of photosynthetically produced hydrocarbons that could substitute for petroleum hydrocarbons currently being used for chemicals, and for starting materials for many synthetic reactions.

In any case, the energy stored and the organic matter made available from photosynthesis is enormous. Many economic obstacles currently hinder the utilization of this vast potential, and hopefully, these problems can be overcome. Continued research directed toward obtaining a better fundamental understanding of the process will aid in our efforts to tap this vast energy source.

13.11. LOOKING AHEAD

At first glance it would appear that photosynthesis has been so intensively investigated that little remains to be done. However, a closer look into almost any of the topics covered will show that many rewarding results await future investigators.

We need to learn much more about photochemical reaction centers. What is it that allows a particular chl that we call P700 or P680 to function as the photochemical traps of photosystems 1 and 2? We have mentioned that extraction of plant material with an organic solvent yields only chl *a,* but in the living plant the chl absorption spectrum is complex, showing it to be associated in the membrane with other substances such as proteins and lipids. We need to obtain a better understanding of the physical and chemical nature of the membrane systems that support the pigments and enzymes that react with each other. These highly ordered structures are a vital requirement for the photosynthetic process to proceed.

Much needs to be learned about how all the individual components in the reaction center can function to perform the primary conversion of light quanta with such high efficiency.

Little is known about the primary acceptors for each of the photosystems. For photosystem 1 we know that the absorbance change seen at 430 nm probably reflects the change in redox state of the primary acceptor. Electron-spin resonance studies indicate that it may be a type of bound ferredoxin. More research is needed on this substance before we can gain a good insight into how it functions in conjunction with the photochemical reaction center.

Even less is known about the acceptor of photosystem 2. C550 is probably only an indicator of the functioning of the primary acceptor for photosystem 2 (Section 13.6.2.1.). We have no good idea of the chemical nature of this acceptor. Likewise, the photochemical reaction center of photosystem 2 "P680," is even less understood.

The mechanism of O_2 evolution will be a challenge to unravel with its changes in the "S" states that are probably caused by some type of manganese-containing compound.

These are just a few of the outstanding problems that remain to be answered. As is usually the case, the answering of one question leads to many more. We have a long way to go before we understand nature's elaborate process of converting solar energy into stable compounds that are vitally important to all life.

13.12. EXPERIMENTS

13.12.1. The Separation and Identification of the Major Leaf Pigments

In this experiment, the major photosynthetic pigments are extracted from leaves, separated from each other, and identified by their absorption and fluorescence characteristics as well as color and position on the chromatogram.

Homogenize about 2 g of leaves and a "pinch" of calcium carbonate in about 10 ml of chilled (refrigerator) 100% acetone in a Waring blender or in a mortar and pestle. Filter the homogenate on a Buchner funnel.

The chromatographic separation of the pigments can be conveniently done on chromatography paper (about 4 cm wide and 20 cm long) or on a cellulose thin layer plate such as Eastman Kodak Chromatogram Sheet for thin layer chromatography (13255 Cellulose without fluorescent indicator, No. 6064, 20 × 20 cm).

Prepare the chromatography plate or paper by allowing 8% vegetable oil in petroleum ether (50–105°C) to run until the front is about 1 cm from the edge. Dry the plate or paper thoroughly.

Streak the acetone extract with a small pipet or medicine dropper repeatedly along the edge of the chromatography plate that is free of the vegetable oil. Let the acetone evaporate between each application. Perform this operation in dim light to avoid light degradation of the pigments, and repeat this procedure until a dark band is obtained.

Develop the chromatogram in a methanol, acetone, water solvent (15:5:1) in darkness. For this purpose the solvent can be put in the bottom of a 25-cm high thin-layer chromatography tank or other suitably sized beaker or glass container. The end of the chromatography paper containing the green band of pigments can be dipped into the solvent (but not so deep as to cover the green band) and held upright by hanging the paper or leaning the chromatography plate against a wire taped in place across the top of the tank or beaker. Cover the tank with a glass plate or the beaker with a Petri plate, and develop the chromatogram in darkness until good separation is obtained (about 45–75 min).

The chromatogram may be examined for fluorescence of the pigments under long-wavelength UV radiation (360 nm). What pigments are fluorescent? Explain why the fluorescence is red.

Note the colors and relative amounts of the green and yellow bands. The pigment bands can be identified by their position on the chromatogram,[83] and can be isolated by scraping off or cutting out the band and eluting the pigment in about 10 ml of acetone.

If a suitable spectrophotometer is available the absorption spectra of the isolated pigments can be measured. If desired, the pigments can be transferred from acetone to other solvents by means of a separatory funnel, and spectra compared to published curves.[6]

(The details of this experiment were provided courtesy of Dr. Joseph Berry and Ms. Susan Reed.)

13.12.2. Hill Reaction With DCIP

Wash and blot dry about 10 g of spinach leaves. Tear off and discard the large central midribs, and cut the remaining parts of the leaves into several small pieces so that they can be ground up in a chilled mortar and pestle with about 60 ml of chilled $0.4M$ sucrose, $0.05M$ KH_2PO_4-Na_2HPO_4 buffer pH $7.2^{[84]}$ and $0.01M$ KCl.

Filter the green juice through a layer of "Miracloth" (Cal Biochem) or through 8 layers of cheesecloth, and centrifuge at about $200g$ for 2 min. Centrifuge the supernatant at about $1000g$ for 10 min (a clinical centrifuge can be used at full speed for this spin). Discard the supernatant and resuspend the pellet (choroplasts) in about 10 ml of the buffer solution. Choroplast preparation should be performed at $\sim 0°C$.

The reduction of DCIP can be measured by adding about 0.5 ml of the chloroplast suspension and 0.5 ml of a $5 \times 10^{-4}M$ stock solution of DCIP (0.145 mg DCIP/ml water) and enough of the buffer solution used to prepare the chloroplasts to bring the volume to 10 ml. These solutions can be added to colorimeter tubes or to test tubes and exposed to a 100-W bulb at a distance of about 15 cm. They may also be conveniently illuminated by placing the samples in the beam from a slide projector. To serve as the dark control, one tube can be covered with aluminum foil to exclude light. All the tubes can be put in a large beaker of water to maintain the desired temperature (usually about 20°C).

The disappearance of the blue color results as a consequence of the reduction of DCIP by the electrons that are obtained from the oxidation of water by the operation of photosystem 2 of photosynthesis (Section 13.6.2.2.). The Hill reaction can be measured quantitatively by determining the change in optical density at 600 nm in a spectrophotometer. The extinction coefficient of DCIP at this wavelength is 21.8 mM^{-1} cm^{-1}[85] (i.e., at 600 nm a 1-mM solution of DCIP in a 1-cm pathlength cuvet will produce an optical density of 21.8). If the concentration of chlorophyll is determined,[86] then the activity of chloroplasts can be expressed as micromoles of DCIP reduced per milligram chl per hour.

The effect of an "uncoupler" of phosphorylation on electron transport (Section 13.7.2.) can be demonstrated if a substance such as methylamine (neutralized with HCl) is added to the reaction mixture to a concentration of about 20 mM.

The effect of the herbicide DCMU (Fig. 13-5) on the Hill reaction can be investigated by adding this compound to the reaction mixture before illumination. For this purpose a 10^{-3} M stock solution of DCMU can be made by dissolving 0.23 mg/ml of DCMU in ethyl or methyl alcohol. Adding 0.1 ml of this solution to the 10 ml of the reaction mixture should give strong inhibition of the Hill reaction.

The effect of temperature on the Hill reaction can be investigated by running the reaction in water baths held at various temperatures, and, if a means of measuring the relative or absolute light intensity is available (photocell, light meter or thermopile), the light curve (at one or different temperatures) showing the rate of DCIP reduction as a function of light intensities can be investigated. It is also possible to determine the effect of different wavelengths of light on the Hill reaction. For this purpose, a slide projector can be adapted to accept square colored glass or interference filters. If heat production is not too great, it is possible to use Kodak Wratten gelatin filters mounted in glass to provide actinic light of appropriate wavelengths.

ACKNOWLEDGMENT

This paper is the Department of Plant Biology Publication No. 557 of the Carnegie Institution of Washington.

13.13. REFERENCES

1. R. K. Clayton, Photochemical reaction centers and photosynthetic membranes, *Adv. Chem. Phys.* **19**, 353–378 (1971).
2. R. K. Clayton, Primary processes in bacterial photosynthesis, *Annu. Rev. Biophys. Bioeng.* **2**, 131–156 (1973).
3. P. Loach and J. J. Katz, Primary photochemistry of photosynthesis, *Photochem. Photobiol.* **17**, 195–208 (1973).
4. M. D. Kamen, *Primary Processes in Photosynthesis,* Academic Press, New York and London (1963).
5. R. K. Clayton, *Molecular Physics in Photosynthesis,* Blaisdell, New York and London (1965).
6. L. P. Vernon and G. R. Seely (eds.), *The Chlorophylls,* Academic Press, New York and London (1966).
7. A. San Pietro, F. A. Greer, and T. J. Army (eds.), *Harvesting The Sun–Photosynthesis In Plant Life,* Academic Press, New York and London (1967).
8. E. Rabinowitch and Govindjee, *Photosynthesis,* Wiley, New York (1969).
9. R. K. Clayton, *Light and Living Matter,* Vol. 1: *The Physical Part,* McGraw–Hill, New York (1970).
10. R. P. F. Gregory, *Biochemistry of Photosynthesis,* Wiley-Interscience, London (1971).
11. D. O. Hall and K. K. Rao, *Photosynthesis, Studies in Biology,* No. 37, Edward Arnold, London (1972).
12. R. K. Clayton, *Photosynthesis: How Light is Converted to Chemical Energy,* Addison-Wesley Module in Biology No. 13, Addison-Wesley, Reading, Mass. (1974).
13. Govindjee and R. Govindjee, The absorption of light in photosynthesis, *Sci. Am.* **231**, 68–82 (1974).
14. Govindjee (ed.), *Bioenergetics of Photosynthesis,* Academic Press, New York and London (1975).
15. A. L. Lehninger, *Biochemistry,* Chap. 34, Worth, New York (1970).
16. D. H. Kenyon and G. Steinman, *Biochemical Predestination,* McGraw–Hill, New York (1969).
17. A. I. Oparin, The origin of life and the origin of enzymes, *Adv. Enzymol.* **27**, 347–380 (1965).
18. S. W. Fox, Self-ordered polymers and propagative cell-like systems, *Naturwissenschaften* **56**, 1–9 (1969).

19. J. H. C. Smith and C. S. French, The major and accessory pigments in photosynthesis, *Annu. Rev. Plant Physiol.* **14,** 181–224 (1963).

20. J. S. Brown, Forms of chlorophyll *in vivo, Annu. Rev. Plant Physiol.* **23,** 73–86 (1972).

21. Govindjee and B. Z. Braun, Light absorption, emission, and photosynthesis, in: *Algal Physiology and Biochemistry* (W. D. P. Stewart, ed.), pp. 346–390 University of California Press, Berkeley, Calif. (1974).

22. H. Y. Yamamoto, T. O. M. Nakayama, and C. O. Chichester, Studies on the light and dark interconversions of leaf xanthophylls, *Arch. Biochem. Biophys.* **97,** 168–173 (1962).

23. D. Siefermann and H. Y. Yamamoto, Light-induced de-epoxidation of violaxanthin in lettuce chloroplasts. V. Dehydroascorbate, a link between photosynthetic electron transport and de-epoxidation, *Proc. 3rd Int. Congr. Photosyn.* (M. Avron, ed.), pp. 1991–1998 (1974).

24. K. Sauer, Primary events and the trapping of energy, in: *Bioenergetics of Photosynthesis* (Govindjee, ed.), pp. 115–181, Academic Press, New York (1975).

25. D. E. Fleischman and B. C. Mayne, Chemically and physically induced luminescence as a probe of photosynthetic mechanisms, *Curr. Top. Bioenerg.* **5,** 77–105 (1973).

26. J. Lavorel, Luminescence, in *Bioenergetics of Photosynthesis (Govindjee, ed.), pp. 223–317, Academic Press, New York (1975).*

27. G. Hoch and R. S. Knox, Primary processes in photosynthesis, *Photophysiology* **3,** 225–251 (1968).

28. R. S. Knox, Excitation energy transfer and migration: Theoretical considerations, in: *Bioenergetics of Photosynthesis* (Govindjee, ed.), pp. 183–221, Academic Press, New York (1975).

29. T. Förster, Zwischenmolekulare Energiewanderung und Fluoreszenz, *Ann. Physik.* **2,** 55–75 (1948).

30. T. Förster, Delocalized excitation and excitation transfer, in: *Modern Quantum Chemistry,* Part III, *Action of Light and Organic Molecules* (O. Sinanoglu, ed.), pp. 93–137, Academic Press, New York (1965).

31. D. C. Fork and J. Amesz, Spectrophotometric studies of the mechanism of photosynthesis, *Photophysiology* **5,** 97–126 (1970).

32. D. C. Fork and J. Amesz, Action spectra and energy transfer in photosynthesis, *Annu. Rev. Plant Physiol.* **20,** 305–328 (1969).

33. F. T. Haxo, Wavelength dependence of photosynthesis and the role of accessory pigments, in: *Comparative Biochemistry of Photoreactive Systems* (M. B. Allen, ed.), Academic Press, New York and London (1960).

34. J. Myers, Enhancement studies in photosynthesis, *Annu. Rev. Plant Physiol.* **22,** 289–312 (1971).

35. J. Myers and C. S. French, Relationships between time course, chromatic transients, and enhancement phenomena of photosynthesis, *Plant Physiol.* **35,** 963–969 (1960).

36. R. Hill and F. Bendall, Function of the two cytochrome components in chloroplasts: A working hypothesis, *Nature* **186,** 136–137 (1960).

37. L. N. M. Duysens, J. Amesz, and B. M. Kamp, Two photochemical systems in photosynthesis, *Nature* **190,** 510–511 (1961).

38. B. Kok and W. Gott, Activation spectra of 700 mμ absorption change in photosynthesis, *Plant Physiol.* **35,** 802–808 (1960).

39. C. Bonaventura and J. Myers, Fluorescence and oxygen evolution from *Chlorella pyrenoidosa, Biochim. Biophys. Acta* **189,** 366–383 (1969).

40. N. Murata, Control of excitation transfer in photosynthesis. I. Light-induced change of chlorophyll *a* fluorescence in *Porphyridium cruentum, Biochim. Biophys. Acta* **172,** 242–251 (1969).

41. N. Murata, Control of excitation transfer in photosynthesis. IV. Kinetics of chlorophyll *a* fluorescence in *Porphyra yezoensis, Biochim. Biophys. Acta* **205,** 379–389 (1970).

42. N. Murata, Control of excitation transfer in photosynthesis. II. Magnesium ion-dependent distribution of excitation energy between two pigment systems in spinach chloroplasts, *Biochim. Biophys. Acta* **189,** 171–181 (1969).

43. N. Murata, Control of excitation transfer in photosynthesis. V. Correlation of membrane structure to regulation of excitation transfer between two pigment systems in isolated spinach chloroplasts, *Biochim. Biophys. Acta* **245,** 365–372 (1971).

44. N. Murata, H. Tashiro, and A. Takamiya, Effects of divalent metal ions on chlorophyll *a* fluorescence in isolated spinach chloroplasts, *Biochim. Biophys. Acta* **197**, 250–256 (1970).

45. T. Hiyama and B. Ke, Difference spectra and extinction coefficients of P700, *Biochim. Biophys. Acta* **267**, 160–172 (1972).

46. T. Hiyama and B. Ke, A new photosynthetic pigment "P430": Its possible role as the primary electron acceptor of photosystem I, *Proc. Natl. Acad. Sci. USA* **68**, 1010–1013 (1971).

47. T. Hiyama and B. Ke, A further study of P430: A possible primary electron acceptor of photosystem I, *Arch. Biochem. Biophys.* **147**, 99–108 (1971).

48. A. J. Bearden and R. Malkin, Primary photochemical reactions in chloroplast photosynthesis, *Q. Rev. Biophys.* **7**, 131–177 (1975).

49. J. W. M. Visser, K. P. Rijgersberg, and J. Amesz, Light-induced reactions of ferredoxin and P700 at low temperatures, *Biochim. Biophys. Acta* **368**, 235–246 (1974).

50. D. O. Hall, R. Cammack, and K. K. Rao, Non-haem iron proteins, in: *Iron in Biochemistry and Medicine* (A. Jocob and M. Worwood, eds.), Academic Press, New York and London (1974).

51. W. L. Butler, Primary photochemistry of photosystem II of photosynthesis, *Accounts Chem. Res.* **6**, 177–184 (1973).

52. G. Döring, G. Renger, J. Vater, and H. T. Witt, Properties of the photoactive chlorophyll a-II in photosynthesis, *Z. Naturforsch. [B]* **24**, 1139–1143 (1969).

53. R. H. Floyd, B. Chance, and D. DeVault, Low temperature photoinduced reactions in green plants and chloroplasts, *Biochim. Biophys. Acta* **226**, 103–112 (1971).

54. L. N. M. Duysens and H. E. Sweers, Mechanism of two photochemical reactions in algae as studied by means of fluorescence, in: *Microalgae and Photosynthetic Bacteria*, pp. 353–372, University of Tokyo Press, Tokyo (1963).

55. D. B. Knaff and D. I. Arnon, Spectral evidence for a new photoreactive component of the oxygen-evolving system in photosynthesis, *Proc. Natl. Acad. Sci. USA* **63**, 963–969 (1969).

56. K. Erixon and W. L. Butler, The relationship between Q, C550 and cytochrome b_{559} in photoreactions at $-196°C$ in chloroplasts, *Biochim. Biophys. Acta* **234**, 381–389 (1971).

57. K. Erixon and W. L. Butler, Light-induced absorbance changes in chloroplasts at $-196°C$, *Photochem. Photobiol.* **14**, 427–433 (1971).

58. S. Okayama and W. L. Butler, Extraction and reconstitution of photosystem II, *Plant Physiol.*, **49**, 769–774 (1972).

59. R. Malkin and D. Knaff, Effect of oxidizing treatment on chloroplast photosystem II reactions, *Biochim. Biophys. Acta*, **325**, 336–340 (1973).

60. P. Joliot, G. Barbieri, and R. Chabaud, Un nouveau modèle des centres photochimiques du système II, *Photochem. Photobiol.* **10**, 309–329 (1969).

61. B. Kok, B. Forbush, and M. McGloin, Cooperation of charges in photosynthetic O_2 evolution. 1. A linear four step mechanism, *Photochem. Photobiol.* **11**, 457–475 (1969).

62. G. M. Cheniae, Photosystem II and O_2 evolution, *Annu. Rev. Plant Physiol.* **21**, 467–498 (1970).

63. M. Schwartz, The relation of ion transport to phosphorylation, *Annu. Rev. Plant Physiol.* **22**, 469–484 (1971).

64. M. Avron and J. Neumann, Photophosphorylation in chloroplasts, *Annu. Rev. Plant Physiol.* **19**, 137–166 (1968).

65. G. Hind and A. T. Jagendorf, Separation of light and dark stages in phosphorylation, *Proc. Natl. Acad. Sci. USA* **49**, 715–722 (1963).

66. A. T. Jagendorf, Mechanism of phosphorylation, in: *Bioenergetics of Photosynthesis* (Govindjee, ed.), pp. 413–492, Academic Press, New York (1975).

67. P. Mitchell, Coupling of phosphorylation to electron and hydrogen transfer by a chemi-osmotic type of mechanism, *Nature* **191**, 144–148 (1961).

68. P. Mitchell, Chemiosmotic coupling in oxidative and photosynthetic phosphorylation, *Biol. Rev.* **41**, 445–502 (1966).

69. G. Hind and A. T. Jagendorf, The effect of uncouplers on the conformational and high energy states of chloroplasts, *J. Biol. Chem.* **240**, 3202–3209 (1965).

70. A. T. Jagendorf and E. Uribe, ATP formation caused by acid–base transition of spinach chloroplasts, *Proc. Natl. Acad. Sci. USA* **55**, 170–177 (1966).

71. R. A. Dilley, Coupling of ion and electron flow in chloroplasts, *Curr. Top. Bioenerg.* **4**, 237–271 (1971).

72. H. Baltscheffsky and M. Baltscheffsky, Energy conversion reactions in bacterial photosynthesis, *Curr. Top. Bioenerg.* **4**, 273–325 (1971).

73. D. C. Fork and J. Amesz, Light-induced shifts in the absorption spectrum of carotenoids in red and brown algae, *Photochem. Photobiol* **6**, 913–918 (1967).

74. D. C. Fork, Light-induced shifts in the absorption spectrum of carotenoids and chlorophyll *b* in the green alga *Ulva, Carnegie Inst. Year Book* **72**, 374–376 (1973).

75. H. T. Witt, Coupling of quanta, electrons, fields, ions and phosphorylation in the functional membrane of photosynthesis, *Q. Rev. Biophys.* **4**, 365–437 (1971).

76. H. T. Witt, B. Rumberg, W. Junge, G. Döring, H. H. Stiehl, J. Weikard, and Ch. Wolff, Evidence for the coupling of electron transfer, field changes, proton translocation and phosphorylation in photosynthesis, *Prog. Photosyn. Res.* **3**, 1361–1373 (1969).

77. D. Branton, Structure of the photosynthetic apparatus, *Photophysiology* **3**, 197–224 (1968).

78. C. J. Arntzen and J.-M. Briantais, Chloroplast structure and function, in: *Bioenergetics of Photosynthesis* (Govindjee, ed.), pp. 51–113 Academic Press, New York (1975).

79. J. A. Bassham, Kinetic studies of photosynthetic carbon reduction cycle, *Annu. Rev. Plant Physiol.* **15**, 101–120 (1964).

80. M. D. Hatch and C. R. Slack, Photosynthetic CO_2 fixation pathways, *Annu. Rev. Plant Physiol.* **21**, 141–162 (1970).

81. O. Björkman and J. Berry, High efficiency photosynthesis, *Sci. Am.* **229**, 80–93 (1973).

82. M. Calvin, Solar energy by photosynthesis, *Science* **184**, 375–381 (1974).

83. T. W. Goodwin (ed.), *Chemistry and Biochemistry of Plant Pigments,* Academic Press, New York and London (1965).

84. A. Dunn and J. Arditti, *Experimental Physiology,* Holt, Rinehart, and Winston, New York (1968).

85. J. McD. Armstrong, The molar extinction of 2,6-dichlorophenol indophenol, *Biochim. Biophys. Acta* **86**, 194–197 (1964).

86. D. I. Arnon, Copper enzymes in isolated chloroplasts. Polyphenoloxidase in *Beta vulgaris, Plant Physiol.* **24**, 1–15 (1949).

14

Bioluminescence

14.1. INTRODUCTION

For most people, bioluminescence is represented by the flash of the firefly or the ''phosphorescence'' that occurs on agitating the surface of ocean water. Indeed, because of the abundance of material, the firefly's bioluminescence reaction has received an intensive study with the result that this system is the archetype of the variety of enzymatic processes that produce light in many bioluminescent organisms, ranging from marine bacteria to large luminous beetles from South America. What is usually understood by the term bioluminescence is a cold light emission of high efficiency, which is used by the organism for some survival purpose, although in many cases the purpose may still be conjectural. Also, a growing number of biological reactions have been found to emit light at a very low level, and this low level emission is called ''biological chemiluminescence.'' This, and the fact that bioluminescence is so widespread among many phyla (although rarely do many members of a phylum possess this property) has led to the suggestion that the ability to produce light arose very early in biochemical evolution, and that the efficient light-production ability was a secondary adaptation of biological chemiluminescence, which enabled the organism to compete more effectively within its biological niche.

Bioluminescence occurs in many terrestial forms but is most common in the

John Lee • Department of Biochemistry, University of Georgia, Athens, Georgia

sea, particularly in the deep ocean where almost all species are luminescent. In addition to the firefly, there exist luminous fungi ("foxfire"), glowworms and freshwater snails *(Latia)* found only in New Zealand, another type of beetle (Coeleoptera), the "railroad worm" found in South America, and earthworms. In the ocean, protozoa are mainly responsible for the so-called "phosphorescence," but the largest bioluminescent group is the coelenterates, i.e., the soft corals (Anthozoa), jellyfish (Hydrozoa), and comb-jellies (Ctenophores). There are many types of luminous fish some of which derive their luminescence from a culture of symbiotic bacteria. There are marine worms *(Chaetopterus, Balanoglossis, Odontosyllis),* a clam *(Pholas),* crustacea *(Cypridina),* squid *(Watasenia),* shrimp *(Holophorus),* and echinoderms (sea stars and sea urchins).

The color of the emitted light ranges from the red of the railroad worm through the deep-blue characteristic of most of the marine creatures. Thus, the first questions that might arise in the mind of an investigator are: Are all these bioluminescence systems the same or similar? What is the reaction and molecule responsible for the light emission? Why are the colors different? The spectral emission of the bioluminescence from the organism *(in vivo)* is first measured, particularly important being the position of the spectral emission maximum (λ_B). Attempts are then made to obtain extracts in solution that will react to produce the same bioluminescence λ_B with high efficiency, as measured by the quantum yield of bioluminescence Φ_B, i.e., the number of photons produced per molecule of reactant or product in the reaction. With these measurements, and after extensive efforts to obtain chemically pure components, the reaction leading to light emission can be characterized in a chemical sense. Since a chemiluminescence reaction is one that proceeds with sufficient release of energy to produce a molecule in the system in an electronically excited state, *bioluminescence can be viewed as an enzyme-catalyzed chemiluminescence.*

To return to the first of the above questions, it was discovered by Boyle in the seventeenth century that oxygen was required for the bioluminescence of bacteria and fungi, and it is now known that oxygen in some form is involved in all bioluminescence systems. At the end of the last century, Dubois found that two components could be extracted from the firefly light organ, one with hot water and the other with cold water. When mixed they would give a brief flash of the bioluminescence light. The cold-water extract, which was heat-labile, he named *luciferase,* and the hot-water extract, *luciferin.* He made a similar observation with the clam *Pholas,* but found that a cross-reaction between clam luciferin and firefly luciferase, and vice versa, did not occur. In general, it has been found that the luciferins and luciferases of bioluminescent members of a class usually cross-react; between classes it is less common, and seldom if the relationship is more distant. It is now known that this is a consequence of the chemical structure of luciferin, which is quite different, for example, between the *Cypridina* and firefly, and therefore the luciferin from the one is not a substrate for the luciferase of the other species.

All the bioluminescence reactions appear to be of the enzyme-substrate type, but of differing complexity; some involve three or four substrates, and others even require a system of three or four enzymes.

14.2. CHEMILUMINESCENCE REACTIONS[1,2]

Many chemical reactions may proceed with the release of sufficient free energy to excite one of the participants into an excited electronic state that can emit light. If this participant is a product of the reaction, the process is called "direct chemiluminescence":

$$A + B \rightarrow C^* \tag{14-1}$$
$$C^* \rightarrow C + h\nu_1 \tag{14-2}$$

The asterisk means that C is in an excited singlet state. The spectral distribution of chemiluminescence corresponds to the fluorescence of C [Eq. (14-2)]. The excitation energy, however, may end up on another molecule, D, which is not a chemical participant. This is called "indirect-" or "sensitized-chemiluminescence":

$$A + B + D \rightarrow C + D^* \tag{14-3}$$
$$D^* \rightarrow D + h\nu_2 \tag{14-4}$$

The emission in this case is the fluorescence of D, which is otherwise chemically unchanged by the process. There are some cases of sensitized chemiluminescence in which the nature of the primary excited species, C, is established or strongly inferred, and the excitation of D is by one of the energy-transfer processes discussed in Section 2.7. Electron-transfer interactions between excited molecules are also possibilities. In most instances, the nature of C has not been established, and one of these is the reaction of oxalyl chloride and hydrogen peroxide in the presence of a dye, which is one of the most efficient chemiluminescence reactions known. This is the reaction used in the commercially available "light-stick."

It will be remembered that an exothermic reaction between two molecules in the gas phase cannot occur without the presence of a third body, needed to carry away the excess energy. Gas reactions at low pressures are thus adiabatic, i.e., the energy does not immediately escape from the system, and many of them are chemiluminescent if the energy released exceeds the first electronic excited level of a product. In any event, the energy may at least populate ground vibrational states of the product and infrared chemiluminescence results.

In solution the vibrational interaction with the medium is so predominant that the release of energy cannot be adiabatic, unless the process is so rapid that fast cooling by collision with solvent molecules does not compete with the energy release. Electron transfer is one such process, i.e., a one-electron oxidation between a donor (D) and acceptor (A) of such a potential difference that the first electronic excited singlet energy of one is exceeded. A good example of this is the reaction of the radical anion of diphenylanthracene, DPA^-, with its own radical cation, DPA^+. This is called "annihilation," and it is found that about

10% of the annihilations put the product into its excited state:

$$DPA^+ + DPA^- \rightarrow 0.2\ DPA^* + 1.8\ DPA \qquad (14\text{-}5)$$

There is a second process, concerted bond cleavage of an intermediate dioxetanone [I, Eq. (14-6)], which has received a lot of attention recently as a possible chemiluminescence excitation mechanism. Briefly the idea was that, because most efficient chemiluminescence reactions involve oxygen, the key intermediate in the process could be a highly energetic species, I, which, by concerted bond cleavage, would be the source of a large energy release, and by the requirement for symmetry conservation in the bond rearrangement, would necessarily place one of its decomposition products in an excited state [Eq. (14-6)].

$$(I)$$

$$R \underset{O-O}{\overset{R}{\underset{\bigg|}{\bigg|}}}\hspace{-0.3em}\diagup\hspace{-0.3em}{\overset{O}{}} \longrightarrow \overset{R}{\underset{R}{\diagdown}}{=}O\ +\ CO_2 \qquad (14\text{-}6)$$

The prediction remained qualitative, since the dioxetane ring is heterocyclic and strict application of the Woodward–Hoffman symmetry conservation rules was not possible, but experimenters soon found, that instead of the expected excited singlet-state products, the decomposition of dioxetanones produced a high yield of triplet state and very little singlet state. If a dioxetanone that decomposed to a highly fluorescent ketone could be synthetized, it would be predicted to have a high chemiluminescence efficiency.

The quantum yield is an important characteristic of chemiluminescence reactions just as it is for photochemical reactions (Section 3.6.). In order to preserve the Second Law of Photochemistry, i.e., the correspondence between one photon and one molecule, we define the chemiluminescence quantum yield, $\Phi_C(A)$, or bioluminescence quantum yield $\Phi_B(A)$, as the number of photons emitted divided by the number of molecules of A reacting. Note that this is the inverse of the definition of quantum yield when photons are entering the system.

The Φ_C for direct chemiluminescence is the product of two efficiencies, one for the fluorescence of the emitter C, $\Phi_F(C)$ [Eq. (14-2)], and the other for the efficiency for putting the product into its excited state, Φ_E [Eq. (14-1)]:

$$\Phi_C(A) = \Phi_E \cdot \Phi_F(C) \qquad (14\text{-}7)$$

We have already seen in Section 2.4.3. how a number of competing processes reduce Φ_F below unity. For reasons that are not understood, Φ_E is found to be much less than one for most chemiluminescence reactions in solution, but in the gas phase, and also surprisingly for some bioluminescence reactions for which reliable information is at hand, Φ_E can be very nearly unity. Thus, any mechanism proposed for excitation in bioluminescence has to account for this important observation.

The Φ_E obviously cannot exceed unity, and this means that $\Phi_F(C) < \Phi_C(C)$. Not only then does the fluorescence spectrum of a proposed emitter, C, have to be the same as the chemiluminescence spectrum, but its fluorescence yield has to be sufficient to satisfy this inequality. In addition, the measurement of Φ_C is one method of determining the stoichiometry of the reaction. Clearly, for Eq. (14-1),

$$\Phi_C(A) = \Phi_C(B) = \Phi_C(C)$$

but if instead the reaction were

$$2A + B \rightarrow C^* \qquad (14\text{-}8)$$

then

$$2\Phi_C(A) = \Phi_C(B) = \Phi_C(C)$$

14.3. GENERALIZATIONS ABOUT BIOLUMINESCENCE REACTIONS[3,4]

The present state of knowledge of the details of six or more bioluminescence systems, the chemical structure of the luciferins (LH_2) and the characteristics of the luciferase enzyme (E), suggests a general sequence of reactions that lead to the emission of light. Only one of these is the true chemiluminescence reaction, but they all have to be untangled from the total system before this one can be isolated for study.

The term "bioluminescence system" will be used to describe the combined sequence of reactions used. Each bioluminescence system may contain all or only a few of the general reactions.

After the initial binding of the substrates to luciferase, the first step in the sequence may be an activation of the luciferin,

$$E\text{---}LH_2 + A \rightarrow E\text{---}LH_2B + C \qquad (14\text{-}9)$$

where A is a cofactor, for instance adenosine triphosphate (ATP) in the case of firefly bioluminescence.

Cypridina bioluminescence does not require activation, and its first reaction, the second in our generalized scheme, is oxygenation:

$$E\text{---}LH_2A + O_2 \rightarrow E\text{---}LAHOOH \qquad (14\text{-}10)$$

Equation (14-11), then, is the true chemiluminescence reaction, the excitation:

$$E\text{---}LAHOOH + M \rightarrow E\text{---}LAO^* + H_2O + M \qquad (14\text{-}11)$$

where M is a second cofactor that may be needed, like Ca^{2+} in the bioluminescent jellyfish extract (aequorin) reaction (see below), and H_2O is one likely product, or CO_2 for the firefly; or M might be another substrate that is oxidized to MO_2 in the process. Light emission results from deexcitation of E—LAO*.

Finally, the organism might want to use the enzyme again, i.e., enzyme "turnover":

$$E—LAO \rightarrow E + LO + A \qquad (14\text{-}12)$$

Apparently the firefly does not need to do this *in vivo* since it has all the luciferase it needs for a lifetime of one-time reactions. *In vitro*, firefly luciferase and most of the other types of luciferases can be made to react more than one time.

It is clear that to understand a bioluminescence reaction in chemical terms the nature of the oxygenated species produced by Eq. (14-10) must first be established, and then the excitation process, Eq. (14-11), can be guessed. In the following sections we shall examine a number of bioluminescence systems, and show what sequence of reactions are involved, and where understanding of the excitation reaction has progressed up to this time.

14.4. FIREFLY BIOLUMINESCENCE[1]

Firefly luciferin (LH_2) is D(−)-2-(6'-hydroxy-2'-benzothiazolyl)-Δ^2-thiazo-line-4-carboxylic acid (II):

(II)

The reference mark (#) indicates the optically active center, and this turns out to be the "business end" of the molecule. The first step in the sequence is activation by a luciferase-catalyzed reaction of the carbonyl group on LH_2 with ATP to form the luciferyl adenosine monophosphate derivative, LH_2AMP, and release of pyrophosphate, PP:

$$E—LH_2 + ATP \rightarrow E—LH_2AMP + PP \qquad (14\text{-}13)$$

The L(+) isomer of luciferin also carries out this reaction but its adenylate, L(+)-LH_2AMP, does not produce light on reaction with luciferase and oxygen.

Oxygenation proceeds in several steps, the first possibly being Eq. (14-14):

$$E—LH_2AMP + O_2 \rightarrow E—LHOOHAMP \qquad (14\text{-}14)$$

and the structure of this intermediate has been proposed to be

(III)

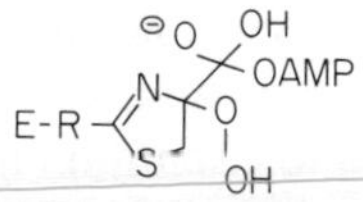

where R stands for the benzothiazolyl group in II. The next step is excitation, which overall is:

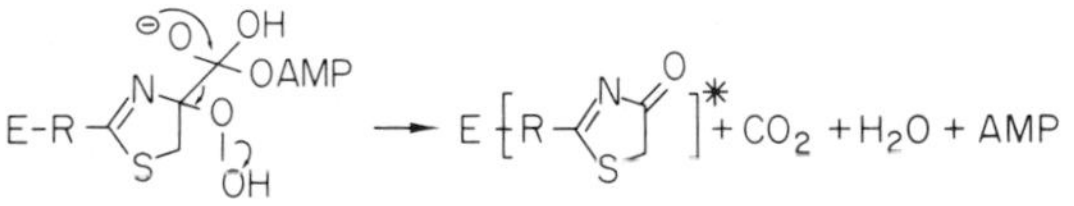

$$(14\text{-}15)$$

(IV)

On the other hand, it is also proposed by others that the oxygenated intermediate is a dioxetanone, formed by the direct attack of molecular oxygen on luciferyl adenylate [Eq. (14-16)].

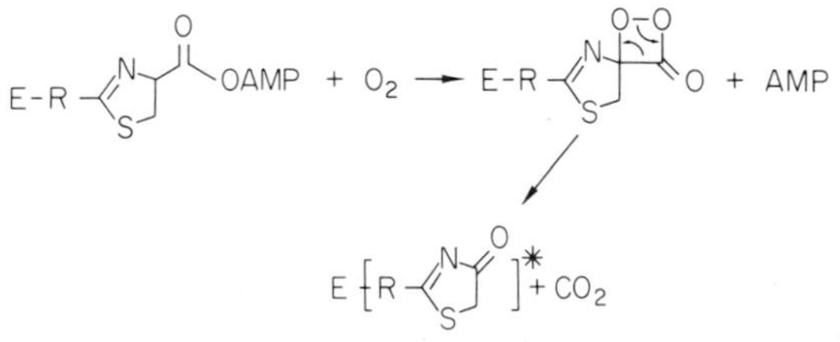

$$(14\text{-}16)$$

The breakdown of such a dioxetanone formed from oxygen-18 labeled O_2 would produce labeled CO_2 and product IV. With the chemiluminescence reaction of luciferin in strong basic dimethylsulfoxide the product IV is so labeled, but in the case of the bioluminescence reaction carried out in water, the CO_2 is not labeled unless the H_2O is labeled with oxygen-18, in which case one of the oxygens of CO_2 now comes out labeled. This is why structure II is proposed for the first step of the bioluminescence reaction, since this involves both molecular oxygen and OH^- addition from the solvent. On the basis of similar data an analogous mechanism is proposed for the coelenterate reaction, Eq. (14-18), in the next section.

The $\Phi_B(LH_2)$ has been found to be 0.8, and $\Phi_B(O_2)$ and $\Phi_B(CO_2)$ are about the same. This means that Φ_E must be about unity, but, unfortunately, product IV is too unstable to have its fluorescence measured accurately.

The *in vivo* and *in vitro* bioluminescence emission spectra are the same but the maximum λ_B is species-dependent. Figure 14-1 shows three spectra representing the range of λ_B observed, 552–575 nm. When the luciferase from a particular species is reacted with synthetic firefly luciferin, the *in vitro* λ_B is the same as the *in vivo* λ_B for that species. Therefore, the environment in the active site on the enzyme must provide a perturbation on the energy level of the emitting molecule (see Section 2.5.), and the degree of this perturbation is characteristic of the species of luciferase used.

Firefly luciferin can be made to chemiluminesce if it is reacted with oxygen in an aprotic solvent, i.e., one like dimethylsulfoxide, which cannot provide any protons for hydrogen bonding. If the solution is kept very dry and made very basic, the chemiluminescence is yellow-green, with practically the same emission spectrum as the bioluminescence. In a weaker base, a red emission of lower efficiency is found. The yellow-green emission (λ_C 562 nm) comes from V, a dianion that in solution may add a proton at a rate in competition with the radiative rate to form the red-emitting species (λ_C 615 nm), VI[2]:

(V)

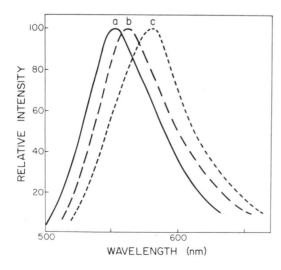

(VI)

Thus, the bioluminescence is believed to come from V. Luciferase prevents its protonation unless the pH is lowered below 6 (Fig. 14-2). Raising the temperature or adding heavy-metal ions (Zn^{2+} or Cd^{2+}) also induces the red-emission band, and obviously these are all conditions that cause the enzyme to interfere with the molecule's ability to lose its proton before emitting light.[5]

Fig. 14-1. Firefly bioluminescence spectra *in vivo*. (a) *Photurus pennsylvanicus;* (b) *Photinus pyralis;* (c) *Purophorus plagiothalamus.*

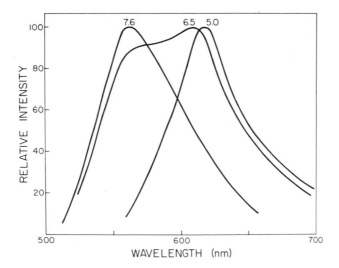

Fig. 14-2. Effect of pH on the firefly bioluminescence emission spectra *in vitro*. The λ_B 616 nm band is normalized to unity, but has an absolute contribution less than half that of the λ_B 562 nm band. The pH values are indicated on each spectrum.

Before we consider other bioluminescence systems, let us first summarize what we have learned about the firefly reaction. The luciferin molecule has had its structure accurately determined, and is found to react with ATP and O_2 on luciferase to produce a certain product, along with CO_2. This product has a fluorescence emission spectrum that looks like the bioluminescence spectrum. This product fluorescence also has an acid–base transition between two emitting species, as does the chemiluminescence of luciferin in dimethylsulfoxide, which if not too basic is a red emission spectrum like that seen in the bioluminescence when the pH is lowered or the temperature raised. Luciferases from different firefly species give slightly different yellow spectral emission maxima, which suggests that they each provide a slightly different perturbing environment. We shall encounter these same or similar phenomena when we deal with the other bioluminescence systems.

14.5. *CYPRIDINA* BIOLUMINESCENCE[1,4]

This system has been studied with extracts of a small ostracod, *Cypridina hilgendorfii,* a crustacean that is particularly abundant in Japanese waters. The same system is also found in many species of luminous fish. When stimulated, the ostracod secretes the luciferin and luciferase from separate glands into the surrounding seawater, so it can be said that the *in vivo* and *in vitro* reactions are the same.

Three reactions comprise this system: oxygenation, Eq. (14-10): excitation, Eq. (14-11); and turnover, Eq. (14-12). The structure of *Cypridina* luciferin is

VII, which we will write in the general form VIII for comparison to the luciferin involved in coelenterate bioluminescence.

(VII)   (VIII)

The overall bioluminescence reaction on the *Cypridina* luciferase (E) is

$$VII + O_2 \longrightarrow \text{(IX)} + CO_2 \tag{14-17}$$

(R_1 = indoyl; R_2 = arginyl; R_3 = isobutyl; see VII). The Φ_B is 0.3 and the spectrum is a featureless blue emission with λ_B at 460 nm, corresponding exactly to the fluorescence spectrum of IX bound to E.[6] In aqueous solution, free IX is not fluorescent. Thus, E provides an environment inhibiting nonradiative loss of the excited state energy.

In basic dimethylsulfoxide, VII and compounds of the general structure VIII (R_1 = phenyl or H; R_2 = methyl or H; R_3 = methyl) react with O_2 to produce a bright green chemiluminescence. For bioluminescence, however, R_1 and R_2 must be as in VII, although R_3 may have other alkyl substituents. The green chemiluminescence is believed to come from X, the anion of IX[1]

(X)

Thus, the luciferase in the *Cypridina* reaction must prevent *deprotonation* before emission occurs, in contrast to firefly luciferase that prevents protonation.

There is still some uncertainty about the nature of the chemical steps in the overall reaction [Eq. (14-17)], and of the oxygen-18 labeling patterns. In the *Cypridina* bioluminescence, one of the oxygen atoms in the CO_2 comes from molecular oxygen, as would be expected for a dioxetanone type of intermediate in the excitation reaction.[1]

14.6. COELENTERATE BIOLUMINESCENCE[7]

Coelenterates are a diverse group of organisms made up of two main subphyla, the Cnidaria and Ctenophra. Classes within Cnidaria are the Anthozoa, such as the soft coral, *Renilla reniformis* (''sea pansy''), the Hydrozoa (for example, jellyfish) such as *Aequorea aequorea,* and the Scyphozoa. The species mentioned are the bioluminescence systems that have received most study. The bioluminescence systems of all the coelenterates involve all four of the general reactions shown in Eqs. (14-9)–(14-12), and, although the chemical mechanism is the same for all of them, they differ between the classes of the Cnidaria in the steps of the activating reactions, and between the subphyla as to the nature of the excitation. In fact, these systems now require us to introduce more complications into the simple picture represented by Eqs. (14-9)–(14-12), and to introduce, for the first time, the phenomenon of ''sensitized bioluminescence'' [cf. Eq. (14-3)].

Coelenterate luciferins have the general structure VIII. For the sea pansy, *Renilla,* the luciferin has a structure XI (R_4 is not yet known), and for *Aequorea,* XII.

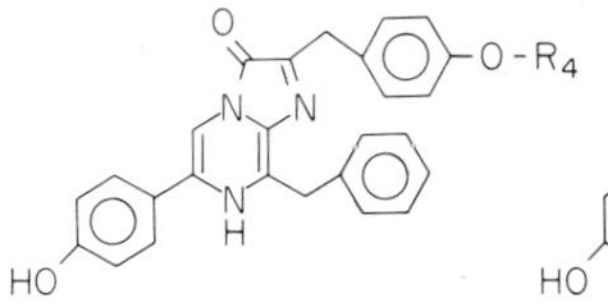

(XI) 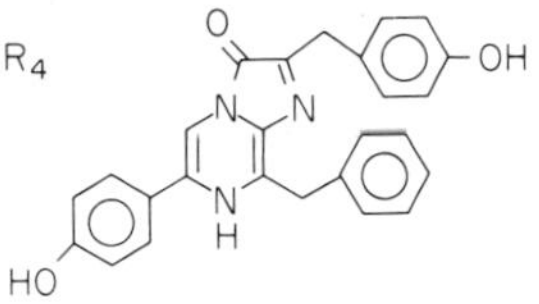   (XII)

The marked resemblance to *Cypridina* luciferin, VII, should be noted.

The light reaction produces X and CO_2. Labeling patterns with oxygen-18 O_2 and H_2O indicate that one oxygen atom of the CO_2 is derived from H_2O, and not from molecular oxygen. This is the same as found for the firefly reaction, and so a similar mechanism has been proposed.

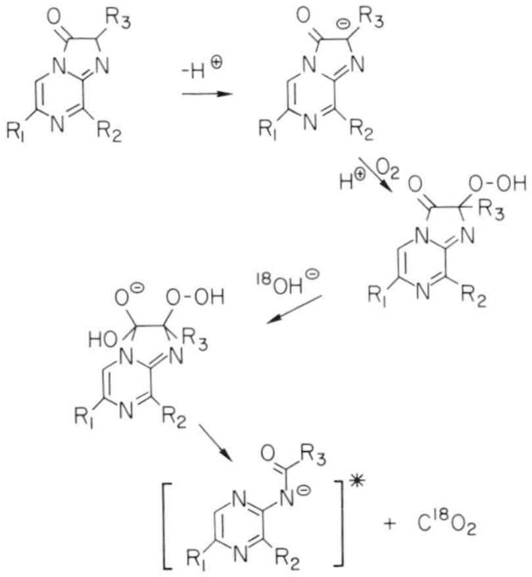

$$(14\text{-}18)$$

(X)

This mechanism has been worked out with *Renilla* luciferin and these intermediates are still speculative.[1] The anion, X, has been proposed as the emitter from studies of chemiluminescence in the aprotic solvent, dimethylformamide. The λ_C is 480 nm, as compared to the *in vitro* λ_B at 490 nm. Both are broad structureless spectra. In basic dimethylformamide, the λ_C is 538 nm, which corresponds to the dianion fluorescence. The fluorescence maximum of the neutral species is λ_F 402 nm and of the monoanion, λ_F 480 nm, Φ_F 0.06. In aqueous solution none of these species is fluorescent. Thus, the chemiluminescence reaction must also produce the excited monoanion directly. For both the chemiluminescence in dimethylformamide and the bioluminescence on the luciferase, protonation of the excited monoanion is slower than its rate of radiative emission. The Φ_B is 0.05, much higher than Φ_C, 0.001. Thus, in addition to preventing protonation of the excited state before emission, another function of luciferase must be to cause Φ_E to approach unity.

The *in vivo* bioluminescence of the Cnidaria, however, is a narrow structured emission, with λ_B 509 nm, identical to the fluorescence of a "green fluorescent protein" that can be purified from extracts of these organisms. The *in vitro* bioluminescence and the green fluorescence spectrum are compared in Fig. 14-3. The green fluorescent protein which has Φ_F 0.7, has no luciferase activity, but if present in the *in vitro* reaction at a concentration of more than 10^{-5} M, the bioluminescence emission corresponds exactly to the green fluorescence and the *in vivo* λ_B, and the Φ_B is raised over 0.2. Thus, it is evident that the green fluorescent protein *sensitizes* the bioluminescence.

No green fluorescent protein is found in extracts of Ctenophores, and the *in vivo* bioluminescence is a broad structureless emission, λ_B ~490 nm, red-shifted by 3–9 nm over the *in vitro* λ_B.[8]

In contrast to their apparent similarity in chemical mechanism, there is an important difference in the nature of the activation and oxygenation steps between Anthozoa on the one hand, and Hydrozoa and Ctenophores on the other. The Anthozoa give an oxygen-dependent luciferin–luciferase reaction, whereas the others show no oxygen dependence, and until recently, no separate luciferin and luciferase, but instead a protein complex called "photoprotein," which emits light on the addition of Ca^{2+}.

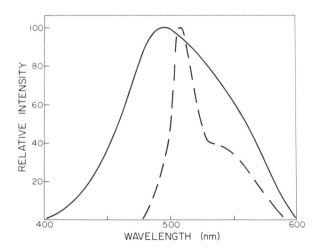

Fig. 14-3. Bioluminescence emission spectrum from the reaction of *Renilla* luciferase *in vitro* (solid line), and the fluorescence emission of the "green fluorescent protein."

Coelenterate bioluminescence can be summarized by the following scheme:

$$LH_2S + PAP \xrightarrow{E_S} LH_2 + PAPS \tag{14-19}$$

$$E_B\text{—}LH_2 \xrightarrow{Ca^{2+}} E_B + LH_2 \tag{14-20}$$

$$E\text{—}LH_2 + O_2 \rightarrow E\text{—}LHOOH \text{ ``photoprotein''} \tag{14-21}$$

Anthozoa $\downarrow$ $\downarrow$ $\begin{array}{c} + Ca^{2+} \\ Hydrozoa \end{array}$

$$E\text{—}LO^* + CO_2 \tag{14-22}$$
$$E\text{—}LO^* + E_G \rightarrow E_G^* + E\text{—}LO \tag{14-23}$$

Luciferyl sulfate, LH_2S (LH_2 is XI) and a sulfokinase, E_S, have been obtained from *Renilla*. The activating reaction [Eq. (14-19)] requires the cofactor, 3′,5′-diphosphoadenosine, PAP, and liberates the corresponding sulfate, PAPS. Luciferyl sulfate is probably the storage form in the animal since free luciferin is rapidly autoxidized in aqueous solution. Luciferyl sulfate has been found in both subphyla and classes of coelenterates. A luciferin-binding protein, E_B, has also been purified from *Renilla;* E_B releases LH_2 on addition of Ca^{2+} [Eq. (14-20)]. This process probably has a function in the control of the light flash in the whole animal. Thus the sulfokinase and binding protein both function, in a sense, in activating reactions.

The next steps are the luciferase-catalyzed oxygenation [Eq. (14-21)] and excitation [Eq. (14-22)]. The luciferases from Anthozoa and Hydrozoa must have different properties in the light reaction, however, since in the latter this bound oxygenated intermediate is extractable and stable, and requires Ca^{2+} for the light step; in Anthozoa, the intermediate is not stable, does not require Ca^{2+}, and reacts immediately to give light. If *Aequorea* luciferin (XII) is reacted with *Renilla* luciferase, the same result ensues, showing that this property of stabilization of the intermediate is determined by the luciferase itself, and not by the luciferin.

The next step [Eq. (14-23)] is the sensitization that occurs in the presence of the green fluorescent protein, E_G. The mechanism for this process is unknown, but may require a prior association between E and E_G.

If Ca^{2+} is added to a mixture of *Renilla* luciferin bound to E_B, and luciferase, the light observed (Fig. 14-4, soluble) rises in a few tenths of a second and falls very slowly, not very like the *in vivo* flash (Fig. 14-4) obtained on stimulating the animal. This discrepancy is explained by the observation that the light-giving organ, or photocyte of *Renilla,* is found to be made up of much smaller structures called *luminelles,* and each luminelle in turn contains hundreds of *lumisomes,* membrane enclosed vesicles about 0.2 μm in diameter, containing E_B, LH_2, E, and E_G. On adding Ca^{2+}, the lumisomes flash much more like the *in vivo* flash (Fig. 14-4). The enzymes are therefore believed to be packaged in the

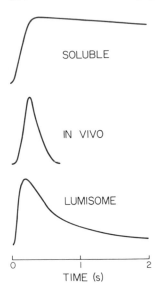

0   1   2
TIME (s)

Fig. 14-4. Light flash from *Renilla* bioluminescence by the addition of Ca^{2+} at zero time. See discussion in the text.

lumisome so that the transient concentration of Ca^{2+} is controlled in such a manner as to produce the *in vivo* kinetics.

Thus, we see in this organism, which lies lower in the evolutionary scale than the firefly, a complex multienzyme package for the bioluminescence system, together with a sensitization of the emission process. The firefly has "learned" to place the activation and light reaction on the same luciferase molecule, and relies solely on a direct but highly efficient chemiluminescence. We shall now look at the next lower form of life, the bacteria.

14.7. BACTERIAL BIOLUMINESCENCE[1,3]

The luminous bacteria are mostly of marine origin, and are found both free-living and symbiotic with certain fish. Although the *in vitro* reaction has received much more study than the *in vivo* one, the same components probably react in both systems. If luciferase is purified from cell extracts, the light reaction requires reduced flavin mononucleotide ($FMNH_2$), oxygen, and an aliphatic aldehyde (RCHO) of carbon chain length longer than seven. In contrast to the other bioluminescence reactions already discussed, no carbon dioxide is produced in the bacterial system; instead, the aldehyde is oxidized directly to its corresponding acid, e.g., decanal yields decanoic acid. Other identified products are FMN and H_2O_2.

There is at present insufficient agreement about the steps in the reaction mechanism to be able to divide it into activation and oxygenation parts. If the $FMNH_2$ and O_2 are mixed with the luciferase, the light response is still obtained if the RCHO is added at a later time. It is surmised that certain intermediates are formed by the luciferase-catalyzed reaction of $FMNH_2$ and O_2 [Eq. (14-24), (14-25)]:

$$E + FMNH_2 \leftrightharpoons E\text{—}FMNH_2 \qquad (14\text{-}24)$$
$$E\text{—}FMNH_2 + O_2 \rightarrow X \qquad (14\text{-}25)$$

X represents one or several intermediates, and free H_2O_2 also appears at this point. If the RCHO addition is delayed too long, the light-producing ability diminishes, and this is interpreted as resulting from a breakdown of X to produce more H_2O_2 and FMN [Eq. (14-26)]:

$$X \rightarrow E + H_2O_2 + FMN \qquad (14\text{-}26)$$

The rate of breakdown is strongly dependent on the temperature and the species of bacterium from which the luciferase is isolated. Typically, the time for loss of half the light-producing activity of X is in the range of 5–20 s at room temperature, but may become as long as an hour at 0°C.

The excitation reaction is

$$X + RCHO \rightarrow Y^* + RCOO^- \qquad (14\text{-}27)$$

where the RCHO has the role of a luciferin. Y^* stands for the molecule that emits the bioluminescence, and for this there are as many suggestions as there are investigators. Several proposals assume that the flavin plays an important role in the emitting process. One proposed emitter is $FMNH^+$ (the cation of FMN), another is a substituted reduced flavin intermediate, and a third is a degraded FMN structure. The last proposal is more difficult to reconcile with the fact that, at the end of the reaction, the luciferase "turns over" and releases FMN back into solution. The $FMNH^+$ hypothesis predicts that this will happen readily:

$$\begin{array}{c} E\text{—}(FMNH^+)^* \rightarrow E\text{—}FMNH^+ + h\nu \\ \downarrow \\ E + FMN + H^+ \end{array} \qquad (14\text{-}28)$$

Counter to these proposals, it has been reported the FMN may be separated from X, and that the remainder still retains activity for light production. Also, a protein-bound chromophore has recently been separated and purified from bacterial extracts, and this chromophore has a fluorescence identical to the bioluminescence. It has been suggested that it plays the role of a sensitizer.

The values for Φ_B reflect the stoichiometry of the reaction. It has been shown that $\Phi_B(RCHO) = 2\Phi_B(FMNH_2)$, and this would imply a stoichiometric requirement of two $FMNH_2$ for every RCHO used, but this is also in dispute. The absolute values for Φ_B depend on the type of luciferase used, and for $\Phi_B(RCHO)$, range from a high of 0.17 to a low of 0.04.

The bioluminescence emission spectra are broad and structureless. *In vivo*

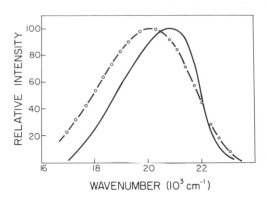

Fig. 14-5. Bioluminescence emission from *Photobacterium phosphoreum* bacteria *in vivo* (solid line), and *in vitro* (dashed line). The circles are the fluorescence of FMNH$^+$ in a rigid solvent at 77 K.

the λ_B ranges from 472 to 505 nm depending on the type of bacteria, but the λ_B *in vitro* all cluster around 490 nm. An example of the *in vivo* to *in vitro* shift is shown in Fig. 14-5; the reason for the shift is not known but is probably some environmental perturbation. FMN is the only fluorescent product of the reaction, and is formed quantitatively from the FMNH$_2$. The emission spectra, however, are quite blue-shifted from the fluorescence of FMN in free solution (Fig. 14-6), and this cannot be explained as a perturbation due to enzyme binding, since most enzymes that bind FMN usually quench the fluorescence, and produce only a small blue shift (less than 10 nm). Also, flavin fluorescence, in strongly perturbing environments like ether–isopentane–acetone at 77 K, exhibits a clear vibrational structure that is not evident in the bioluminescence emission (Fig. 14-6).

All the suggestions for the nature of the emitting molecule are based on a degree of similarity between their fluorescence and the bioluminescence. Figure 14-5 shows, e.g., that the fluorescence of FMNH$^+$ in an appropriately rigid solvent is identical to the *in vitro* bioluminescence. A requirement of the

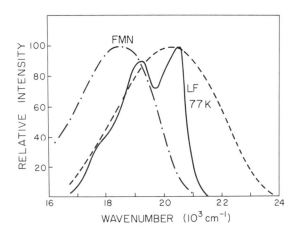

Fig. 14-6. Fluorescence spectra of FMN in water 25°C (dash-dot line), of lumiflavin (LF) in a glass at 77 K (solid line), and of bioluminescence from the *in vitro* reaction of bacterial luciferase (dashed line).

FMNH$^+$ hypothesis is that the luciferase must prevent the *loss* of the proton before the radiative process, and also hold the molecule in a rigid conformation, a condition necessary for efficient fluorescence of FMNH$^+$. This same property of impeding proton transfer has already been noted for firefly and *Renilla* luciferases.

14.8. PEROXIDE SYSTEMS[3,9]

A number of bioluminescence systems have been found to function with hydrogen peroxide rather than molecular oxygen. These are *Pholas dactylus* (clam), *Diplocardia longa* (earthworm), *Balanoglossus* (marine "acorn" worm), *Chaetopterus* (a marine segmented worm), and some echinoderms. Some conclusions reached through study of the *Pholas* system, in particular, probably apply to all these systems; for one it is very suggestive that superoxide anion, O_2^-, is the active intermediate and is produced by luciferase reacting with oxygen or peroxide alone, or in combination. This is a very interesting possibility in view of the fact that there is a great deal of current interest in an enzyme called superoxide dismutase, which is found in all aerobic cells, and whose function appears to be to remove O_2^- by rapidly catalyzing the dismutation [Eq. (14-29)] thereby preventing this

$$2 O_2^- + 2H^+ \rightarrow H_2O_2 + O_2 \qquad (14\text{-}29)$$

species from producing undesirable side reactions. Some of these luciferases must catalyze the reverse of the reaction in Eq. (14-29), or may even produce O_2^- from H_2O_2 or O_2 alone, similar to the reaction of molecular oxygen with Fe^{2+} complexed to a ligand like pyrophosphate:

$$Fe^{2+} + P_2O_7^{4-} + O_2 \rightarrow Fe^{3+} + P_2O_7^{4-} + O_2^- \qquad (14\text{-}30)$$

These luciferases are also found to have peroxidase-like properties, e.g., catalyzing the hydrogen peroxide oxidation of ascorbate. The luciferin can be made to chemiluminesce with other sources of O_2^-, like horseradish peroxidase, FMNH$_2$, etc. Apparently this chemiluminescence has the same quantum yield and emission spectrum as the bioluminescence.[9]

The *Pholas* luciferin is bound to a protein (MW 50,000) and is inactivated if the protein integrity is interfered with, e.g., by mild digestion. The chemical structure of the luciferin is not known. This protein is analogous to the luciferin-binding protein (E_B) in *Renilla*.

The bioluminescence system is comprised of two reactions, an activation [Eq. (14-31)], which is just the production of "active oxygen," O_2^-, by the luciferase

$$E\text{---}Cu^+ + O_2 \rightarrow E\text{---}Cu^{2+} + O_2^- \qquad (14\text{-}31)$$

followed by excitation

$$O_2^- + E_B\text{—}LH_2 \rightarrow E_B\text{—}LO^* + H_2O \tag{14-32}$$

The bioluminescence emission is broad and structureless, λ_B 490 nm, Φ_B 0.03.

The luciferase, E—Cu, is a copper glycoprotein (MW 300,000), and it is speculated that the Cu undergoes a one-electron oxidation in the production of O_2^-. Most types of superoxide dismutases are Cu–Zn enzymes, and are believed to act in a similar manner.[9]

The other peroxide systems have been less well investigated than *Pholas*. *Chaetopterus* luciferin is also protein-bound, and, like *Pholas,* can be made to chemiluminesce with Fe^{2+}, hydrogen peroxide, and oxygen.[4] The *Balanoglossis* and earthworm systems behave similarly, except that the luciferins are small molecules, e.g., the earthworm luciferin has a molecular weight around 300. Its luciferase is also a copper protein.[10]

14.9. LOW-LEVEL LUMINESCENCE. BIOLOGICAL CHEMILUMINESCENCE[1,11,12]

Although we have been discussing the terms light emission, chemiluminescence, and quantum yield, we have not yet considered the important device that has made the accurate measurement of these quantities possible. This is the photomultiplier tube (Section 1.4.3.), a vacuum tube device containing a photosensitive cathode coating on the inside of the front window, which emits an electron on absorption of a photon, the electron then being attracted by a voltage gradient to a series of electroemissive dynodes. These dynodes, made of a Ag/Mg alloy, have a low work function, and when one electron hits the surface of the dynode it will knock off another two or more electrons. These are attracted by a positive voltage to the next dynode where they multiply again, and so on down the line of 10 or more dynodes until out of the anode comes a burst of 10^6 or more electrons, a negative pulse of current, corresponding to that initial impinging photon.

The photomultiplier serves as the "eye" of a television camera, and was developed primarily for this purpose, and for scintillation counting in nuclear physics in the early 1940s. Since that time, photomultipliers have been developed to be more and more sensitive to light and to have wide spectral sensitivity. At the present time it is possible to make a photomultiplier device that can detect less than 100 photons/s coming from a chemical reaction contained in a volume of about 1 cm³. This means that for a fast reaction with a Φ_C of below 10^{-17}, light emission would be measurable. It has been observed that a great many exothermic chemical reactions, and a number of biological systems, do indeed emit light at a very low level.

The human eye is also a very good light detector, comparable in sensitivity, but not accuracy, to the photomultiplier. Weak light emission from a number of chemical reactions has been known, therefore, for hundreds of years. Many investigations, particularly by groups of Russian workers during the 1930s, have shown that a variety of living objects also give weak visible and ultraviolet luminescence.[11] They used Geiger tubes for detection, and this luminescence is sometimes referred to as mitogenetic or Gurvich radiation, after its discoverer. This luminescence is usually associated with rapidly growing or respiring cells— onion root tips, dividing yeast cells, white blood cells (leukocytes), liver mitochondria and microsomes, contracting skeletal and heart muscle, etc.[11,12]

In order to distinguish this low-level luminescence from what we have been previously discussing, it is called "biological chemiluminescence." Unlike most cases of bioluminescence, where the light serves some definite biological function, communication in the fireflies for instance, biological chemiluminescence has no apparent advantage to the organism, and may just be the result of a very minor energy wastage through an inevitable Maxwell–Bolztmann probability of populating an excited state of a product molecule in a very exergonic reaction. It could be, on the other hand, that it is a very efficiently chemiluminescent reaction, but occurring to only a minor extent. Oxygen and radical intermediates appear to be involved in biological chemiluminescence.

Bioluminescence and biological chemiluminescence can be better compared in terms of the light emission from the functional biological unit, i.e., from a single bacterium or the organelle of a higher organism. Luminous bacteria each typically emit 10^2–10^3 photons/s, while a lumisome from *Renilla* will produce a flash of peak intensity about 10^2 photons/s, and a single dinoflagellate which probably contains thousands of organelles, when stimulated will give more than 10^8 photons per individual. In contrast, the rapidly respiring mitochondria or leukocytes undergoing phagocytosis, have a light level far less than one photon per second per organism, on the average.[12] Actually the distinction between bioluminescence and biological chemiluminescence is blurred by the fact that the so-called "dark" mutants of luminous bacteria are only dark because they are giving off 1000 or more times less light than the wild type, i.e., at levels that could be regarded as that of biological chemiluminescence.

Not much is known about the mechanism of biological chemiluminescence, but since the oxidation of lipids and other unsaturated hydrocarbons gives rise to a chemiluminescence of low efficiency, it is proposed that such a process is probably the basis for biological chemiluminescence. The luminescence requires molecular oxygen, and can often be stimulated by the addition of hydrogen peroxide. A mechanism based on the chain oxidation of a hydrocarbon is

$$R\cdot + O_2 \rightarrow ROO\cdot \tag{14-33}$$
$$ROO\cdot + RH \rightarrow ROOH + R\cdot \tag{14-34}$$
$$ROO\cdot + ROO\cdot \rightarrow (RO)^* + RO + O_2 \tag{14-35}$$

The RO* is an excited ketone, and the fluorescence yields of these are extremely low, resulting in the very low overall yield of light.

14.10. ORIGIN AND FUNCTION OF BIOLUMINESCENCE[4,13-17]

It is generally believed that life arose on the primitive earth in an atmosphere that was extremely reducing, i.e., containing methane and other hydrocarbons, ammonia, hydrogen, etc. Oxygen concentrations were probably one thousand times below that of the present time, and were maintained there by photolytic dissociation of H_2O by highly energetic UV radiation ($\lambda < 180$ nm). Living organisms must have obtained free energy by oxidation, first, of the easily oxidizable compounds; let us call one AH_2

$$AH_2 + B \rightarrow A + BH_2 \qquad (14\text{-}36)$$

thus accumulating BH_2 with a less reducing potential, and so on up the energy scale. At some point they would have had to make use of the hydrocarbons that are very difficult to oxidize, but this could have been achieved through use of the extremely reactive oxygen radicals or hydroxyl radicals ($HO\cdot$):

$$FMNH_2 + O_2 \rightarrow FMNH\cdot + O_2^- + H^+ \qquad (14\text{-}37)$$
$$2O_2^- + 2H^+ \rightarrow H_2O_2 \qquad (14\text{-}38)$$
$$O_2^- + H_2O_2 \rightarrow OH^- + O_2 + HO\cdot \qquad (14\text{-}39)$$
$$O_2^- + RH \rightarrow HO_2^- + R\cdot \qquad (14\text{-}40)$$
$$HO\cdot + RH \rightarrow H_2O + R\cdot \qquad (14\text{-}41)$$

Such reactions can give rise to biological chemiluminescence, as discussed in the last section.[13]

Another theory of the origin of bioluminescence is based on the toxicity of oxygen to primitive anaerobes.[14] With the progression of evolution, and particularly after the advent of photosynthesis based on water, oxygen levels started to rise in the environment. This change was uncomfortable for the anaerobes, since they would produce toxic amounts of H_2O_2 and O_2^- by reactions such as Eq. (14-36). Survival would have been enhanced, however, by mutants that were able to reduce O_2 directly to water without making the reactive intermediates. This would have required getting rid of a lot of excess energy that could have been conveniently radiated off as light. Bioluminescence reactions all work well at very low oxygen concentrations, well below the levels required to support the growth of aerobes. This fact, however, supports either theory equally well.

It could not have been until the development of visual processes that there was evolutionary pressure to develop the high light efficiency reactions of bioluminescence. The first stage of the process would have been to use a highly fluorescent acceptor molecule to efficiently sensitize the chemiluminescence process. We see this mechanism operative in the bacterial and coelenterate systems. Higher on the evolutionary scale, selection was made of reactions that would produce highly fluorescent product molecules *directly* in their excited singlet state. This is the mechanism of the firefly, *Cypridina,* and other higher forms.

From the point of view of function, the firefly has received most study and is best understood. The flash is a sexual signaling, and really consists of an elaborate series of flashes (Fig. 14-7), a "Morse code" that is characteristic of the species.[15] The flying male flashes and may be answered after a definite time interval, about 2 s, by the female. The female of one species is insectivorous, and is able to lure its prey by imitating the female response appropriate to the prey species.

It has recently been shown that starting with a mixed population of luminous and nonluminous dinoflagellates and their predators, under controlled laboratory conditions, that luminous types show an increase in survival over nonluminous types. The luminous flash probably induces an avoidance response in the predator, conferring survivorship on the species that possess this property.[16] Other escape flashes are evident in shrimp, squid, and the *Cypridina,* but have received little controlled study.

Some luminous fish have a light organ that is a symbiotic culture of luminous bacteria. Some fish have even adapted a lid to move over the light organ to create a flashing of the otherwise continuous emission. Other fish appear to contain a luminous system that is the same as in *Cypridina,* i.e., the luciferins and luciferases cross-react. The luminescence probably serves as a mating signal, or as a lure.[17]

In the Australasian glowworm, which is the larva of a diptera (fly), in contrast to the American firefly, which is a coeleoptera (beetle), the luminescence has a lure function. The larva hangs in the middle of a web, and emits a blue glow (λ_B 490 nm). Small winged insects are attracted to the glow, are ensnared in the web, and devoured.

Bioluminescence therefore appears to be a secondarily adapted biological chemiluminescence, functioning in ways for other species similar to those described above, to the advantage of the possessor. H. H. Seliger has summarized the functions of bioluminescence in all species by the "four P's": *Predation, Protection, Prenuptial* (firefly communication), and *Perfidy* (the insectivorous firefly).

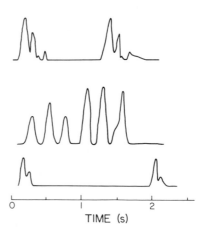

Fig. 14-7. Flash sequence from live fireflies. Top: *Photinus gracilobus,* middle: *Photinus melanuris,* lower: *Photonus nothus.* (Redrawn from reference 14.)

TIME (s)

14.11. APPLICATIONS OF BIOLUMINESCENCE REACTIONS[1,18]

The application of bioluminescence reactions to provide illumination is impractical due to the cost of preparations and the difficulty of storing enzymes for long periods. Some efficient chemiluminescence reactions, however, are widely used as emergency light sources. A really important application of bioluminescence reactions has been the detection and analysis of cofactors that are involved or can be coupled to the bioluminescence reaction, e.g., the analysis of ATP with the firefly reaction.[19] Of great potential still to be fully exploited, are methods for analyzing for reduced nicotine adenine dinucleotide (NADH) and FMN with the bacterial bioluminescence reaction,[18,20] and the detection of Ca^{2+} with the photoprotein, aequorin.[21]

For very low amounts of ATP, the firefly assay is practically the only available technique. If a limiting quantity of ATP is added to a solution of firefly luciferin and luciferase, a flash of light is produced, which, if detected by a photomultiplier linked to a current meter and recorder, produces a curve like that shown in Fig. 14-8. A straight line results when the maximum light intensity (I_0) is plotted against the amount of ATP added, and this is the calibration line (Fig. 14-9). For an unknown sample, the ATP content can be calculated from its I_0. This method has been used to find out how much ATP is present in nerve cells, plant seeds, and bacteria, and modification of the procedure by enzyme coupling enables the assay of other important cofactors like AMP, cyclic AMP, and pyrophosphate. Enzymes that use ATP can be determined, e.g., creatine phosphokinase. One of the most important applications has been to the assay of biomass via the ATP content, to which it is directly related.[1] In a hospital laboratory, e.g., a blood or urine specimen can be rapidly measured for infectious levels of bacteria in this way.

Of wider application than the obvious assay of FMN with the bacterial bioluminescence reaction, is the measurement of NADH. There is an enzyme present in crude extracts of bioluminescent bacteria that can reduce FMN with NADH.

$$NADH + H^+ + FMN \rightarrow FMNH_2 + NAD^+ \qquad \text{(dehydrogenase)} \qquad (14\text{-}42)$$
$$FMNH_2 + O_2 + RCHO \rightarrow light \qquad \text{(luciferase)} \qquad (14\text{-}43)$$

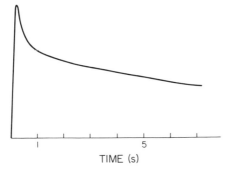

TIME (s)

Fig. 14-8. Light flash from firefly bioluminescence *in vitro* obtained by injecting ATP into a mixture of firefly luciferase and luciferin. For 10^{-9} g of ATP the flash height maximum (I_0) corresponds to about 10^{10} photons/s.

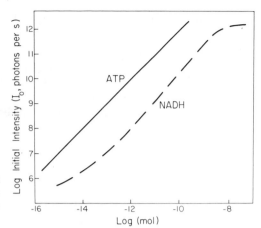

Fig. 14-9. Calibration lines for the estimation of ATP with the firefly bioluminescence system, and NADH with the bacterial bioluminescence system.

Figure 14-9 shows the I_0 obtained with such a mixed enzyme system. Many enzymes of clinical interest are assayed by their reaction with NADH, but the bioluminescence procedure remains yet to be developed for this work.[18]

The reaction for Ca^{2+} using aequorin is able to detect less than 10^{-13} mol. Although specific for Ca^{2+}, other ions such as Sr^{2+} are able to trigger the reaction but only at 100-times higher concentration. Many physiological processes involve rapid changes in Ca^{2+} concentrations, and the aequorin reaction has found use, therefore, in studying such things as Ca^{2+} ion transients in muscular contraction.[22]

Finally, an application of a different nature has been made by the use of image intensifier tubes, which, when coupled to a microscope, reveal the localization and time responses of the bioluminescence in whole organisms. This has enabled the visualization of nerve impulses across single dinoflagellates, and to microscopically localize the photocytes in some species of coelenterates.[23]

14.12. EXPERIMENTS

A number of simple demonstrations can be carried out without access to elaborate laboratory facilities. The firefly reaction can be used to demonstrate the ATP assay, the oxygen requirement, and the pH effect on the spectra.[14,19] With the luminous bacteria, an oxygen effect and the NADH reaction can be demonstrated. The light emission can be seen by the dark-adapted (10 min) eye, but for more quantitative measurements some sort of photometer should be used, such as a fluorometer with the excitation lamp turned off, or a scintillation counter operated in the out-of-coincidence mode. Quantitative experiments can be developed if the student has access to a refrigerated centrifuge and a pH meter.

Fireflies may be easily captured, or dried whole specimens may be obtained from several chemical supply companies (Worthington Chemical Corp., Freehold, New Jersey; Sigma Chemical Co., St. Louis, Missouri). Take about 10 dried fireflies, cut off the tails, and grind them with about 1 g of clean sand and 3 ml of buffer solution. This solution is made by dissolving 0.4 g glycine and 0.1 g ammonium bicarbonate in 100 ml of distilled water. It should have a final pH of about 7.6. Grinding should be done for about 10 min, and then the solution should be filtered (or better if available, centrifuged). The filtrate is the crude extract. All these procedures should be carried out under ice-cold conditions. Divide the extract into about three portions in test tubes. Make up a solution of ATP by dissolving 0.1 g in 100 ml of buffer, and add to this 0.01 g $MgSO_4$ (or other soluble Mg salt). All the light reactions must be carried out at room temperature. To a portion of extract, rapidly add 1 ml of the ATP solution. A flash of yellow light will be observed.

Bubble nitrogen gas for about 5 min into a 1-ml sample of ATP solution and also a second portion of the firefly extract, and then pour the solutions together rapidly while still bubbling with nitrogen. The nitrogen removes oxygen from the solutions and only after agitation for a few seconds in air will there be enough oxygen returned to induce the light emission. Add one drop of 10% concentrated acetic acid to 1 ml of the ATP solution, and then add this to the last portion of extract. A weak red glow should be seen (see Fig. 14-2).[14]

A culture of luminous bacteria can be purchased from the American Type Culture Collection, 12301 Parklawn Drive, Rockville, Md. 20852; ATCC 11040, *Photobacterium phosphoreum,* grows well. Alternatively a wild strain can be isolated from the surface of a decaying fish or squid. Standard microbiological techniques can be used to transfer and grow the bacteria on a Petri plate containing sterile media of the following composition (100 ml): sodium chloride 2.5 g, sodium phosphate tribasic 0.5 g, potassium phosphate monobasic 0.2 g, ammonium phosphate dibasic 0.05 g, magnesium sulfate 0.01 g, Bacto-peptone ("Difco") 1 g, glycerol 0.3 ml, Bacto-Agar 1 g. The final pH of this mixture should be around 7.3. About 3 days after the inoculation, the surface of the plate should be covered with the growth of the luminous bacteria. If a stream of nitrogen is passed into the Petri plate, keeping the lid as tightly closed as possible, the light will dim perceptibly, and then return when reexposed to air.[17]

Bacterial luciferase may be purchased from the above-mentioned chemical companies. Make a solution of reduced nicotine adenine dinucleotide (NADH) by dissolving 0.1 g in 10 ml of buffer, another solution of flavin mononucleotide (FMN) by dissolving 0.05 g in 1000 ml of water, and then a suspension of dodecanal in water by taking 1 drop of the pure dodecanal and shaking it vigorously in 10 ml of buffer for 1–2 min to obtain a cloudy suspension. To 2 mg of bacterial luciferase dissolved in 1 ml of buffer in a test tube, add one drop of the dodecanal suspension, then 1 drop of the FMN. In a darkened room, add to this 1 drop of the NADH solution. A blue flash will be seen, not as bright as the firefly light, but it then settles down to a steady glow that continues for many minutes.

14.13. REFERENCES

1. M. J. Cormier, D. M. Hercules, and J. Lee (eds.), *Chemiluminescence and Bioluminescence,* Plenum Press, New York (1973).

2. E. H. White, J. D. Miano, C. J. Watkins, and E. J. Breaux, Chemically produced excited states, *Angew. Chem. [Engl.]* **13,** 229–243 (1974).
3. M. J. Cormier, J. Lee, and J. E. Wampler, Bioluminescence: Recent advances, *Annu. Rev. Biochem.* **44,** 255–272 (1975).
4. F. H. Johnson and Y. Haneda (eds.), *Bioluminescence in Progress,* Princeton University Press, Princeton. N.J. (1965).
5. L. J. Bowie, R. Irwin, M. Loken, M. DeLuca, and L. Brand, Excited-state proton transfer and the mechanism of action of firefly luciferase, *Biochemistry* **12,** 1852–1857 (1973).
6. O. Shimomura, F. H. Johnson, and T. Masugi, Cypridina bioluminescence: Light-emitting oxyluciferin-luciferase complex, *Science* **164,** 1299–1300 (1969).
7. M. J. Cormier, K. Hori, and J. M. Anderson, Bioluminescence in coelenterates, *Biochim. Biophys. Acta* **346,** 137–164 (1974).
8. W. W. Ward and H. H. Seliger, Properties of mnemiopsin and berovin calcium activated photoproteins from the ctenophores *Mnemiopsis* sp. and *Beroe ovata, Biochemistry* **13,** 1500–1509 (1974).
9. A. M. Michelson and M. F. Isambert, Bioluminescence. XI. Pholas dactylus system. Mechanism of luciferase, *Biochemie* **55,** 618–634 (1973).
10. R. Bellisario, T. E. Spencer, and M. J. Cormier, Isolation and properties of luciferase, a non-heme peroxidase, from the bioluminescent earthworm, *Diplocardia longa, Biochemistry* **11,** 2256–2266 (1972).
11. G. M. Barenboim, A. N. Domanskii, and K. K. Turoverov, *Luminescence of Biopolymers and Cells* (translated from Russian), Plenum Press, New York (1969).
12. R. C. Allen, R. L. Stjernholm, and R. H. Steele, Evidence for the generation of an electronic excitation state in human polymorphonuclear leukocytes and its participation in bacteriacidal activity, *Biochem. Biophys. Res. Commun.* **47,** 679–684 (1972).
13. H. H. Seliger, Origin of bioluminescence, *Photochem. Photobiol.* **21,** 335–361 (1975).
14. H. H. Seliger and W. D. McElroy, *Light: Physical and Biological Action,* Academic Press, New York (1965).
15. J. E. Lloyd, Bioluminescent communication in insects, *Annu. Rev. Entomol.* **16,** 97–122 (1971).
16. W. E. Esaias and H. C. Curl, Effects of dinoflagellate bioluminescence on copepod ingestion rates, *Limnol. Oceanogr.* **17,** 901–906 (1972).
17. E. N. Harvey, *Bioluminescence,* Academic Press, New York (1952).
18. P. E. Stanley, Analytical bioluminescence assays using the liquid scintillation spectrometer. A review, *Liquid Scintillation Counting* (M. A. Crook and P. Johnson, eds.), Vol. 3, pp. 253–271, Heyden, London (1974).
19. B. L. Strehler, Bioluminescence assay: Principles and practice, *Methods Biochem. Anal.* **16,** 99–181 (1968).
20. E. Chappelle and G. L. Picciolo, Assay of flavine mononucleotide (FMN) and flavine adenine dinucleotide (FAD) using the bacterial luciferase reaction, *Methods Enzymol.* **18,** 381–385 (1971).
21. F. H. Johnson and O. Shimomura, Preparation and use of aequorin for rapid microdetermination of calcium ions in biological systems, *Nature New Biol.* **237,** 287–288 (1972).
22. J. R. Blinks, F. G. Prendergast, and D. G. Allen, Photoproteins as biological calcium indicators, *Pharmacol. Rev.* **28,** 1–93 (1976).
23. G. T. Reynolds, Image intensification applied to biological problems, *Q. Rev. Biophys.* **5,** 295–347 (1972).

15

New Topics in Photobiology

15.1. INTRODUCTION

The highlighting of new topics in photobiology, or of new advances in older subspecialties of photobiology, is meant to call to the attention of the reader those areas of photobiology which, in this author's view, offer unusual opportun-

Kendric C. Smith • Department of Radiology, Stanford University, School of Medicine, Stanford, California

ities for exciting research in the coming years. Additional examples are also given in the Conclusion sections of the preceding chapters.

Section 15.2., on misconceptions about light, may at first seem out of place in a chapter entitled "New Topics in Photobiology," but it is not. Several of the misconceptions about the properties of light have held back progress in photobiology for many years. For example, if one thinks, and incorrectly so, that light does not penetrate the human body, then there would seem to be no point in investigating the effects of light on man in addition to studies on vision, skin cancer induction, and vitamin D synthesis. Once this misconception is dispelled, however, the study of the effects of light on man (e.g., activation of enzymes, alteration of hormone levels, etc.) becomes one of the most exciting "new topics" in photobiology.

Photobiology is a fascinating field. One can approach it from at least two different view points: (1) one can simply study the effects of light on biological systems; both the beneficial and the detrimental effects; or (2) having learned something about the effects of light on biological systems, one can then use radiation as a selective tool in research or in the treatment of disease. Some examples of both approaches are given below.

15.2. COMMON MISCONCEPTIONS ABOUT LIGHT

There exist, even among scientists and physicians, several common misconceptions about the properties of light. They can be considered to be in the category of "old wives' tales." At first, they might seem amusing, but when one realizes that they are responsible for holding back progress in the science of photobiology, they no longer seem humorous.

1. *False: Visible light is not as photochemically reactve as UV radiation because the energy of the photons of visible light is too low.* This misconception has probably arisen because biological systems are generally more easily inactivated by UV radiation than by visible light. This result is not due to the property of the light, but rather to the property of the light-absorbing molecules whose photochemical alteration leads to the inactivation of the biological system.

Thus, since DNA is the most important molecule in a cell, and since it can be altered by the absorption of UV radiation, a cell is most readily killed by UV radiation. If, for example, the electronic structure of DNA were such that it absorbed blue light rather than UV radiation, then cells would be most easily killed by blue light. Therefore, it should be remembered that visible light is just as "photochemically active" as UV radiation, under the appropriate conditions.

2. *False: Visible light is natural and therefore it is safe.* This is a variation of the above misconception. Because something is natural should convey no concept of safety to the thinking person. Poisonous snakes and plants are natural, but they certainly are not "safe."

Visible light, or even UV radiation is perfectly safe if it is not absorbed. Safety is not an intrinsic property of the wavelength of the light. Blue light is "safe" for nucleic acids, since it is not absorbed by nucleic acids, but it is not safe for bilirubin (see Chapter 7). Under appropriate conditions, "natural" visible light is quite detrimental (Section 15.7.3.).

3. *False: We need not be concerned about the biological effects of visible light because it does not penetrate human tissues.* This point arises from the misconception that because one cannot see through his (her) hand, that light is not transmitted through the hand. Actually a considerable amount of light is transmitted through a human hand, as will be documented and discussed in Section 15.3.1. Since light is transmitted and absorbed in human tissues, it is most appropriate to wonder about its biological effects.

4. *False: Water is a good filter for UV radiation.* This point is most dramatically refuted by the observation that even on a cloudy day, when most of the heat of the sun, and much of the visible light are filtered out by the clouds, one can still get badly sunburned, because the amount of solar UV radiation is only reduced by about 50%.[1] Data on the transmission of solar radiation into natural waters have recently been reviewed.[2]

5. *False: Since the amount of solar UV radiation (below 320 nm) reaching the surface of the earth is less than 1% of the total solar radiation reaching the earth, it can be ignored when considering the biological effects of sunlight.* This misconception arises from the common practice of neglecting small percentage errors in homogeneous samples; however, sunlight is not a homogeneous sample. Each wavelength of light (UV through visible) has the same chance of producing a photochemical change if absorbed, yet, because all living cells contain DNA, UV radiation is orders of magnitude more effective in killing cells than is visible light, in the absence of added photosensitizers. This point is documented by the action spectrum for the germicidal effectiveness of the various wavelengths of light (Section 15.7.1.). Although less than 1% of the total energy output of the sun is at wavelengths less then 320 nm,[3] it accounts for ~99% of the deleterious biological effects of sunlight (Section 15.7.1.). Clearly, this 1% cannot be ignored.

15.3. ABSORPTION SPECTRA OF OPAQUE OBJECTS

Opacity is a relative term, based upon the rather limited spectral range of human visual acuity. This range extends from about 380 to 700 nm, with a peak of efficiency around 550 nm (Fig. 1-2), thus, an object that transmits light poorly between 380 and 700 nm is considered to be opaque. Similarly, however, if an object readily transmits light at wavelengths below 380 nm or above 700 nm, it would still be considered opaque, since our eyes do not respond to these wavelengths.

This rather limited view of opacity, based upon human visual acuity, has prejudiced people's thinking about possible biological effects of light on opaque systems, and has slowed progress in certain areas of photobiology, especially the photobiology of humans.

15.3.1. Light Is Transmitted through Human Tissues

Since we cannot ordinarily see through the human body, a common misconception has arisen that light does not penetrate the human body except through the eyes. When confronted with this incorrect statement, I simply ask if that person played the childhood game of putting the lighted end of a flashlight in his (her) mouth while standing before a mirror in a dark room. Such a trick dramatically exemplifies that light does penetrate living tissue. It is mainly red light (670–760 nm) that is transmitted.

An example of the transmission of light by the human hand may be found on the cover of the March 1972 issue of *Scientific American*. The space between the fingers of a hand was carefully masked, and a flash lamp was fired on one side of the hand; a color photograph taken on the other side by the *transmitted light* yielded a picture of a red hand.

A more artistic example of light passing through a hand comes from a painting by Georges de La Tour (1593–1652) that hangs in the Louvre. In this painting, entitled Saint Joseph the Carpenter, a child shields the flame of a candle with her hand, and the light is shown to be transmitted through the tissue of the fingers.

The absorption spectrum of "opaque" biological samples may be recorded on a very sensitive spectrophotometer with the sample in close juxtaposition with the photocathode.[4] The absorption spectrum of a human hand (Fig. 15-1)

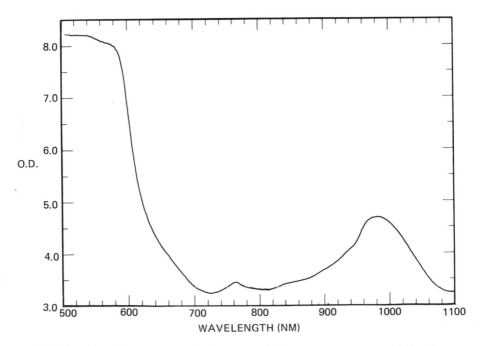

Fig. 15-1. The absorption spectrum of a human hand. The spectrum was recorded with a very sensitive spectrophotometer with the hand in close juxtaposition with the photocathode. (Unpublished data of Karl H. Norris.)

demonstrates that it is rather transparent to light of wavelengths between 650 and 900 nm. An optical density of 3.5 corresponds to a percent transmission (%T) of about 0.05%. Therefore, whether or not we see red light being transmitted through living tissue depends, in part, upon the incident intensity (I_0) of the light [$\%T$ = I(transmitted)/I_0].

A similar trick with a flash lamp and a camera can be used to demonstrate hydrocephaly in babies,[5] a condition in which part of the space normally occupied by the brain is replaced with cerebrospinal fluid. Such transillumination of the heads of babies is routinely done by physicians with a flashlight in a dark room. Where fluid replaces brain tissue, light is easily transmitted from one side of the head through the other.

When light is absorbed, photochemistry can occur! Therefore, since light can penetrate deeply into the tissues of man, it is appropriate to ask, what are the biological consequences?

Extraocular photoreceptors are known in lower animals and birds (Chapter 9), but the possibility of such photoreceptors in man is just beginning to be studied. If there are extraocular photoreceptors in man, this would greatly affect our ideas concerning artificial lighting, which is now optimized only for vision.

15.3.2. Light Is Transmitted through Plant Tissues

Absorption spectroscopy on opaque objects has been a valuable tool for plant sciences. Not only has it been used as a tool in research in photosynthesis, photoperiodism, and photomorphogenesis,[6] but also in agriculture for measuring the interior quality of fruits and vegetables, and for sorting eggs.[7]

15.3.3. Photoacoustic Spectroscopy

The technique of photoacoustic spectroscopy is based on the experiments of Alexander Graham Bell and others in the 1880s. These experiments showed that any energy absorbed from periodically interrupted light illuminating a gas in an enclosed cell produced pressure fluctuations that could be detected as audible sound. The effect has been thoroughly studied for gases, but the analogous effect with solids has not been studied in detail until recently.

For the photoacoustic spectroscopy of solids, the sample is placed inside a cell containing a suitable gas, and a sensitive microphone. The sample is then illuminated with periodically interrupted monochromatic light. Any light absorbed by the solid is converted, in whole or in part, into heat by nonradiative deexcitation processes within the solid. The resulting periodic heat flow from the solid absorber, as it is cyclically heated by the periodically interrupted light, to the surrounding gas creates pressure fluctuations within the cell that are detected by the microphone. The analog signal from the microphone can then be recorded as a function of the wavelength of the incident light. The photoacoustic spectrum corresponds, qualitatively at least, to the optical absorption spectrum of the solid, provided that the nonradiative processes dominate in the dissipation of the absorbed light energy.

The principal advantage of photoacoustic spectroscopy is that it enables one to obtain spectra similar to optical absorption spectra on any type of solid or semisolid material, whether it is crystalline, powder, amorphous, smear, gel, etc. Furthermore, since only the absorbed light is converted to sound, light scattering, a very serious problem when dealing with many solid materials (e.g., biological samples) by conventional spectroscopic techniques, presents no difficulty in photoacoustic spectroscopy.

Photoacoustic spectroscopy has been used to study membrane-bound cytochrome P450, green leaves, opaque crystals of myoglobin, and hemoglobin in whole blood. It should find use in the study of hard tissues such as teeth and bone. It has been used to detect and identify spots on thin-layer chromatography plates, and to demonstrate the uptake of an antibacterial agent in guinea pig skin. It can provide, in minutes, information about the compounds present in complex biological systems, and requires only milligrams of material, and no special sample preparation. Photoacoustic spectroscopy should find wide use, in the near future, in the study of the absorption spectra of opaque biological materials.[8]

15.4. EFFECTS OF ENVIRONMENTAL LIGHTING ON ANIMALS, INCLUDING MAN

Light (its intensity, wavelength distribution, and cyclicity) is probably one of the most important elements in our environment, yet its effect on man has been little studied. A considerable amount of research has been done on the effects of temperature (arctic and desert expeditions) and of pressure (high altitude and deep-sea expeditions) on man, but little effort has been expended to determine the effects of environmental lighting on man, except for the study of circadian rhythms (i.e., the effects of the cyclicity of light; see Chapter 8).

Also, little research has been done on the effects of environmental lighting on laboratory animals. The following statement was made by Dr. Fouts in his overview of the recent symposium entitled "Environmental and Genetic Factors Affecting Laboratory Animals: Impact on Biomedical Research" (ref. 9, p. 1164).

> Lighting has been mentioned, but primarily in terms of duration and cycling with respect to sacrifice time. Does it matter whether light is bright or dim, from incandescent or fluorescent lights? What wavelengths of room light have biologic activity? A lot of work has been done on selected aspects of these problems (e.g., pineal gland amine modulation by light), but more generalized effects have not often been studied in sufficient detail (e.g., drug metabolism, absorption, distribution, excretion and reactivity of non-CNS receptors). . . . Thus, this field of environmental and genetic factors affecting the response of laboratory animals represents an exciting area of biological research for future generations. Its appeal lies in the following: (1) Small changes in environmental conditions and genetic constitution can produce large alterations in the responses of the animal; (2) Only a portion of the environmental and genetic conditions that cause such changes in laboratory animals have been identified, or once identified, adequately quantified; and (3) Systematic investigations of such variables can result not

only in closer correlation between animal models and physiological or disease states existing in man but also in discovery of new biological principles.

One can only hope that the concern expressed by Dr. Fouts over the effects of environmental lighting on experimental animals will become a *general concern* of scientists who use experimental animals, because this may give us valuable information that can be used in designing environmental lighting for man.

Anecdotal effects of different lighting regimens on plants, animals, and man have been reported.[10] It would be of value if these observations could be confirmed and expanded.

15.5. IMMUNOLOGY IN PHOTOBIOLOGY

Currently, the field of immunology is enjoying an unprecedented popularity. This has been stimulated by developments in organ transplant techniques, the discovery of autoimmune diseases, and the belief that immunological techniques may offer ways of curing cancer.

Immunologists have recently made some significant contributions to photobiology. The application of immunological techniques to problems in photobiology should be a very profitable field in the coming years.

15.5.1. Immunological Detection of Thymine Photoproducts *in Vivo*

The use of procedures of immunology to detect and localize damaged DNA in histological sections of the skin of UV-irradiated animals (including man) is a powerful new technique. For example, rabbit antiserum containing antibodies specific for UV-irradiated DNA (the antigen) will react with damaged DNA in tissue sections from the skin of UV-irradiated animals. These specific antigen–antibody complexes are then visualized microscopically after reacting them with fluorescein-conjugated sheep antiserum to rabbit γ-globulin (indirect immunofluorescence technique).

The antiserum appears to be highly specific for thymine photoproducts and does not cross-react with native DNA.[11,12] It has been used to detect the presence of UV-induced DNA damage in human skin[13] and cultured human cells.[12] The kinetics of the repair of UV-induced DNA damage to human cells in culture have been followed with this technique.[12] A rough action spectrum has been obtained for the production of UV-induced nuclear damage in the skin of mice.[14]

An exciting potential of this technique is to quantitate the depth of penetration of the different wavelengths of solar radiation that produce damage to DNA. Presumably such studies would be equally applicable to plants as well as animals, and should provide valuable information relevant to the problem of possible adverse reactions if the stratospheric ozone layer is perturbed by man.

A sensitive radioimmuneassay (RIA) has been developed for the serologic estimation of thymine dimers in bacterial and mammalian cells following UV irradiation and postirradiation repair.[15] Thymine dimers formed by 5 J m^{-2} were easily detectable, which makes it somewhat more sensitive than chromatographic techniques.

15.5.2. Immunology of UV Radiation-Induced Skin Cancer

Most skin tumors induced in mice by chronic UV irradiation are highly antigenic. Usually these tumors (mainly fibrosarcomas) are immunologically rejected when transplanted to normal, genetically identical (syngeneic) mice, although they will grow in immunosuppressed mice (e.g., thymectomized at 3–4 weeks of age and given 450 rads of whole-body X-radiation 2 weeks later). This raises the important question of why these highly antigenic tumors are able to persist and grow progressively in their original hosts.

This question was recently investigated in an exciting series of experiments on the immunologic parameters of UV-induced carcinogenesis.[16] While normal syngeneic mice rejected grafts of UV-induced tumors, animals that had been UV-irradiated dorsally did not reject ventrally implanted grafts. Therefore, in addition to its carcinogenic action at the site of irradiation, chronic UV treatment also induces a *systemic alteration* that prevents immunologic rejection of these tumors. Thus, a new effect of UV irradiation on animals has been described that may have far-reaching implications upon our attitudes toward excessive exposure to sunlight.

In addition, it raises the interesting possibility (yet to be tested) that this UV-induced systemic alteration may also affect the balance between host and tumor during viral or chemical carcinogenesis. This possibility certainly warrants investigation.

15.6. LIGHT ACTIVATION OF ENZYMES

Light-activated enzymes may constitute an important class of photoreceptors for controlling biological functions. It has been proposed that the photoactivation of an enzyme, which catalyzes the conversion of a small substrate molecule to a product effector, initiates a train of successive enzyme activations through the interaction of allosteric control enzymes. Such a cascade would account for the amplification needed from light signal to biological response. Negative effectors can also be accommodated in the scheme.[17]

The light activation of enzymes certainly constitutes a "new topic" in photobiology, which should develop rapidly over the next few years. Six mechanisms by which light can affect enzyme reactions are described below.

15.6.1. Photoactivation of Enzyme–Substrate Complexes

When cyclobutane-type pyrimidine dimers are formed in the DNA of cells by UV irradiation, the photoreactivating enzyme, present in most cells, combines with these lesions. If these enzyme–substrate complexes are then illuminated with light at about 320–410 nm, the dimers are split to monomers, thus reversing, by a combination of visible light and an enzyme, the damage produced by UV radiation. The photoreactivating enzyme (photolyase) is the only known example where light plays a primary role in an enzymatic process at the level of the enzyme–substrate complex (Section 5.4.2.).

15.6.2. Photoactivation by Producing a Conformational Change in the Enzyme

Urocanase is an important enzyme in the pathway for the conversion of histidine to glutamic acid. Urocanase from *Pseudomonas putida* is activated by near-UV radiation (~360 nm). This is true for cells, cell extracts, and for the enzyme after purification by gel electrophoresis. Photoactivation of cell extracts does not require oxygen, and is not dependent on temperature (0–40°C). The action spectrum for photoactivation shows a major peak at 275 nm, a secondary peak at 320 nm, and has some resemblance to the absorption spectrum of α-ketobutyrate, the coenzyme of urocanase.[18] The action spectrum for inactivation of the active enzyme, however, corresponds to the absorption spectrum of a simple protein, suggesting that tyrosine or tryptophan may be the chromophore for inactivation. A comparison of purified active and inactive forms of the enzyme suggest that urocanase undergoes a slight conformational change upon activation.[19] The active enzyme slowly loses activity when stored in the dark, but can be reactivated by another exposure to light. Light is not needed for the enzyme reaction itself, however.

15.6.3. Photoactivation of the Enzyme by the Removal of a Blocking Compound from an Essential Amino Acid

Papain, as usually obtained commercially, must be activated chemically by the addition of dithiothreitol, a reducing agent. The activation reaction requires the splitting of a mixed disulfide group involving the active cysteinyl residue at position 25 in papain. Papain can also be activated by UV irradiation.[20] UV radiation is known to split disulfide bonds, in fact, disulfide bonds are the most photochemically reactive groups in proteins (Table 5-1). When inactive papain is irradiated *under nitrogen* at $\lambda > 250$ nm, activation of papain occurs for a time, but, with longer irradiations, inactivation occurs. Only inactivation occurs if partially active papain is UV-irradiated in air-saturated solutions. Flash photolysis studies lead to the conclusion that the activation of papain by UV irradiation involves the photolysis of a mixed disulfide, and inactivation involves the photoionization of an essential tryptophan residue.[20]

This particular activation reaction probably has little biological relevance, but is offered as a possible mechanism for enzyme control by light, i.e., the photochemical removal of a blocking group at an active site. For such a reaction to be of biological importance, it would need to be produced by visible light. An example of such a reaction may be the photoactivation of NADP-malate dehydrogenase in spinach and barley leaves. The purified enzyme is activated by dithiothreitol *in vitro,* but is activated by visible light *in vivo.*[21] It will be of interest to elucidate the mechanism of this photoactivation reaction.

15.6.4. Photoactivation of an Enzyme by Photoproducing Its Substrate

When rod outer segments, prepared from retinas, are incubated with Mg^{2+} and ATP, rhodopsin is phosphorylated. The reaction is markedly stimulated by light. The light could be acting by directly stimulating the activity of the kinase, by altering the concentration of some cofactor of the enzyme, or by altering the conformation of the rhodopsin so that it becomes a substrate for the kinase. The protein kinase has been isolated, and was found to act only on photobleached rhodopsin, and not on unbleached rhodopsin or on histones, phosvitin, or casein.[22] Thus, photobleached rhodopsin appears to be the unique substrate for this kinase.

The photoactivation of enzymes by the photoproduction of their substrates also occurs when cells are UV-irradiated. The photoproducts thus produced in DNA are then acted upon by the many DNA repair enzymes (see Chapter 5).

15.6.5. Photoactivation of Enzymes by the Isomerization of a Photochromic Inhibitor

An enzymatic process, itself insensitive to light, can be made subject to photoregulation through the use of a light-sensitive enzyme inhibitor. For example, a specific inactivator or chymotrypsin, *p*-azophenyldiphenylcarbamyl chloride, exists as two geometric isomers, *cis* and *trans,* which are interconvertible by light. The *trans* isomer is converted to *cis* by irradiation at 320 nm; the *cis* isomer is converted to *trans* by irradiation at 420 nm. The *cis* isomer is five times more effective as an enzyme inhibitor than is the *trans* isomer; thus, the degree of inactivation of the enzyme can be regulated by light.[23] A photochromic reagent has also been used to control acetylcholinesterase activity.[24]

While no naturally occurring photochromic enzyme inhibitor has yet been isolated, photochromic substances play major roles in vision (i.e., retinal), and in photomorphogenesis (i.e., phytochrome).

15.6.6. Enhancement of Enzyme Activity by Visible Light Irradiation of the Substrate in Crystalline Form

It has been reported that the irradiation of enzyme substrates in crystalline form with visible light (~400 footcandles at 546 nm) induces enhancements in enzyme reactions. This effect is cyclical, and occurs only at discrete irradiation times. The minimum irradiation time (seconds) inducing the first enhancement, and the time interval between subsequent enhancements (seconds), are dependent upon the particular enzyme.[25] This phenomenon is unique, and quite controversial. It awaits general confirmation by others, and a plausible hypothesis that can be tested.

15.7. NEAR-ULTRAVIOLET AND VISIBLE RADIATION (DETRIMENTAL EFFECTS)

15.7.1. Germicidal Action Spectrum between 220 and 700 nm: An Explanation

The action spectrum for the killing of *E. coli,* shown in Fig. 15-2, indicates that radiation from 220 to 700 nm, (i.e., far-UV, near-UV, and visible radiation) can kill bacteria, although the relative germicidal efficiencies of the different wavelengths vary by as much as several orders of magnitude.

The most effective range, i.e., between 220 and 300 nm, has been correlated with the absorption spectrum of DNA.[26] The direct absorption of radiation to raise the nucleic acid bases from their ground states (S_0) to their first excited singlet states (S_1), is the important reaction over this wavelength range.

Although of very low probability (i.e., a "forbidden transition"), it is possible to populate the triplet state of a molecule by the direct absorption of light in the ground state (Section 2.3.6.). For the nucleic acids, this would occur over the range of 300–380 nm.[27] The probability of such a transition (i.e., $S_0 \rightarrow T_1$) is about 10^{-6} compared to the transition, $S_0 \rightarrow T_1$. This probably explains why it takes about 10^6 more energy at 365 nm than at 254 nm to produce the same amount of thymine dimers in cells.[28] (*N.B.*: The value of $\sim 10^{-6}$ for the ratio of $(S_0 \rightarrow T_1)/(S_0 \rightarrow T_1)$ may be calculated from the ratio of the radiative lifetimes for phosphorescence/fluorescence, which are proportional to the absorption coeffi-

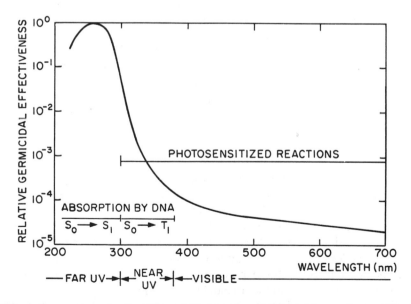

Fig. 15-2. Action spectrum for the killing of *Escherichia coli* (adapted from reference 26), and an indication of the wavelength ranges for different types of photochemical reactions. See the text for further details.

cients; i.e., the probability of excitation is proportional to the probability of deexcitation).

From about 300 to 700 nm, photosensitized reactions (Chapter 4) probably play the most important role in inactivating bacteria. Such inactivation would result from chromophores other than the nucleic acids absorbing the light energy, but this energy, in many cases, would subsequently be transmitted to the nucleic acids, resulting in cell killing.

Oxygen-dependent killing occurs over the range of 313–650 nm, while oxygen-independent killing occurs out to about 460 nm.[29] The oxygen-dependent reactions would be classified as photodynamic reactions (Chapter 4); possible natural photosensitizers include riboflavin, vitamin K, and flavin mononucleotide (FMN).

Oxygen-independent reactions could occur via triplet–triplet energy transfer (Section 3.8. and reference 27) from ketones such as acetone.[30] DNA addition reactions,[31] induced by endogenous photosensitizers and visible light, have not been studied, but extensive studies have been performed on exogenous photosensitizers such as furocoumarin (Section 4.4.1.).

Thus, while the action spectrum for the killing of bacteria mimics the absorption spectrum of DNA in the region of 220–300 nm, the shape of the action spectrum from 300 to 700 nm will depend upon the nature and amount of endogenous or exogenous photosensitizers present in the system.

15.7.2. Near-UV Radiation (300–380 nm)

Near-UV radiation is absorbed only marginally by proteins and DNA, yet it produces many of the deleterious effects usually ascribed to far-UV radiation (i.e., wavelengths below 300 nm), such as killing, mutation induction, and growth delay, although requiring much higher radiation fluences.[32,33] Of these effects, the one induced by the lowest fluences of near-UV radiation is the inhibition of growth [i.e., either a decrease in the initial rate of growth or a complete, but temporary, cessation of growth ("growth delay")].

15.7.2.1. Growth Delay

Bacterial studies have shown that growth delay has a narrow action spectrum centered at 340 nm. Earlier suggestions that quinones might be the chromophores, and oxidative phosphorylation the cellular target have not been confirmed. Recently, it has been shown that fluences of near-UV radiation capable of inducing a large growth delay in *E. coli* B/r result in a complete (but temporary) cessation of net RNA synthesis, the effects on protein and DNA synthesis being less severe. The action spectrum for the inhibition of RNA synthesis is the same as that for growth delay, and fits very closely the absorption spectra of tRNAs that contain the unusual base 4-thiouracil. *In vitro* studies have shown that near-UV irradiation produces an internal cross-link between the thiouracil at position 8 and a cytidine at position 13 of the tRNA; this cross-link reduces the biological activity of some tRNAs. The fluences required to inactivate tRNA are similar to those required to produce an inhibition in bacterial growth. Thus, it seems established that near-UV-induced growth delay in *E. coli* is due to the photochemical alteration of tRNA molecules that contain 4-thiouracil.[34]

In other organisms, other mechanisms for light-induced growth delay have been observed, e.g., inactivation of cytochrome a_3 in cells by blue light (Section 15.7.3.).

15.7.2.2. Lethal and Mutagenic Effects

The lethal and mutagenic effects of near-UV radiation have been reviewed.[29,33] Two molecular mechanisms for cellular inactivation by near-UV radiation are discussed in Section 15.7.1. Near-UV radiation can produce thymine dimers in the DNA of bacteria,[28] and can produce DNA chain breaks and/or alkali labile bonds.[35] The near-UV irradiation of tryptophan in solution produces the lethal and mutagenic agent hydrogen peroxide,[36] probably through the action of electrons photoejected from tryptophan.

15.7.3. Visible Light

Much of this book has been on the beneficial effects of visible light (e.g., vision, photomorphogenesis, photosynthesis), but visible light can also be detrimental. It is mutagenic in bacteria, and requires the presence of oxygen; thus mutagenesis by visible light is probably produced by a photodynamic mechanism.[37] Visible light is also lethal (Fig. 15-2), especially to repair-deficient bacteria, and the synergistic interaction of different wavelengths of visible light appears to be important in this regard (Section 15.8.1.).

Chromosomal aberrations are induced,[38] and DNA synthesis is inhibited[39] in pig kidney cells by far-red (750 nm) light. Green-yellow (546–579 nm) light represses the growth of liquid cell cultures of Ginkgo pollen and of monolayer cultures of HeLa cells, but this effect can be negated by irradiation with red (640–675 nm) light.[40] Blue light inhibits cell division reversibly in microsporocytes of *Lilium* and *Trillium,* presumably by photochemically altering cytochrome a_3 needed for respiration.[41] Blue light also produces free radicals in wool.[42] Other effects of visible light have recently been reviewed.[43]

The few effects described above are in the *absence* of added photosensitizers. The effects of visible light in the presence of added sensitizers is the subject of Chapter 4.

15.7.4. Conclusions

The photochemically active portion of sunlight (far-UV, near-UV, and visible light) produces both beneficial and detrimental effects on biological systems. Organisms have had to develop ways of making use of the portions of the solar spectrum that are essential to their survival while protecting themselves from the detrimental wavelengths of radiation, which differ for different organisms. The latter has been accomplished by preventing undue exposure through avoidance reactions and shielding (e.g., fur, shells, pigment), and by developing effective repair systems (Section 5.4.).

Most published work on the effects of UV radiation on cells has made use of the inexpensive and nearly monochromatic germicidal lamps emitting at 254 nm. However, most biological systems never encounter such radiation, except in experimental situations. Since sunlight contains a considerable amount of near-UV radiation, and most biological systems are exposed to sunlight, it makes sense to be concerned about the biological effects of near-UV radiation. Furthermore, many of the photochemical changes produced by near-UV radiation differ from those produced by 254 nm radiation. We need to know the effects of near-UV radiation on plants and animals in the presence and absence of added sensitizers (e.g., environmental pollutants). We need more information on the synergistic interaction of near-UV radiation with radiation at other wavelengths (see Section 15.8.). In short, we need to know more about the environment in which we live.

So much attention has been directed toward the detrimental effects of far-UV radiation that very little effort has been expended to elucidate possible beneficial effects, except for vitamin D synthesis. The reverse situation is true for visible radiation, i.e., so much attention has been directed toward the beneficial effects of visible light, that, relatively speaking, very little effort has been expended on the study of the deleterious effects of visible light in the absence of added photosensitizers.

It is important to know the intrinsic sensitivity of all types of organisms to the various wavelengths of radiation present in sunlight, for only with this information can we hope to make correct decisions about stopping human activites that may affect the quality and quantity of sunlight. A current problem is the possible perturbation of the atmospheric ozone layer by the exhaust emissions of supersonic aircraft, and by chlorofluorocarbons from spray cans and leaks from refrigeration systems.

A decrease in the ozone layer would affect the amount of radiation reaching the earth in the region of 290–320 nm. It would have little if any effect at longer wavelengths. On the other hand, aerosols (dust) and water droplets (clouds) can cause a significant reduction of both UV and visible radiation.[44] There is a desperate need for data on the effects of enhanced amounts of radiation in sunlight between 290 and 320 nm on key species such as plankton, and on major food crops.

15.8. THE NONADDITIVE EFFECTS OF DIFFERENT WAVELENGTHS OF RADIATION (SYNERGISM AND ANTAGONISM)

When a biological system is exposed to two different wavelengths of radiation, the observed effect is frequently not the summation of the effects of the individual wavelengths. Most frequently one observes a synergistic effect (i.e., the effect of two wavelengths of radiation given together is greater than the sum of the two effects when given independently). Sometimes the effects are antagonistic (i.e., the opposite of synergistic), and sometimes there is no interaction at all. It is especially important to keep in mind the possibility of synergism or antagonism when one uses a polychromatic source of radiation, whether it be an artificial source or natural sunlight.

15.8.1. Synergism

Recombination-deficient bacteria are also deficient in several pathways of DNA repair, and are very sensitive to killing by near-UV radiation and by radiation from ordinary white fluorescent lights. However, when an action spectrum for killing was run on recombination-deficient bacteria, they were not particularly sensitive to any *single* wavelength of near-UV radiation. These results suggest that the special sensitivity of these mutant bacteria to broad spectrum near-UV radiation may be due to synergistic effects of different wavelengths of near-UV radiation.[45]

An example of synergism between closely related wavelengths of near-UV radiation is the inactivation of transforming DNA. While radiation at 334 nm caused little or no inactivation of transforming DNA, it produced a marked synergistic effect on the inactivation by 365 nm radiation.[46] While the molecular basis of this synergism is not known, it is probably due to chromophores bound to the DNA.

Numerous papers have been published on the synergistic interaction of UV radiation (254 nm) and ionizing radiation. Recently it has been demonstrated that one basis for this synergism is the selective inhibition, by UV radiation, of one particular system for the repair of X-ray-induced DNA single-strand breaks.[47]

Near-UV radiation (365 nm) has also been shown to act synergistically with ionizing radiation by interfering with DNA repair.[48]

Far-red radiation has been reported to interact synergistically with X-rays in the production of chromosomal abberations in *Vicia faba*; however, this has now been shown to be an artifact caused by a far-red light-induced mitotic delay. Far-red irradiation does not alter the number of chromosome aberrations, it merely shifts the time of arrival of cells at metaphase where they can be scored.[49]

The production of erythema and histological changes in human skin by UV-B (290–320 nm) is enhanced by additional exposure to fluences of UV-A (320–400 nm) that cause no skin changes.[50]

A synergistic response that is receiving considerable attention by those interested in the effects of ionizing radiation on mammalian cells is the synergism between heat (hyperthermia) and ionizing radiation (for a review, see reference 51). When mammalian cells are heated to 40–43°C, either before or after X-irradiation, the cells are much more easily killed by the ionizing radiation. There is currently great hope of using this synergistic response in the treatment of cancer. Techniques using ultrasound or microwaves are being developed for the localized deposition of heat within the tumor.

It is of interest that under the conditions where heat is found to be synergistic with X-irradiation, no synergistic effect was observed for 254 nm irradiation.[51] This is consistent with recent results for bacteria, where only a minor synergistic effect of heat (52°C) was found for 254 nm irradiation, but a marked synergistic effect was found for 334 nm > 365 nm > 405 nm irradiations.[52] This latter synergistic response is particularly interesting in view of the older observations that heat enhanced the production of UV-induced (290–320 nm) skin cancer in experimental animals (reviewed in reference 53).

15.8.2. Antagonism

Perhaps the best known case of antagonism between different wavelengths of radiation is the red–far-red effect on phytochrome-mediated responses; red light promotes the effect, and far-red inhibits (Chapter 11). Thus, far-red light is antagonistic to the biological effects of red light.

Another example, described in Section 5.2.4., is that irradiation of DNA at 280 nm produces thymine dimers, and subsequent irradiation at 240 nm will split the dimers back to monomers. This situation requires very high fluences of UV radiation so that a photochemical equilibrium between the formation and splitting of dimers is reached. At 280 nm, the equilibrium favors the formation of dimers, while at 240 nm, the equilibrium is shifted toward monomer formation.

While the above examples are purely photochemical, there is another type of antagonism that appears to be more biological in nature. There is synergism between X-rays and 254 or 365 nm radiation when the two radiations are given in rapid succession (Section 15.8.1.); however, if an hour or two of metabolism in growth medium is permitted between the two irradiations, then antagonism is observed, i.e., the cells show a much increased resistance to the second irradiation. This is true for 365 nm irradiation prior to X-irradiation,[54] and for X-irradiation prior to 254 nm irradiation or X-irradiation.[55] The possible radiation induction of DNA repair enzymes is currently receiving considerable attention.[56] Alternative explanations for the radiation-induced enhancement of resistance to radiation have been discussed.[55]

15.9. USING LIGHT AS A SELECTIVE TOOL

A classical approach to understanding how a biological system functions, or in determining the limits of its adaptability, is to perturb the system, e.g., by changes in nutrition, temperature, etc., and see how it responds. As we have seen in the preceding chapters, light has also been a useful tool for perturbing biological systems in order to learn how they function. However, after sufficient knowledge has been gained about the photochemical responses of biological systems, one can design experiments in which light is used as a tool for studying specific biological or biochemical problems.

Example 1. It was discovered in the author's laboratory some years ago that the UV irradiation of cells causes the cross-linking of proteins and nucleic acids. These studies were concerned with gaining a better understanding of the lethal effects of UV radiation. Recently, the concept that nucleic acid and proteins, which are in close proximity, can be covalently cross-linked by UV radiation, has been applied to the study of the sites of close association between nucleic acids and proteins in enzyme–substrate complexes. The question has been asked, where on a tRNA molecule does its aminoacyl tRNA synthetase "sit"? In a beautiful series of experiments, Schimmel and co-workers[57] have answered this question.

For each amino acid there is at least one specific synthetase and tRNA. The enzymes must attach each amino acid without error to its cognate tRNA in order to prevent altered proteins from being synthesized. The tRNAs are single-stranded polynucleotide chains composed of about 75–85 bases, and are folded into the same basic hydrogen-bonded cloverleaf structure. A key question is what is the molecular basis for specificity in synthetase–tRNA interactions. One approach to answering this question is to determine those regions on a tRNA molecule that are, and those that are not, in close proximity to the surface of the enzyme. The enzyme–tRNA complex is UV-irradiated to form RNA–protein cross-links. The irradiated and unirradiated complexes are then treated with T1 ribonuclease, which cleaves RNA only near guanine residues. Fourteen distinct RNA fragments are isolated from unirradiated samples, however, after UV irradiation, three of the fragments are missing from the chromatogram, indicating that they have been cross-linked to the enzyme by UV irradiation. The nature of the missing fragments tells the location on the tRNA of the attachment sites of the enzyme.

In some cases, a synthetase will attach its amino acid to the wrong tRNA species. An example is yeast valine tRNA synthetase, which can attach valine to *E. coli* isoleucine tRNA. The question is whether this noncognate yeast enzyme binds to the *E. coli* tRNA by a similar or by a different topological pattern compared to the cognate enzyme. Using the photochemical cross-linking technique, two of the three attachment sites were found to be identical with the cognate complex, but one was different.

Other photochemical studies into the structural relationships of specific nucleic acid–protein complexes have been reviewed.[57] All known DNA addition reactions, both chemically and radiation-induced, have been reviewed with special reference to their importance in the fields of aging, carcinogenesis, and radiation biology.[31]

Example 2. 5-Bromouracil (BrUra) is an analog of thymine. Because the bromine atom is about the same size as the methyl group of thymine (5-methyluracil), cells can be made to incorporate BrUra into their DNA in place of thymine (Thy). BrUra is more photochemically reactive than Thy, and cells containing this analog are much more easily killed by UV (and X) irradiation than are normal cells. Such results were used years ago to help prove that DNA is the target for the lethal effects of both UV and ionizing radiation. BrUra has another useful property; its absorption spectrum extends to about 330 nm, while that of

Thy extends only to about 310 nm. Thus, cells containing BrUra can be inactivated by irradiation at 313 nm, while normal cells are little affected. This selectivity of inactivation has been a useful tool in molecular biology; only a few examples will be given.

Puck and Kao[58] enriched cultures of mammalian cells for spontaneous mutants that require a specific nutrient by growing them in a medium lacking the specific nutrient, but containing 5-bromodeoxyuridine. Under these conditions, the mutant cells would not grow or incorporate BrUra into their DNA, but the other cells would. The cultures were then heavily irradiated with long-wavelength UV radiation to kill the normal cells, and the surviving mutant cells (i.e., those that did not contain BrUra) were tested for their nutritional deficiency.

The selective photolysis of BrUra-containing DNA has also been used to determine the presence of repair "patches" in the DNA of mammalian cells after UV irradiation. Subsequent to UV irradiation, repair is allowed to proceed in medium containing 5-bromodeoxyuridine; then the cells are irradiated at 313 nm. If the cells in question are able to perform excision repair, BrUra will be present in the patches, and 313 nm irradiation will produce breaks in the DNA at the site of these patches. These breaks can be measured on alkaline sucrose gradients. If the cells lack the ability to perform excision repair, no BrUra will be incorporated, and no single-strand breaks will be produced by the 313 nm irradiation.[59] The photochemistry, photobiology, and uses of BrUra in molecular biology have been reviewed.[60]

Example 3. The selective fluorescence of cells after exposure to a suitable dye has been utilized recently in developing several very useful techniques.

Cell Sorter. One problem in cell biology is obtaining pure populations of cells from a given tissue (e.g., T and B lymphocytes). Frequently, different cell types will take up different amounts of a dye (e.g., fluorescein) on their surface, and will therefore show different amounts of fluorescence. An apparatus has been devised that will sort cells on the bases of their fluorescence properties, and thus allow the isolation of closely related types of animal cells at rates up to 5000 cells per second.[61,62] A droplet containing a cell with the appropriate fluorescence is given either a negative, positive or no charge. The charged droplets will be deflected in an electric field, and will fall into different collecting flasks. This new technique holds much promise for investigations in cell biology.

Cytofluorimeter. Another problem in cell biology is determining the relative numbers of cells of a pure cell population that are in different stages of their growth cycle. If one uses a fluorescent dye that selectively binds to DNA (e.g., ethidium bromide), the fluorescence of the cells is then proportional to their DNA content. Cytofluorimeters are now available that can determine the fluorescence properties of individual cells in a flowing stream of cells. The data are then presented as a distribution of the number of cells that have different amounts of DNA.[62] Those cells in G_1 have one complement of DNA, those in G_2-M have two complements, and those in S phase (i.e., DNA synthesis phase) contain an intermediate amount of DNA. This technique has proven valuable in studies on the effects of drugs and of radiation on cells in culture.

Both the cell sorter and cytofluorimeter use a laser to excite the fluorescence. Other biological, photochemical, and spectroscopic applications of lasers have recently been reviewed.[63]

15.10. REFERENCES

1. F. Urbach (ed.), *The Biologic Effects of Ultraviolet Radiation,* p. 363, Pergamon Press, N.Y. (1969).
2. R. C. Smith and J. E. Tyler, Transmission of solar radiation into natural waters, in: *Photochemical and Photobiological Reviews* (K. C. Smith, ed.), Vol. 1, pp. 117–155, Plenum Press, New York (1976).
3. M. Luckiesh, *Applications of Germicidal, Erythemal and Infrared Energy,* Van Nostrand, New York (1946).
4. K. H. Norris and W. L. Butler, Techniques for obtaining absorption spectra on intact biological samples. *IRE Trans. Bio-Med. Electron.* **8,** 153–157 (1961).
5. D. R. Laub, D. J. Prolo, W. Whittlesey, and H. Buncke, Jr., Median cerebrofacial dysgenesis, *Calif. Med.* **112,** 19–21 (1970).
6. W. L. Butler, Absorption spectroscopy *in vivo*: Theory and application. *Annu. Rev. Plant Physiol.* **15,** 451–470 (1964).
7. K. H. Norris, Measuring and using light transmittance properties of plant materials, in: *Electromagnetic Radiation in Agriculture,* pp. 64–66, Illuminating Engineering Society, New York (1965).
8. A. Rosencwaig, Photoacoustic spectroscopy—A new tool for investigation of solids, *Anal. Chem.* **47**(6), 592A–604A (1975).
9. E. S. Vesell and C. M. Long (co-chairman), Environmental and genetic factors affecting laboratory animals: Impact on biomedical research, *Fed. Proc.* **35,** 1123–1165 (1976).
10. J. N. Ott, *Health and Light* (The Effects of Natural and Artificial Light on Man and Other Living Things), Devin-Adair, Old Greenwich, Conn. (1973).
11. P. G. Natali and E. M. Tan, Immunological detection of thymidine photoproduct formation *in vivo, Radiat. Res.* **46,** 506–518 (1971).
12. C. J. Lucas, Immunological demonstration of the disappearance of pyrimidine dimers from nuclei of cultured human cells, *Exp. Cell Res.* **74,** 480–486 (1972).
13. E. M. Tan and R. B. Stoughton, Ultraviolet light-induced damage to desoxyribonuclcic acid in human skin, *J. Invest. Dermatol.* **52,** 537–542 (1969).
14. E. M. Tan, R. G. Freeman, and R. B. Stoughton, Action spectrum of ultraviolet light-induced damage to nuclear DNA *in vivo, J. Invest. Dermatol.* **55,** 439–443 (1970).
15. E. Seaman, H. van Vunakis, and L. Levine, Serologic estimation of thymine dimers in the deoxyribonucleic acid of bacterial and mammalian cells following irradiation with ultraviolet light and postirradiation repair, *J. Biol. Chem.* **247,** 5709–5715 (1972).
16. M. L. Kripke and M. S. Fisher, Immunologic parameters of ultraviolet carcinogenesis, *J. Natl. Cancer Inst.* **57,** 211–215 (1976).
17. D. H. Hug, D. Roth, and J. K. Hunter, Photoactivation of an enzyme and biological photoreception: An hypothesis, *Physiol. Chem. Phys.* **3,** 353–360 (1971).
18. D. Roth and D. H. Hug, Photoactivation of urocanase in *Pseudomonas putida*: Action spectrum, *Radiat. Res.* **50,** 94–104 (1972).
19. D. H. Hug and D. Roth, Photoactivation of urocanase in *Pseudomonas putida*: Purification of inactive enzyme, *Biochemistry* **10,** 1397–1402 (1971).
20. J. F. Baugher and L. I. Grossweiner, Ultraviolet inactivation of papain, *Photochem. Photobiol.* **22,** 163–167 (1975).
21. H. S. Johnson, NADP-malate dehydrogenase: Photoactivation in leaves of plants with Calvin Cycle photosynthesis, *Biochem. Biophys. Res. Commun.* **43,** 703–709 (1971).

22. M. Weller, N. Virmaux, and P. Mandel, Light-stimulated phosophorylation of rhodopsin in the retina: The presence of a protein kinase that is specific for photobleached rhodopsin, *Proc. Natl. Acad. Sci. USA* **72**, 381–385 (1975).

23. H. Kaufman, S. M. Vratsanos, and B. F. Erlanger, Photoregulation of an enzymic process by means of a light-sensitive ligand, *Science* **162**, 1487–1489 (1968).

24. J. Bieth, N. Wassermann, S. M. Vratsanos, and B. F. Erlanger, Photoregulation of biological activity by photochromic reagents, IV. A model for diurnal variation of enzymic activity, *Proc. Natl. Acad. Sci. USA* **66**, 850–854 (1970).

25. S. Comorosan, The measurement process in biological systems: A new phenomenology, *J. Theor. Biol.* **51**, 35–49 (1975).

26. F. L. Gates, A study of the bactericidal action of ultraviolet light. III. The absorption of ultraviolet light by bacteria, *J. Gen. Physiol.* **14**, 31–42 (1930).

27. A. A. Lamola, M. Gueron, T. Yamane, J. Eisinger, and R. G. Shulman. Triplet state of DNA, *J. Chem. Phys.* **47**, 2210–2217 (1967).

28. R. M. Tyrrell, Induction of pyrimidine dimers in bacterial DNA by 365 nm radiation, *Photochem. Photobiol.* **17**, 69–73 (1973).

29. R. B. Webb, Lethal and mutagenic effects of near-ultraviolet radiation, in: *Photochemical and Photobiological Reviews,* Vol. 2, (K. C. Smith, ed.), pp. 169–261, Plenum Press, New York (1977).

30. H.-D. Menningmann and A. Wacker, Photoreactivation of *Escherichia coli* B_{s-3} after inactivation by 313 nm radiation in the presence of acetone, *Photochem. Photobiol.* **11**, 291–296 (1970).

31. K. C. Smith (ed.), *Aging, Carcinogenesis, and Radiation Biology* (The Role of Nucleic Acid Addition Reactions), Plenum Press, New York (1976).

32. J. Jagger, Growth delay and photoprotection induced by near-ultraviolet light, *Res. Prog. Org. Biol. Med. Chem.* **3**, 383–401 (1972).

33. R. B. Webb, Photodynamic lethality and mutagenesis in the absence of added sensitizers, *Res. Prog. Org. Biol. Med. Chem.* **3**, 511–530 (1972).

34. T. V. Ramabhadran and J. Jagger, Mechanism of growth delay induced in *Escherichia coli* by near ultraviolet radiation, *Proc. Natl. Acad. Sci. USA* **73**, 59–63 (1976).

35. R. M. Tyrrell, R. D. Ley, and R. B. Webb, Induction of single-strand breaks (alkali-labile bonds) in bacterial and phage DNA by near UV (365 nm) radiation, *Photochem. Photobiol.* **20**, 395–398 (1974).

36. J. P. McCormick, J. R. Fischer, J. P. Pachlatko, and A. Eisenstark, Characterization of a cell-lethal product from the photooxidation of tryptophan:hydrogen peroxide, *Science* **191**, 468–469 (1976).

37. R. B. Webb and M. M. Malina, Mutagenic effects of near ultraviolet and visible radiant energy on continuous cultures of *Escherichia coli, Photochem. Photobiol.* **12**, 457–468 (1970).

38. S. A. Gordon, A. N. Stroud, and C. H. Chen, The introduction of chromosomal aberrations in pig kidney cells by far-red light, *Radiat. Res.* **45**, 274–287 (1971).

39. C. H. Chen and S. A. Gordon, Inhibition of ^{3}H-thymidine incorporation in pig kidney cells by far-red light, *Photochem. Photobiol.* **15**, 107–109 (1972).

40. R. M. Klein and P. C. Edsall, Interference by near ultraviolet and green light with growth of animal and plant cell cultures, *Photochem. Photobiol.* **6**, 841–850 (1967).

41. H. Ninnemann and B. Epel, Inhibition of cell division by blue light, *Exp. Cell Res.* **79**, 318–326 (1973).

42. A. Shatkay and I. Michaeli, EPR study of wool irradiated with blue light, *Photochem. Photobiol.* **15**, 119–138 (1972).

43. N. I. Krinsky, Cellular damage initiated by visible light, in: *The Survival of Vegetative Microbes* (T. G. R. Gray and J. R. Postgate, eds.) (Soc. Gen. Biol. Symp. No. 26), pp. 209–239, Cambridge University Press, London (1976).

44. P. Halpern, J. V. Dave, and N. Braslaw, Sea-level solar radiation in the biologically active spectrum, *Science* **186**, 1204–1208 (1974).

45. D. Mackay, A. Eisenstark, R. B. Webb, and M. S. Brown, Action spectra for lethality in recombinationless strains of *Salmonella typhimurium* and *Escherichia coli, Photochem. Photobiol.* **24**, 337–343 (1976).

46. M. J. Peak, J. G. Peak, and R. B. Webb, Synergism between different near-ultraviolet wavelengths in the inactivation of transforming DNA, *Photochem. Photobiol.* **21**, 129–131 (1975).

47. K. D. Martignoni and K. C. Smith, The synergistic action of ultraviolet and X radiation on mutants of *Escherichia coli* K-12, *Photochem. Photobiol.* **18**, 1–8 (1973).

48. R. M. Tyrrell, The interaction of near UV (365 nm) and X-radiations on wild-type and repair deficient strains of *Escherichia coli* K-12: Physical and biological measurements, *Int. J. Radiat. Biol.* **25**, 373–390 (1974).

49. S. Wolf and H. E. Luippold, Mitotic delay and the apparent synergism of far-red radiation and X-rays in the production of chromosomal aberrations, *Photochem. Photobiol.* **4**, 439–445 (1965).

50. I. Willis, A. Kligman, and J. H. Epstein, Effects of long ultraviolet rays on human skin: Photoprotective or photoaugmentative? *J. Invest. Dermatol.* **59**, 416–420 (1972).

51. W. C. Dewey, L. E. Hopwood, S. A. Sapareto, and L. E. Gerweck, Cellular responses to combinations of hyperthermia and radiation, *Radiology* **123**, 463–474 (1977).

52. R. M. Tyrrell, Synergistic lethal action of ultraviolet–violet radiations and mild heat in *Escherichia coli, Photochem. Photobiol.* **24**, 345–351 (1976).

53. J. H. Epstein, Ultraviolet carcinogenesis, *Photophysiology* **5**, 235–273 (1970).

54. R. M. Tyrrell, RecA⁺-dependent synergism between 365 nm and ionizing radiation in log-phase *Escherichia coli*: A model for oxygen-dependent near-UV inactivation by disruption of DNA repair, *Photochem. Photobiol.* **23**, 13–20 (1976).

55. K. C. Smith and K. D. Martignoni, Protection of *Escherichia coli* cells from ultraviolet and X-irradiation by prior X-irradiation: A genetic and physiological study, *Photochem. Photobiol.* **24**, 515–523 (1976).

56. E. M. Witkin, Ultraviolet mutagenesis and inducible DNA repair in *Escherichia coli, Bacteriol. Rev.* **40**, 869–907 (1976).

57. P. R. Schimmel, G. P. Budzik, S. S. M. Lam, and H. J. P. Shoemaker, *In Vitro* studies of photochemically cross-linked protein-nucleic acid complexes. Determinations of cross-linked regions and structural relationships in specific complexes, in: *Aging, Carcinogenesis and Radiation Biology* (K. C. Smith, ed.), pp. 123–148, Plenum Press, New York (1976).

58. T. T. Puck and F.-T. Kao, Genetics of somatic mammalian cells. V. Treatment with 5-bromodeoxyuridine and visible light for isolation of nutritionally deficient mutants, *Proc. Natl. Acad. Sci. USA* **58**, 1227–1234 (1967).

59. J. D. Regan, R. B. Setlow, and R. D. Ley, Normal and defective repair of damaged DNA in human cells: A sensitive assay utilizing the photolysis of bromodeoxyuridine, *Proc. Natl. Acad. Sci. USA* **68**, 708–712 (1971).

60. F. Hutchinson, The lesions produced by ultraviolet light in DNA containing 5-bromouracil, *Q. Rev. Biophys.* **6**, 210–246 (1973).

61. L. A. Herzenberg, R. G. Sweet, and L. A. Herzenberg, Fluorescence-activated cell sorting, *Sci. Am.* **234**, 108–117 (1976).

62. P. M. Kracmer, L. L. Deaven, H. A. Crissman, J. A. Steinkamp, and D. F. Petersen, On the nature of heteroploidy, *Cold Spring Harbor Symp. Quant. Biol.* **38**, 133–144 (1974).

63. M. W. Berns, Biological, photochemical, and spectroscopic application of lasers, *Photochemical and Photobiological Reviews,* Vol. 2 (K. C. Smith, ed.), pp. 1–37, Plenum Press, New York (1977).

Index